Ultra-Low Voltage Nano-Scale Memories

SERIES ON INTEGRATED CIRCUITS AND SYSTEMS

Anantha Chandrakasan, Editor
Massachusetts Institute of Technology
Cambridge, Massachusetts, USA

Published books in the series:

Ultra-Low Voltage Nano-Scale Memories
Kiyoo Itoh, Masashi Horiguchi, and Hitoshi Tanaka
2007, ISBN 978-0-387-33398-4

Routing Congestion in VLSI Circuits: Estimation and Optimization
Prashant Saxena, Rupesh S. Shelar, and Sachin S. Sapatnekar
2007, ISBN 978-0-387-30037-5

Ultra-Low Power Wireless Technologies for Sensor Networks
Brian Otis and Jan Rabaey
2007, ISBN 978-0-387-30930-9

Sub-threshold Design for Ultra Low-Power Systems
Alice Wang, Benton H. Calhoun, and Anantha Chandrakasan
2006, ISBN 0-387-33515-3

High Performance Energy Efficient Microprocessor Design
Vojin Oklibdzija and Ram Krishnamurthy (Eds.)
2006, ISBN 0-387-28594-6

Abstraction Refinement for Large Scale Model Checking
Chao Wang, Gary D. Hachtel, and Fabio Somenzi
2006, ISBN 0-387-28594-6

A Practical Introduction to PSL
Cindy Eisner and Dana Fisman
2006, ISBN 0-387-35313-5

Thermal and Power Management of Integrated Circuits
Arman Vassighi and Manoj Sachdev
2006, ISBN 0-398-25762-4

Leakage in Nanometer CMOS Technologies
Siva G. Narendra and Anantha Chandrakasan
2005, ISBN 0-387-25737-3

Statistical Analysis and Optimization for VLSI: Timing and Power
Ashish Srivastava, Dennis Sylvester and David Blaauw
2005, ISBN 0-387-26049-8

Ultra-Low Voltage
Nano-Scale Memories

Edited by

KIYOO ITOH
Hitachi, Ltd.
Tokyo, Japan

MASASHI HORIGUCHI
Renesas Technology Corp.
Tokyo, Japan

and

HITOSHI TANAKA
Hitachi ULSI Systems Co., Ltd.
Tokyo, Japan

 Springer

Kiyoo Itoh
Hitachi, Ltd.
Tokyo, Japan

Masashi Horiguchi
Renesas Technology Corp.
Tokyo, Japan

Hitoshi Tanaka
Hitachi ULSI Systems Co., Ltd.
Tokyo, Japan

Ultra-Low Voltage Nano-Scale Memories

Library of Congress Control Number: 2007920040

ISBN-13: 978-0-387-33398-4   e-ISBN-13: 978-0-387-68853-4

Printed on acid-free paper.

Printed in the United States of America.

9 8 7 6 5 4 3 2 1

springer.com

Preface

Ultra-low voltage nano-scale large-scale integrated circuits (LSIs) are becoming more important to ensure the reliability of miniaturized devices, to meet the needs of a rapidly growing mobile market, and to offset a significant increase in the power dissipation of high-end microprocessor units. Such LSIs cannot not be made without ultra-low voltage nano-scale memories because they need low-power large-capacity memories. Many challenges arise, however, in the process of achieving such memories as their devices and voltages are scaled down below 100 nm and sub-1-V. A high signal-to-noise (S/N) ratio design is necessary in order to cope with both a small signal voltage from low-voltage memory cells and with large amounts of noise in a high-density memory-cell array. Moreover, innovative circuits and devices are needed to resolve the increasing problems of leakage currents when the threshold voltage (V_t) of MOSFETs is reduced and serious variability in speed and leakage occur. Since the solutions to these problems lie across different fields, e.g., digital and analog, and even SRAM and DRAM, a multidisciplinary approach is needed.

Despite the importance of this field, there are few authoritative books on ultra-low voltage nano-scale memories. This book has been systematically researched and is based on the authors' long careers in developing memories, and low-voltage designs in the industry. Ultra-Low Voltage Nano-Scale Memories gives a detailed explanation of various circuits that the authors regard as important because the circuits covered range from basic to state-of-the-art designs. This book is intended for both students and engineers who are interested in ultra-low voltage nano-scale memory LSIs. Moreover, it is instructive not only for memory designers, but also for all digital and analog LSI designers who are at the leading edge of such LSI developments.

Chapter 1 describes the basics of digital, analog, and memory circuits, and low-voltage related circuits. First, the basics of LSI devices, leakage currents, and CMOS digital and analog circuits including circuit models are discussed. Then the basics of memory LSIs, DRAMs, SRAMs, and flash memory are explained, followed by a discussion of memory related issues such as soft errors, redundancy, and error checking and correcting (ECC) circuits. Issues related to voltage, such as the scaling law, power-supply schemes, and trends in power-supply voltages are also described. Finally, various power-supply management issues for future memory and on-chip voltage converters are briefly discussed.

Chapter 2 describes ultra-low voltage nano-scale DRAM cells. First, the trends in DRAM-cells and 1-T-based DRAM-cells are discussed. After that, the design of the folded-data-line 1-T cell is described five ways: in terms of the lowest necessary V_t and word voltage, the signal charge and the signal voltage, noise sources, the gate-over drive of the sense amp, and noise reductions. Open-data-line 1-T cells and state-of-the-art DRAM cells, such as the two-transistor (2-T) DRAM cell, the so-called 'twin cell', as well as a double-gate fully-depleted SOI 2-T cell, and gain cells are also explained.

Chapter 3 describes ultra-low voltage nano-scale SRAM cells. An explanation of the recent trends in SRAM-cell developments, is followed by a discussion of the leakage currents, and the voltage margin of 6-T SRAM cells, as well as their improvements. Finally, the 6-T SRAM cell is compared with the 1-T cell in terms of its voltage margin and soft error immunity.

Chapter 4 describes various circuit techniques that are used to reduce subthreshold leakage currents in RAM peripheral circuits. The basic principles of how to reduce leakage are described, with particular emphasis on the use of gate-source reverse biasing schemes. Various biasing schemes are discussed in detail, followed by applications to RAM cells and peripheral circuits in both standby and active modes.

Chapter 5 deals with the issue of variability in the nanometer era. The main focus is leakage and speed variations that are caused by variations in V_t. Various solutions with redundancy and ECC, layout, controls of internal supply voltages, and new devices such as planar double-gate fully-depleted SOI are discussed.

Chapter 6 describes the reference voltage generators that provide reference voltages for other converters. Various generators such as V_t-referenced, V_t-difference, band-gap generators, voltage trimming circuits, and burn-in test capability are described in detail.

Chapter 7 describes voltage down-converters in terms of their basic design concept, transient characteristics and phase compensation as well as their power-supply rejection ratio. Half-V_{DD} generators are also briefly discussed.

Chapter 8 deals with the circuit configurations of various voltage-up converters and negative voltage generators. Basic voltage converters with capacitors, Dickson-type voltage multipliers, and switched-capacitor-type voltage multipliers are explained and compared. Level monitors are also discussed.

Chapter 9 describes high-voltage tolerant circuit techniques that manage the voltage differences between peripheral circuits as well as between internal circuits and interface circuits of chips operating at a high external voltage.

We are indebted to many people, especially to our research colleagues at the Hitachi Central Research Laboratory, Tokyo who have collaborated with us, and one particular member of the administrative team, Ms. Anzai. They have offered support, advice, and the material needed to complete our work. Without their support this book would not have been possible.

Kiyoo Itoh
Masashi Horiguchi
Hitoshi Tanaka
Tokyo, September 25, 2006

Table of Contents

1
An Introduction to LSI Design

1.1. Introduction

LSI (Large-Scale Integrated Circuit) technologies have rapidly advanced since their advent in the early 1970's [1]. Integration per chip has reached the 1- to 4-Gb level for DRAMs (Dynamic Random Access Memories), the 256-Mb level for SRAMs (Static Random Access Memories), the 8-Gb level for flash memories, and above the 1.3-billion-MOST (Metal-Oxide-Semiconductor Transistor, or MOSFET) level for MPUs (Microprocessor Units) [2]. Such advances have been made possible by the extensive high-density LSI technologies on devices and circuits, lithography and fabrication processes, packaging, and so on. In the nanometer era, however, we are facing many emerging challenges. Of these, realization of ultra-low voltage (sub-1-V) operations is vital because it ensures the reliability of miniaturized devices, meets the needs of the rapidly growing mobile market, and offsets the sky-rocketing increase in the power dissipation of high-end MPUs. Ultra-low voltage operation is especially crucial for embedded memories that dominate chip sizes of the LSIs. Thus, we must solve evolving problems [3] such as the ever-smaller signal-to-noise-ratio (S/N) of memory cells, the ever-increasing leakage with reducing the threshold voltage (V_t) of MOSTs, and variations in speed and leakage enhanced by variations of design parameters such as V_t. Power management with on-chip voltage converters is thus playing a critical role in solving these problems.

In this chapter, the basics of high-density LSI technologies are described and state-of-the-art technologies are summarized with an emphasis on ultra-low voltage nano-scale memory technology. First, the basics of LSI devices are discussed, focusing on MOSTs, silicon substrates, and resistors. Next, the basics of the complementary-MOS (CMOS) digital and analog circuits are described. The basics of memories are also explained, focusing on DRAMs, SRAMs, and flash memories. After that, issues relevant to memory such as soft error, redundancy and the ECC (error checking and correcting) circuit, the scaling law, and power-supply issues are discussed. Finally, the power management design to solve the above-described problems and the roles of on-chip voltage converters are described.

1.2. Basics of LSI Devices

1.2.1. MOST Characteristics

Figure 1.1 shows a schematic cross-section [1] of an n-channel MOST (nMOST). The drain, gate, and substrate voltages measured from the source are called

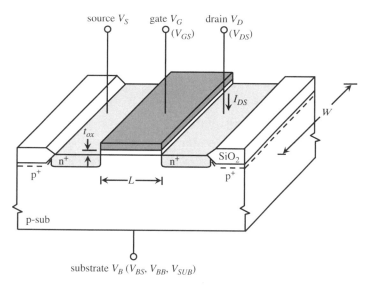

V_{DS}, V_{GS}, and V_{BS} (i.e. backgate bias or substrate bias V_{BB}). V_{BB} is usually negative or 0 V, and is supplied directly from an external source or is internally generated from V_{DD} with an on-chip voltage converter (or V_{BB} generator). When V_{GS} is increased, the drain-source current I_{DS} starts to flow at a certain V_{GS}. This value of V_{GS} is called the MOS threshold voltage V_t. I_{DS} increases with V_{GS} above V_t.

There are basically two different types of MOSTs, nMOST and pMOST, depending on the type of channel carrier. The MOST with an n$^+$ source and drain (Fig. 1.1) is called an n-channel MOST (nMOST, nMOS, NMOST, or NMOSFET). In the nMOST, electrons flow from source to drain through an n-channel; that is, a positive current I_{DS} flows from drain to source. If $I_{DS} = 0$ at $V_{GS} = 0$, we must apply a positive V_{GS} to form the n-channel. This type is known as the enhancement (normally off) nMOST (the E-nMOST). If an n-channel exists at $V_{GS} = 0$, we must apply a negative V_{GS} to deplete carriers in the channel. This type is called the depletion (normally on) nMOST (the D-nMOST). Similarly, we have the p-channel enhancement (normally off) MOST (the E-pMOST) and the depletion (normally on) MOST (the D-pMOST). The transfer characteristics of the four types are shown in Fig. 1.2. Note that for the E-MOST, the nMOST has positive values for both V_{DS} and V_t, while the pMOST has negative values for these voltages. Figure 1.3 shows circuit symbols and voltage relationships for the E-MOST, in which V_{DD} and V_{SS} are external supply voltages. Symbols (a)(or substrate without an arrow) and (c) are used in Chapters 1 to 5 to show nMOST and pMOST, respectively, except in section 1.5, and symbols (b) and (d) are used in Chapters 6 to 9 to show nMOST and pMOST, respectively. The following shows dc characteristics of E-nMOSTs.

Figure 1.4 shows I_{DS} versus V_{DS} for different values of V_{GS} at a given V_{BB}. Three regions of operation can be distinguished, as follows [1].

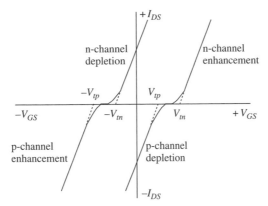

FIGURE 1.2. The transfer characteristics of four types of MOST [1].

The Cutoff Region. This region corresponds to $V_{GS} < V_t$ and $I_{DS} \cong 0$. It is also referred to as the subthreshold region, where I_{DS} in this region is much smaller than its value when $V_{GS} > V_t$. Thus, in many nMOS circuits for $V_{GS} < V_t$, the transistor is considered off and $I_{DS} = 0$. However, the small value of I_{DS} (i.e. the subthreshold current) in the region could affect the performance of the circuit, as in MOS dynamic circuits that operate with charges held at the floating nodes. In ultra-low-voltage operations with a lower V_t, the ever-increasing subthreshold current is an emerging issue, as discussed later.

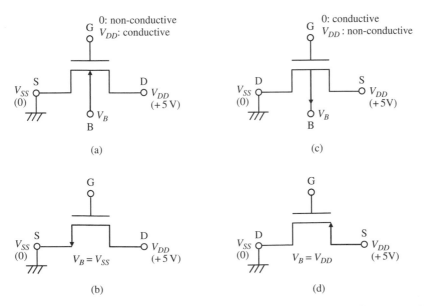

FIGURE 1.3. Circuit expressions and voltage relationships for an E-MOST [1]. (a) and (b) are for nMOST. (c) and (d) are pMOST.

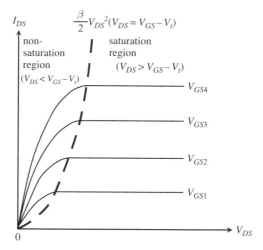

FIGURE 1.4. I_{DS} versus V_{DS} for a given V_{BB} [1].

The Non-Saturated Region. This region is also referred to as the "triode region", in which I_{DS} increase with V_{DS} for a given V_{GS} larger than V_t. The current is given by

$$V_{GS} \geq V_t,$$

$$V_{DS} \leq V_{GS} - V_t,$$

$$I_{DS} = \beta \left\{ (V_{GS} - V_t) V_{DS} - \tfrac{1}{2} V_{DS}^2 \right\} \tag{1.1}$$

where β is the channel conductance, which is explained in (1.4). When $V_{DS} \ll V_{GS} - V_t$, the $I - V$ characteristics can be approximated by a straight-line, $I_{DS} = \beta(V_{GS} - V_t)V_{DS}$, and the transistor is modeled by a resistor of

$$R_{ON} = \frac{1}{\beta (V_{GS} - V_t)}. \tag{1.2}$$

The Saturated Region. In this region I_{DS} becomes constant and independent of V_{DS}. It is determined only by V_{GS}, and is expressed as

$$V_{GS} \geq V_t$$

$$V_{DS} \geq V_{GS} - V_t$$

$$I_{DS} = \tfrac{\beta}{2} (V_{GS} - V_t)^2 . \tag{1.3}$$

The conductance β is given by

$$\beta = \frac{W}{L} \frac{\varepsilon_{OX} \mu}{t_{OX}} = \frac{W}{L} \beta_0 = \frac{W}{L} \mu C_{OX}, \tag{1.4}$$

where β_0 is the conduction factor, ε_{ox} is the permittivity of the gate oxide, t_{OX} is the thickness of the gate oxide, μ is the average surface mobility of carriers

(μ_n in the case of electrons in nMOST, and μ_p in the case of holes in pMOST), C_{OX} is the gate capacitance per unit area, L is the channel length between the n^+ source and drain edges, and W is the channel width perpendicular to L.

The threshold voltage V_t is given by

$$V_t = V_{t0} + \Delta V_t \left(V_{BB} \right), \tag{1.5a}$$

$$V_{t0} = V_{FB} + K\sqrt{2\psi} + 2\psi, \tag{1.5b}$$

$$\Delta V_t \left(V_{BB} \right) = K \left(\sqrt{|V_{BB}| + 2\psi} - \sqrt{2\psi} \right), \tag{1.5c}$$

$$K = \sqrt{2\varepsilon_s qN} \Big/ C_{OX}, \psi = \left(kT/q \right) \ln \left(N/n_i \right) \tag{1.5d}$$

where V_{t0} is V_t for $V_{BB} = 0$, V_{FB} is the flat-band voltage, ψ is the Fermi potential, N is the substrate doping concentration, n_i is the intrinsic carrier density of silicon, ε_s is the permittivity of silicon, q is the magnitude of electric charge, and k is the Boltzmann constant. The body-effect coefficient or substrate-bias effect constant K, which is about 0.1–$1.0 \mathrm{V}^{1/2}$, represents the sensitivity of V_t to V_{BB}. V_t decreases linearly with increasing temperature, as given by

$$V_t \left(T \right) = V_t \left(0 \right) - a \left(T - T_0 \right), \tag{1.6}$$

where $V_t(0)$ is for room temperature ($T_0 = 298\,\mathrm{K}$), and a is 0.5–$5.0\,\mathrm{mV/K}$. Note that the temperature dependence of I_{DS} tends to be canceled with reductions in β_0 and V_t as the temperature increases.

- Increases in V_t in Source-Follower Mode

When the source voltage (and thus source-substrate voltage) changes, V_t changes even at a fixed V_{BB}, as exemplified in Fig. 1.5(a). V_t increases as node A – which is the source in this case – is charged up by applying a sufficiently high voltage to the gate and V_{DD} to node B of an nMOST. Finally, V_t becomes a maximum at a source-substrate voltage of $V_{BB} + V_{DD}$, as follows:

$$V_{t\,\mathrm{max}} = V_{t0} + K \left(\sqrt{|V_{BB}| + V_{DD} + 2\psi} - \sqrt{2\psi} \right). \tag{1.7}$$

On the contrary, when node A is discharged from V_{DD} to $0\,\mathrm{V}$ by fixing node B (which is in turn the source), V_t becomes a minimum, as follows:

$$V_{t\,\mathrm{min}} = V_{t0} + K \left(\sqrt{|V_{BB}| + 2\psi} - \sqrt{2\psi} \right). \tag{1.8}$$

The difference is thus

$$V_{t\,\mathrm{max}} - V_{t\,\mathrm{min}} = K \left(\sqrt{|V_{BB}| + V_{DD} + 2\psi} - \sqrt{|V_{BB}| + 2\psi} \right). \tag{1.9}$$

For example, the difference is as large as $0.67\,\mathrm{V}$ at $V_{BB} = 0\,\mathrm{V}$ for $V_{DD} = 5\,\mathrm{V}$, $K = 0.3\,\mathrm{V}^{1/2}$ and $2\psi = 0.6\,\mathrm{V}$. If $V_{t\mathrm{min}}$ is set to be $0.5\,\mathrm{V}$, as for

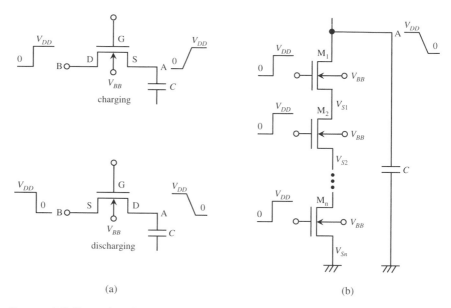

FIGURE 1.5. Examples of source-follower mode. (a) Charging and discharging of node A; (b) discharging of node A by stacked nMOSTs [1].

the usual 5-V V_{DD} designs, V_{tmax} is 1.17 V. Hence, to charge node A to V_{DD}, with elimination of the reduction in V_t, the gate voltage must be higher than 6.17 V $(= V_{DD} + V_{tmax})$, giving the gate oxide a high stress voltage. A higher $|V_{BB}|$ reduces the difference (e.g. 0.31 V at $|V_{BB}| = 3$V), enabling a lower gate voltage (e.g. 5.81 V). DRAM cells provide an example of this, as discussed later.

The discharging of node A by stacked MOSTs, shown in Fig. 1.5(b), is another example. During discharging, the V_t values of all nMOSTs except M_n rise due to their instantaneous source voltages. In particular, the V_t of the upper MOST M_1 is maximized at the beginning of discharging, because $V_{S1} > V_{S2} > \cdots > V_{Sn}$. The raised V_t eventually limits the number of stacked nMOSTs for high speed. A NAND gate provides an example of this.

• Channel Length Modulation

The I_{DS} equation at saturation is derived by simply assuming that I_{DS} at saturation does not change when V_{DS} increases, resulting in a zero slope of the I_{DS} versus V_{DS} characteristics. In the actual MOST characteristics shown in Fig. 1.6, however, I_{DS} increases slightly with V_{DS} even in the saturation region, which can be accounted for by empirically multiplying the voltage-dependent term in (1.3) by the factor $1 + \lambda V_{DS}$ [1, 4]; i.e.

$$I_{DS} = \frac{\beta}{2}(V_{GS} - V_t)^2(1 + \lambda V_{DS}) \tag{1.10}$$

where λ is the channel length modulation parameter $(= 0.1\text{--}0.01\text{V}^{-1})$.

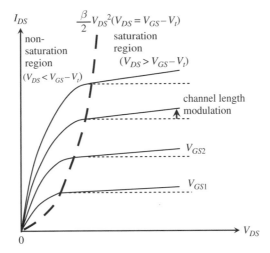

FIGURE 1.6. Actual I_{DS}-V_{DS} characteristics [1].

• Small-Size Effects

The dc equations described above are for a sufficiently large MOST, since they are based on one-dimensional analysis. However, if the channel is short and/or narrow, behavior of actual MOSTs differs from that predicted by the equations. A typical example is the short-channel effects that arise from a two-dimensional potential distribution and high electric fields in the channel region. An excessively short channel of excessively high drain voltage even causes a punch-through, in which the depletion region of the drain junction is punched through to the depletion region of the source junction. The lowering of V_t in a shorter channel, shown in Fig. 1.7(a) [1], is a result of short-channel effects. On the contrary, the increase in V_t in a narrower channel, shown in Fig. 1.7(b),

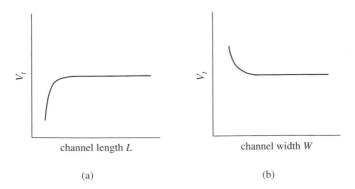

FIGURE 1.7. Small size effects in a MOST [1]. (a) The short-channel effect; (b) the narrow-channel effect.

is a small-size effect, that is related to the depletion region spreading laterally in the substrate along the channel width [1]. For a given size tolerance, such shifts in V_t make the V_t variation more prominent as MOSTs are miniaturized. Therefore, MOST sizes appropriate to circuit operations must be chosen: the use of an excessively short channel must be avoided for differential circuits such as sense amplifiers and comparators, in order to reduce the V_t mismatch (i.e. offset) between paired MOSTs. If the setting of the nominal L and W is changed without changing the W/L ratio, the resultant I_{DS} characteristics can be changed, since the short-channel effects are changed.

- MOS Capacitor

The capacitor components of an E-MOST are the overlap capacitance C_{OV} (= $C_{OX}l_{OV}W$) between the gate and the n^+ regions at the source and drain edges, the gate-source capacitance C_{GS}, the gate-drain capacitance C_{GD}, and the gate-substrate capacitance C_{GB} [4], as shown in Fig. 1.8(a). Figure 1.8(b) shows the gate capacitance C_G versus V_{GS}, excluding the relatively negligible C_{OV}. Here, C_G is the sum of C_{GS}, C_{DG}, and C_{GB}. At $V_{SG} = 0$, C_G is equal to C_{GB}, which

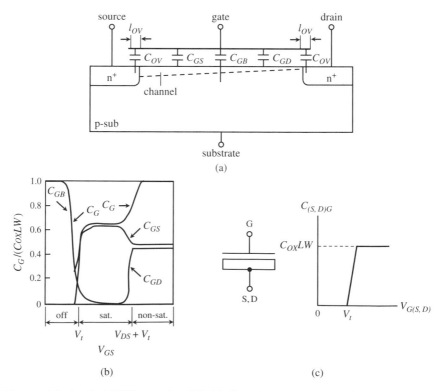

FIGURE 1.8. An E-nMOST capacitor [1]. (a) Capacitor components; (b) C_G versus V_{GS}; (c) the drain-source connected MOS capacitor.

is the gate-oxide capacitance $C_{OX}LW$, because no channel is formed. When V_{GS} is increased, a series connection of the gate capacitance and the depletion-layer capacitance, formed at the silicon surface beneath the gate, is established. Thus C_G is decreased as the depletion capacitance is decreased with V_{GS}. When V_{GS} exceeds V_t, however, C_G increases. In this region C_{GB} becomes zero, because a channel (i.e. an inversion layer or an n-type thin conduction layer) is formed and works as a shield. Instead, C_{GS} appears as the result of a channel formed at the source. In the non-saturation region at $V_{GS} > V_{DS} + V_t$, the channel spreads fully between the source and the drain. Hence, the gate-oxide capacitance, which equals C_G, is equally distributed between C_{GS} and C_{GD}.

When the source (S) and drain (D) are connected to each other, a large capacitance is formed between the S-D terminal and the gate when V_{GS} exceeds V_t, as shown in Fig. 1.8(c).

1.2.2. Bulk MOSTs

In the nMOS DRAM era, in the 1970's and early 1980's, a p-type substrate (Fig. 1.1) with a resistivity of about $10\,\Omega \cdot cm$ was used. A substrate bias voltage V_{BB} of -2 to $-3\,V$ was supplied to the substrate to ensure stable operations and isolation between nMOS memory cells. In the 1980's this was replaced by a double-well (or tub) CMOS structure, shown in Fig. 1.9(a) [1]. In principle, the p-well is unnecessary, because a p-type substrate is used. The double well, however, allows nMOS parameters such as V_t to be controlled by adjusting only the p-well dose concentration. In order to avoid a forward-biasing of the p-n junction, the p-well and p-substrate are back-biased at a voltage lower than any nMOST source voltage, while the n-well is back-biased at another voltage higher than any pMOST source voltage. Since the lowest source voltage is V_{SS} (0 V), and the highest pMOST source voltage is V_{DD}, the p-well voltage is 0 V or a negative voltage, and the n-well voltage is V_{DD} or higher voltage. Recently, a triple-well structure (Fig. 1.9(b)) has been used to protect a memory-cell array from minority carrier injections or to enable ultra-low-voltage operations

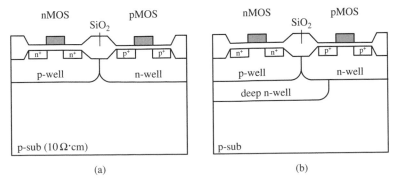

FIGURE 1.9. A double-well (a) and triple-well (b) structures [1].

with reduction of subthreshold currents. Dynamic control of the substrate and gate-source self-backbiasing, discussed in Chapter 4, both need the triple well structure.

CMOS latch-up is a prime concern because it occurs in CMOS structures [1], in which a parasitic thyristor consisting of parasitic npn and pnp bipolar transistors is easily formed. As soon as the thyristor is triggered by a noise spike, it is turned on by positive feedback. As a result, a large current flows from V_{DD} to ground, and melts junctions and metal wiring throughout the chip. For example, in the structure forming an n-well on to a p-substrate shown in Fig. 1.10 [1], an npnp thyristor composed of an nMOS source, a p-substrate, an n-well and a pMOS source is developed. Here, R_W and R_S are the n-well resistance and substrate resistance, respectively. These resistors are represented by lumped devices in the figure, although they are actually distributed. The resultant circuit features positive feedback, and thus it causes latch-up when either of the two transistors is turned on. When a positive noise is input to the Q_1-base, Q_1 is turned on. The resultant collector current creates a voltage drop across R_W, so that the base-emitter of Q_2 is forward-biased. When the bias exceeds a certain voltage, Q_2 is turned on, and the collector current makes the Q_1-base voltage increase further. Thus, the current grows if the product of the current gains of both transistors is larger than unity. Latch-up can be triggered by transient noises, which forward-bias the pn junction, such as local voltage variations in the substrate and wells, caused by capacitive couplings, ringing waveforms on the signal lines, overshoots exceeding V_{DD} and undershoots exceeding V_{SS} to gate protection diodes at the input and output pins, and noises coupled to power supply lines. In order to suppress latch-up, the current gain of parasitic bipolar transistors must be reduced, which is realized by deepening the wells for the vertical transistor (Q_2) and isolating MOSTs from the well edges as much as possible for the lateral transistor (Q_1). Reductions of the parasitic resistances R_W and R_S are also effective for suppression. They are achieved by the use of guard rings (Fig. 1.10), a V_{SS}-supplied p$^+$ guard ring surrounding an nMOST, and a V_{DD}-supplied n$^+$ guard ring surrounding a pMOST. The guard rings also lower the current gains, because they capture minority carriers injected before

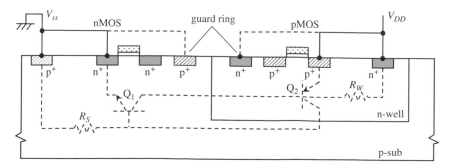

FIGURE 1.10. Parasitic bipolar transistors in a CMOS structure [1].

they can reach the bases. Latch-up immunity and the usefulness of guard rings have been reported as follows [5]. Latch-up occurs more easily with a higher temperature and a wider pulse width of noise. For example, for an npn-parasitic transistor, the latch-up starting base-voltage was lowered from 1.2 V at 25 °C and a 10 ns pulse width to 0.4 V at 125 °C and a 50 ns pulse width. An n^+ guard ring improved the latch-up immunity from 0.4 V to 1.2 V. An additional p^+ guard ring brought about a further improvement, to 2 V.

Guard rings are not so effective for the parasitic transistors formed in the deeper regions of substrate and wells, since they lower only the surface resistances. Therefore, increases in the doping concentrations of the substrate and the wells further improve latch-up immunity with lowered R_S and R_W. Excessive doping, however, increases junction capacitances, degrades the mobilities of MOSTs, and raises V_t. An epitaxial wafer that drastically improves immunity solves the above problems, since it consists of a thin silicon layer with an appropriate doping concentration, grown on to quite a high doping concentration wafer.

1.2.3. SOI MOSTs

Figure 1.11(a) shows the conventional bulk nMOST discussed thus far. There is another type of MOSTs, that is, the SOI (Silicon-On-Insulator) MOST which is isolated from the silicon substrate by a buried oxide (BOX) layer, thus enabling to reduce the junction capacitance of the source and drain. The SOI MOST is categorized as the partially-depleted (PD) SOI MOST and the fully-depleted (FD) SOI MOST, as shown in Figs. 1.11(b) and 1.11(c), respectively. PD-SOI has almost the same MOST size as bulk CMOS, which enables a mask compatible design with the bulk. For example, a PD SOI has been reported to cut logic-gate speed of the bulk CMOS by 25–30% [1, 3]. It also increased the signal voltage of DRAM cells by 25%, and improved the access time of a 64-Kb DRAM test chip by about 35% [1]. Less capacitance in the SOI body favors dynamic V_t control, which is attained by body voltage control. This dynamic V_t control enables unique low-voltage logic circuits, sense amplifiers, and memory cells [1]. Nevertheless, some problems still remain unsolved. In addition to the costly wafer preparation for a thick BOX layer, the floating body effect of a PD-SOI MOST is problematic for some applications. For example, if applied to the transfer MOST in a DRAM cell, the instability of the floating body

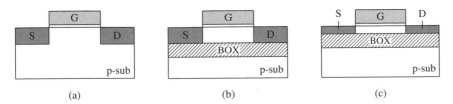

FIGURE 1.11. nMOST structures for bulk (a), PD-SOI (b), and FD-SOI (c).

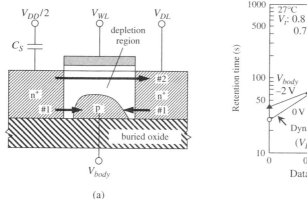
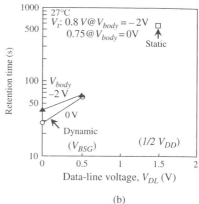

(a)   (b)

FIGURE 1.12. The data-retention characteristics of a SOI DRAM cell [1, 42]. The V_{DL} of 1.5 V (= 1/2 V_{DD}) is for the static mode, while the V_{DL} values of 0 V and 0.5 V are for the dynamic mode and the BSG scheme, respectively.

potential degrades the data-retention characteristics and soft-error immunity, as summarized in [1]. Figure 1.12 shows the degradation mechanism for the data-retention characteristics [42]. The data-retention time is defined as the time until the cell voltage stored by the write operation decays to almost half V_{DD}. This decay stems from the stored charges (holes) lost by pn junction leakage (#1) at the cell storage node. The resulting accumulated holes at the body raise the body potential due to its small capacitance, causing reduced V_t of the cell MOST, which is the origin of the subthreshold leakage current (#2). In the static mode, an SOI DRAM cell achieves a superior data-retention time of 550 s at 27 °C, which is six times longer than that of a bulk memory cell [42]. In this mode, the cell MOST is completely cut off despite the reduced V_t, because the data line remains at a high level of a half V_{DD} (the precharge voltage of the data line), and the word line is maintained at a low level of 0 V during retention. Thus, the long retention time comes from the small area of the pn junction. Unfortunately, the dynamic mode that keeps the data-line voltage at 0 V shortens the data-retention time to 42 s, as shown in the figure. This results from the subthreshold current caused by the reduced V_t. The boosted sense-ground (BSG) scheme described in Chapter 2, which raises the lowest data-line voltage from 0 V to 0.5 V, improves the characteristics to some extent. The floating body also degrades the soft-error immunity of MOSTs, not only in DRAM and SRAM cells, but also in peripheral circuits. For example, after an alpha-particle incidence for the floating body, electrons generated in the floating body diffuse to the source and drain, while holes remain in the floating body region and raise the potential. This potential increase causes a large continuous subthreshold current. The introduction of body contacts [1] to suppress body-potential change is very effective in peripheral circuits. For memory cells, however, it usually increases the memory cell size, especially in DRAMs.

FD-SOI MOSTs have an extremely-thin lightly-doped channel so the MOSTs are fully depleted. Typically, the ratio of the channel length to the channel thickness is more than 3, and an extremely thin silicon channel is thus required. It reduces V_t-variations with fewer short-channel effects and dopant-atom fluctuations in the channel (see Chapters 2–5). Moreover, it reduces soft-error rate because in addition to a small number of collected charges due to an extremely thin silicon layer, most of electron-hole pairs generated by alpha-particle or cosmic-ray irradiations are blocked by the BOX and do not affect the silicon layers (see Fig. 2.27(c)) [49]. It also significantly reduces pn-junction current due to a small junction area (Fig. 2.27(b). Subthreshold current may be reduced for some structures due to a small subthreshold swing (a small S-factor) [1]. Therefore, FD-SOI devices are expected to be suitable most for ultra-low-voltage LSIs, especially for DRAMs, as discussed in Chapter 2. Many FD-SOI structures to increase the drain current have been proposed to date [7], although the V_t must be adjusted by using various gate-materials with different work functions [6], since the poly-silicon gate causes a depleted MOST. Of these structures, the fin FET has been actively developed because it is reportedly the simplest structure to implement [8]. A challenge, however, is to realize multi-V_t without using multi-gate materials from the view point of circuit design. Otherwise circuit design flexibility is lost.

1.2.4. Resistors

Resistors have been widely used for voltage conversions and for eliminating floating nodes [1]. Resistances as high as $100\,k\Omega - 1\,M\Omega$ are needed to suppress the chip standby current to $10\,\mu A$–$100\,\mu A$. They are usually made of poly-silicon because it has no voltage or temperature dependences, and because of ease of use, despite large variations in resistance during volume production. Here, let us evaluate the length of a poly-silicon line with a width of $1\mu m$ and a sheet resistance of $100\,\Omega$/square needed to make a $1\,M\Omega$ resistor. The sheet resistance ρ_s, which is defined by the resistance (Ω/square, or $\Omega/\square$) of a square with thickness t_{int} (Fig. 1.13), is related to the resistivity ρ ($\Omega \cdot cm$) [1] as follows:

$$\rho_S = \rho / t_{int}. \qquad (1.11)$$

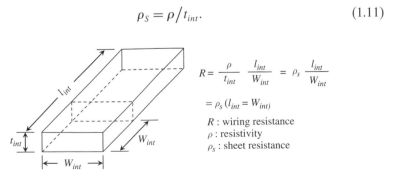

$$R = \frac{\rho}{t_{int}} \frac{l_{int}}{W_{int}} = \rho_S \frac{l_{int}}{W_{int}}$$

$$= \rho_S \ (l_{int} = W_{int})$$

R : wiring resistance
ρ : resistivity
ρ_s : sheet resistance

FIGURE 1.13. Interconnect resistance. Reproduced from [42] with permission; © 2006 IEEE.

The resistance R of a poly-silicon line with a width of W_{int} and a length of l_{int} is

$$R = \rho_S \left(l_{int} / W_{int} \right). \tag{1.12}$$

The length can thus be as long as 10 mm, which is still practical, since the necessary area becomes small relative to the ever-increasing chip size.

Diffused n and p layers could be used for resistors, although they have voltage and temperature dependences and need a large area due to small sheet resistances. A combination of an n-well resistor and a poly-silicon resistor has been proposed for a refresh timer [1]. Different temperature dependences of resistance between the two – large ($+0.6\%/°C$) for the n-well and almost zero for the poly-silicon – enables detection of the junction temperature and accordingly variation of the refresh interval of DRAMs.

1.3. Leakage Currents

The leakage current issue is becoming increasingly important with device and voltage scaling, because it is closely related to data retention characteristics of memory cells, and standby and active currents of chips. The six kinds of leakage currents must be taken into consideration for designing nano-scale LSIs. They are subthreshold current, GIDL (gate-induced drain leakage) current, gate-oxide-tunneling current, substrate current, and forward bias and reverse-biased pn-junction currents. Subthreshold current and GIDL current are developed while the gate is off (i.e., $V_{GS} = 0$). The subthreshold current is a channel-leakage current that flows from the drain to the source due to a weak inversion layer in the channel region. GIDL current, which is very sensitive to the drain-gate voltage (V_{GD}), flows from the drain to the substrate, exemplified by a reduction to one-tenth with reducing V_{DD} from 1.5 V to 1.0 V for 1.3-μm nMOST. On the other hand, gate-oxide-tunneling current and substrate current are developed while the gate is on (i.e., $V_{GS} > V_t$). The forward bias and reverse-biased pn-junction currents always flow as dc currents. Of these currents, the following four leakage currents are significant for low-voltage nanometer LSIs, although GIDL and pn junction currents can shorten the refresh time of DRAM cells and increase the retention current of SRAM cells, despite being small compared with other currents.

1.3.1. Subthreshold Current

The high-speed operations of low-voltage CMOS circuits necessitate reducing V_t, because speed is roughly inversely proportional to $V_{DD} - V_t$. However, when V_t becomes small enough to no longer cut off the MOST, a MOST subthreshold dc current is developed, which increases exponentially with decreasing V_t, as discussed below. To evaluate the current caused by V_t scaling, the definition of V_t must be clarified. There are two kinds of V_t[1] as shown in Fig. 1.14:

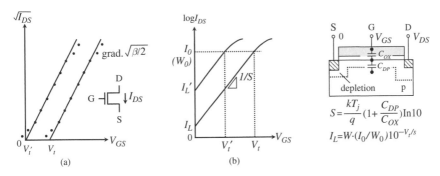

FIGURE 1.14. The definition of V_t [1]. (a) for extrapolation; (b) for constant current.

the extrapolated V_t that is familiar to circuit designers, and constant-current V_t. The extrapolated V_t is defined by extrapolating the saturation current on the $\sqrt{I_{DS}} - V_{GS}$ plane, and by neglecting the tailing current actually developed at approximately V_t. Our major concern is with the subthreshold current, which is developed at $V_{GS} = 0\,\text{V}$. If V_t is high enough, the subthreshold current is negligible. With decreasing V_t, however, the subthreshold current starts to be developed at a V_t higher than expected. This current is not expressed in the definition. Thus, the constant-current V_t is indispensable in evaluating the current. The V_t is defined as a V_{GS} for a given current density on the log I_{DS}-V_{GS} plane. This constant-current V_t is empirically estimated to be smaller than the extrapolated V_t by about 0.2 V to 0.3 V for a current density of $1\,\text{nA}/\mu\text{m}$.

The subthreshold leakage current [1], I_L, is given by

$$I_L = W \cdot \frac{I_0}{W_0} \cdot 10^{-V_t/S} \tag{1.13}$$

$$S \cong \frac{kT_j}{q} \cdot \left(1 + \frac{C_{DP}}{C_{OX}}\right) \cdot \ln 10 \tag{1.14}$$

where W is the gate (channel) width of the MOST, I_0/W_0 is the current density to define V_t, S is the subthreshold slope(or swing), C_{OX} is the gate capacitance, C_{DP} is the depletion-layer capacitance, and T_j is the junction temperature. Obviously, the subthreshold current is sensitive to V_t, S, and T_j, and W.

The subthreshold current degrades data-retention characteristics of RAM cells and increases standby/active currents of peripheral logic circuits in RAMs, as discussed in Chapters 2 and 3. There are several ways to reduce the subthreshold current, as summarized in the following, although the details are explained in Chapter 4.

(1) Increase in V_t. The subthreshold current is quite sensitive V_t. For example, it deceases by one decade with a V_t increase of only 100 mV at 100 °C. Thus, even various reverse-biasing schemes can manage such a small increase in V_t.

(2) Reduction of T_j. This reduces the current with increased V_t and decreased S. Although liquid-nitrogen temperature $(-196\,°C)$ operation [1] has been proposed, it is not suitable for general-purpose CMOS LSIs.

(3) Reduction of the total channel width of low-V_t MOSTs in a chip. In particular, it is crucial for pMOSTs since the subthreshold current of pMOSTs in a usual CMOS logic circuit is larger than that of nMOSTs due to a larger W and a larger S in some devices (see Fig. 4.18).

(4) Reduction of S with new device structures. Unfortunately, however, S is not sensitive to MOST structures. Even the fully-depleted SOI device with reduced C_{DP}/C_{OX} reduces S from about 80mV/decade (dec.) of bulk MOSTs to only about 60mV/dec. at room temperature. It never reduces S to less than 60 mV/dec., because S is always larger than $(kT_j/q)\ln 10 (\cong 60\text{mV}/\text{dec.})$, independent of the values of C_{DP} and C_{OX}, as seen in (1.14).

1.3.2. Gate-Tunneling Current

Reduction of the gate tunneling current is urgent because the gate-oxide thickness (t_{OX}) has been rapidly decreasing to less than 2–3 nm [3], the point where the gate current becomes prominent. The gate tunneling current flows from the gate to the source for nMOST, or from the source to the gate for pMOST, via the channel. Figure 1.15 shows the characteristics [9]. The gate current of nMOST in inversion mode is a major source, which is usually 4 to 10 times larger than that of pMOST for the same size. The gate current is quite sensitive to t_{OX}, while it is less sensitive to the junction temperature (T_j) and gate voltage (V_{GS}). For example, a leakage reduction of one order of magnitude requires a t_{OX} increment of only 2–3 Å, while it requires a V_{GS} reduction of as much as 0.5 V in a low-V_{DD} region, as shown in the figure. Such a large V_{DD} reduction in the sub-1-V

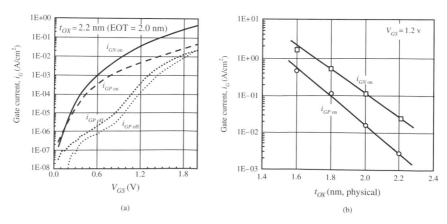

(a)

(b)

FIGURE 1.15. Characteristics of gate tunneling current (i_G) [9]. i_{GN} and i_{GP} are for nMOST and pMOST, respectively.

region is risky in terms of stable operations. Thus, this kind of reduction is the intended responsibility of the process and device designers, and developments of new high-k gate insulator materials are thus in progress. Even so, a few circuit schemes to reduce the leakage have been proposed to date: A thick-t_{OX} high-V_t power switch [10] to shut off the leakage path of the internal thin-t_{OX} core in the standby mode (although it is ineffective for the active mode); reduction of the supply-voltage during standby mode [11, 12](although it is not effective because the leakage is less sensitive to the voltage); circuit reconfiguration and sleep-state assignment techniques for sleep mode [53] (although they involve speed and area penalties); and dynamic control of the gate voltage with a floating gate [9]. Of these schemes, dynamic control [9] is attractive in terms of circuit designs despite an area penalty, since it is applicable to both active and standby modes.

Figure 1.16 shows a gate-leakage-suppressed word driver [9]. For the conventional circuit (a), the word line is fixed to 0 V by turning on the driver M_{n1} during inactive periods. Thus, a large gate tunneling current i_G continues to flow due to a large M_{n1}-channel width that is necessary to quickly drive the heavy word line. For the proposed scheme (b), at word-off timing, N_0 goes to "L", and M_{n0} is off, while N_1 goes to "H", M_{p1} is off, and N_2 goes to "H" because M_{p2} is still on. Thus, the word line is quickly discharged by a large M_{n1}, and M_{p2} is then turned off. Therefore, N_2 is left to a floating "H", so N_2 continues to be discharged by i_G until the resultant reduced V_G finally cuts i_G. After that, the word line is fixed to 0 V with a small level-holder M_{n2}, instead of the large M_{n1}, thus enabling a small i_G.

1.3.3. Substrate Current

As a MOST is scaled down for a fixed V_{DD} the electric field near the drain strengthens. Consequently, electrons flowing from the source to the drain obtain a high energy from the high electric field (so-called hot electrons), and generate electron-hole pairs as a result of impact ionization at the drain, as shown in Fig. 1.17. Some of the generated electrons flow into the drain. The others are

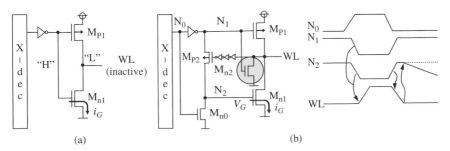

(a)   (b)

FIGURE 1.16. Conventional word driver (a) and gate-tunneling-current-suppressed word driver (b)[9].

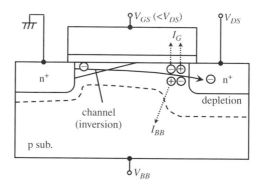

FIGURE 1.17. The mechanism of hot-carrier injection [1].

injected into the gate insulator as gate current (I_G) and trapped there, causing a gradual change of V_t and a decrease in the transconductance of the MOST. On the other hand, some of the generated holes flow into the substrate, resulting in a substrate current (I_{BB}). The I_{BB} of an nMOST is 3 orders larger than that of a pMOST because of a larger impact ionization coefficient and a higher electric field that results from a sharper impurity profile near the drain. Thus, the degradation of nMOST device-parameters is more prominent. The maximum I_{BB}, I_{BBmax}, which is usually developed at $V_{GS} = V_{DS}/2$, is expressed as

$$I_{BB\max} \propto \exp\left(-\gamma/V_{DS}\right),\qquad(1.15)$$

where γ is a constant. The life time (τ_{HC}) of an nMOST, which is defined as the time when V_t degrades by 100mV due to hot electrons, is given by

$$\tau_{HC} \propto \left(I_{BB\max}/W\right)^{-n}/(f \cdot t_{SUB}),\ n = 2.5 - 3.0,\qquad(1.16)$$

where W is the channel width, t_{SUB} is the pulse width of an I_{BB} pulse with amplitude of I_{BBmax}, and f is the pulse frequency. Obviously, the most efficient way to reduce I_{BB} is to lower V_{DS} (i.e., V_{DD}) because I_{BBmax} exponentially reduces with reducing V_{DS}. One-order reduction of I_{BBmax} extends the MOST life time by 3 orders.

1.3.4. pn-Junction Current

The forward body-biasing of MOST bulk-to-drain or bulk-to-source diodes [13, 50–52] is becoming a major concern with device and voltage scaling. It is developed when substrate (well or body) control circuits such as dynamic V_t and gate-well connection circuits are used to achieve high speed or compensate for variations in speed and leakage caused by design parameter variations. However, MOST characteristics during forward body-biasing must carefully be investigated because forward body-biasing always entails an excessively large pn-junction current at high temperature. Figure 1.18 shows a schematic cross

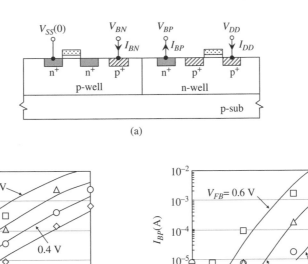

FIGURE 1.18. Forward pn-junction current at $V_{DD} = 1.2$V [13]. Solid line, calculated; Markers, experimental.

section of devices on a 0.13-μm CMOS test chip [13] composed of four 32-bit ALU units (8,000 gates in total), and an experimental dc power-supply current (I_{DD}) and dc n-well current (I_{BP}) for pMOSTs on the chip. The solid lines are for calculations. The same forward bias (V_{FB}) is simultaneously applied to both n- and p-wells with a fixed $V_{DD} = 1.2$ V. The currents increase exponentially with temperature due to the parasitic bipolar transistors. A large drain current is not acceptable for handheld equipment, and a large well current burdens on-chip substrate (i.e. well) generators with excessive pumping currents. For example, for a given well current of 10 μA, the bias must be reduced to 0.4 V at 100 °C. For a drain current $I_{DD} < 10$μA, it must be as shallow as around 0.1 V at 100 °C, as suggested by an extrapolation of the data in the figure. This implies a strong need for compensations for temperature variations.

1.4. Basics of CMOS Digital Circuits

The key issues of LSIs are low power, high speed, ease of design, and a wide voltage margin. Low power, especially for inactive circuits, is crucial for larger-scale integration. This is because almost all circuits in CMOS LSIs are inactive, while only a limited number of circuits are active, thus calling for an extremely low power for inactive circuits. CMOS circuits just meet the requirement. Without CMOS circuits, no LSI chip could have been designed successfully, due to the ever-increasing power with large-scale integration. In this section, basic circuits such as a CMOS inverter and NOR/NAND gates are described. Special circuits, which are between digital and analog circuits, such as a cross-coupled CMOS sense amplifier, level shifter, charge pump, and ring oscillator are also discussed.

1.4.1. CMOS Inverter

Figure 1.19 shows a basic CMOS inverter and its schematic operations [1]. An nMOST discharges the load capacitance quickly and completely, while it charges up slowly, entailing a reduction in V_t. On the other hand, a pMOS charges up quickly without a reduction in V_t, while it discharges slowly, which does entail a reduction in V_t. The CMOS circuit combines the advantages of the two kinds of MOST, enabling the fast charging up by a pMOST and fast discharging by an nMOST, with a full swing of V_{DD}. It features extremely low power (relative to an all nMOST design, for example), since only one of the two devices is conductive, depending on the input voltage, without any floating or boosted node. The simple circuit configuration has another advantage of having no substrate bias effects, because the MOSTs' source-substrate voltages are always fixed. However, the low-power advantage of CMOS circuits is diminished in the ultra-low voltage nano-scale era, where V_{DD} and thus V_t are lowered and subthreshold currents rapidly increase. The threshold of the circuit (V_{TC}) [1], at which the output is in a critical condition at a high or low level, is expressed as

$$V_{TC} = \left(V_{DD} - |V_{tp}| + V_{tn}\sqrt{\beta_R} \right) \Big/ \left(1 + \sqrt{\beta_R} \right), \tag{1.17}$$

$$\beta_R = \beta_n/\beta_p, \beta_n = (W_n/L_n)\mu_n C_{OX}, \beta_p = (W_p/L_p)\mu_p C_{OX}$$

where β_n and β_p are the conductance of nMOST and pMOST, as shown in (1.4). $V_{TC} = V_{DD}/2$ for $\beta_R = 1$ and $|V_{tp}| = V_{tn}$. V_{TC} becomes lower with a larger V_{tp}, while it becomes higher with a larger V_{tn}.

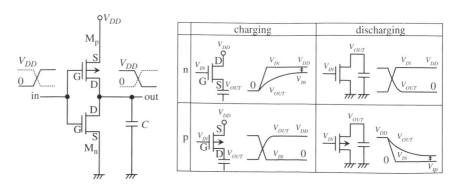

FIGURE 1.19. The CMOS inverter [1].

1.4.2. NOR and NAND Gates

Figure 1.20 shows static and dynamic NOR and NAND gates. The static gates (a, b) are fast due to no additional clock despite a large area, while the dynamic gates (c, d) are small despite a slow cycle due to the need for a precharge clock ϕ_p. In addition, in general, a static gate is suitable for low voltage operations more than a dynamic gate because it does not ever have a floating node. In particular, the operation of the dynamic NOR gate(c) is quite sensitive to V_t. When all inputs are at a low level, the precharged floating-high level at the output tends to be discharged by subthreshold currents from many OR'd nMOSTs if V_t is low, requiring a level keeper (e) [14]. A dynamic NAND gate is relatively immune to subthreshold current due to the stacking effects discussed in Chapter 4. Due to the low power advantage, the NAND gate has been used for address decoders in CMOS memories [1] instead of NOR gates, which were commom in the nMOS era. Discharging of only one of many decoder outputs is responsible for the low power.

1.4.3. Cross-Coupled CMOS Sense Amplifier

The CMOS sense amplifier shown in Fig. 1.21(a) [1] has been used exclusively in modern CMOS DRAMs because of its simple circuit configuration and thus small

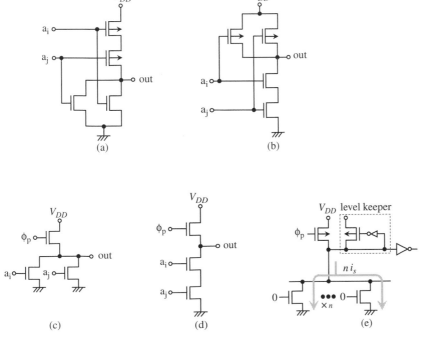

FIGURE 1.20. Static NOR (a) and NAND (b) inverters, dynamic NOR (c) and NAND (d) inverters, and dynamic NOR with level keeper (e).

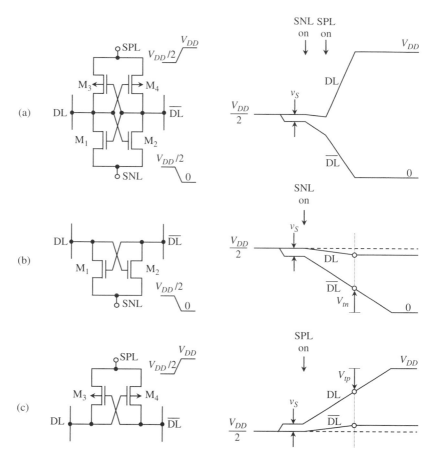

FIGURE 1.21. Operations of CMOS sense amplifier (a), nMOS sense amplifier (b), and pMOS sense amplifier (c). V_{tn} and V_{tp} are threshold voltages of nMOST and pMOST, respectively.

size, low power, and suitability for sensing and rewrite (restoring) operation. It consists of cascaded nMOS and pMOS cross-coupled circuits, each of which works as a sense amplifier. A small signal voltage (v_S) developed on a pair of data lines (DL, $\overline{DL}$) at a floating $V_{DD}/2$ is amplified to a full V_{DD} by the sense amplifier. The amplified voltage is utilized as the rewrite voltage to the DRAM cell. The amplification to V_{DD} results from successive activations of the nMOS amplifier and the pMOS amplifier. Note that a simultaneous activation may cause an increased dc current and an unreliable amplification caused by both offset voltages of the nMOST and pMOS sense amplifiers. The operation becomes more comprehensive if each circuit is explained separately. For example, the nMOS amplifier (Fig. 1.21(b)) amplifies the small signal v_S to $V_{DD}/2$ by activating SNL, as a result of discharging one data line ($\overline{DL}$) to ground with turned-on MOST M_2, and holding another data line (DL) at $V_{DD}/2$ with turned-off MOST M_1. To be more precise, the resultant amplified signal voltage is smaller than $V_{DD}/2$ if V_t is low, because M_1 continues to slightly turn on and to gradually discharge the data line (DL) that should remain floating until $\overline{DL}$ reaches the nMOST's $V_t(V_{tn})$. Similarly the nMOS amplifier, the pMOS amplifier (Fig. 1.21(c)) amplifies the signal voltage to almost $V_{DD}/2$ by activating SPL, as a result of charging up one data line (DL) to V_{DD} and holding another data line ($\overline{DL}$) at $V_{DD}/2$. Thus, a combination of both amplifiers results in a full-V_{DD} amplification, as shown in Fig. 1.21(a).

1.4.4. Level Shifter

A level shifter circuit has been indispensable for low-voltage LSIs in which many internal supply voltages are used. Figure 1.22 shows a typical level shifter [1] to shift from a low-level (V_{DD}) pulse to a high-level (V_{DH} or V_{PP}) pulse. Since either of the cross-coupled pMOSTs may be turned on, depending on the input, there is no ratioed operation with an nMOST. V_{DH} (or V_{PP}) is generated by the charge pump.

1.4.5. Charge Pump

Quasi-dc power supply voltages generated by voltage up-converters are indispensable in low-voltage LSIs that need high-speed, well-controlled raised pulses, as discussed later. The basic idea for the power-supply generation is to use a charge pump with MOS capacitors. Figure 1.23 shows the concept of the charge pump [1]. Nodes N_1 and N_2 are charged to $V_{DD} - V_t$ in the quiescent state. The application of a pulse to P increases N_1 to $(1 + \alpha) V_{DD} - V_t$ where α is the boost ratio, which is determined by C_P and the N_1 parasitic capacitance. Thus, M_0 is turned on and N_1 is charged up. This implies that part of the charge

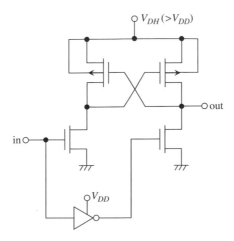

FIGURE 1.22. A level shifter [1].

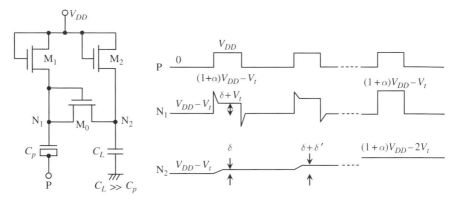

FIGURE 1.23. The charge pump [1].

(i.e. $C_P V_{DD}$) injected to N_1 by C_P is transferred to N_2. However, the charging stops when the voltage difference between N_1 and N_2 reaches V_t. The resultant voltage change δ at N_2 is small because, usually, $C_L \gg C_P$. When the pulse is turned off, N_1 instantaneously drops to below $V_{DD} - V_t$, and then it is charged up again to $V_{DD} - V_t$ by M_1. In this manner, successive pulse applications to P allow N_2 to continue to be charged and to develop a higher voltage. When N_2 reaches $(1 + \alpha) V_{DD} - 2V_t$, however, N_2 stops being charged, thereafter enabling to hold a stable raised voltage on C_L. If there is a pulse load current at C_L, the increased voltage would be degraded. In this case, M_0 would again continue to charge up C_L until the resultant voltage reached a stable level. The stable level is 7.5 V for $V_{DD} = 5$ V, $V_t = 0.5$ V, and $\alpha = 0.8$. The time necessary to reach the stable voltage depends on C_P, C_L, and the pulse frequency. The pulse can be generated by a ring oscillator.

1.4.6. Ring Oscillator

Figure 1.24 shows a ring oscillator consisting of a odd number of inverter stages [1]. It is used for charge-pumping in a voltage up-converter and in substrate-bias generators. In addition, it is indispensable in the control of the DRAM refresh time, via μs-ms interval pulses that are generated with the help of counters. The oscillation frequency f_{OSC} [1] is given by $[(2n + 1)(t_H + t_L)]^{-1}$, where $2n + 1$ is the number of inverter stages, and t_H and t_L are the delay times of each inverter for a high-input and low-input pulse, respectively.

1.5. Basics of CMOS Analog Circuit

1.5.1. Analog Circuits compared with Digital Circuits

Analog circuits deal with continuous signals, while digital circuits deal with a finite set of signals, such as two voltages representing only "1" and "0". An

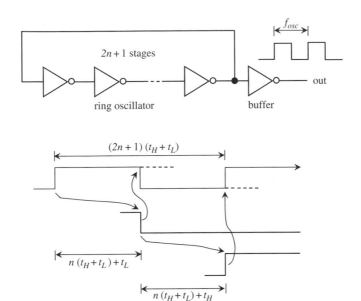

FIGURE 1.24. A ring oscillator [1].

inverter, which is a typical digital circuit, can also be used as an analog circuit (amplifier). Figure 1.25 shows a schematic diagram of an inverter composed of an nMOST driver and an nMOST load, and its input-output characteristics. The transition region B is mainly used in the analog circuit, while the saturation regions (note that this is not the same saturation region as for a single MOST) shown by A and C are used in the digital circuit. A positive DC voltage (bias voltage) V_{BIAS} is applied to the gate of M_D to use the inverter as an amplifier.

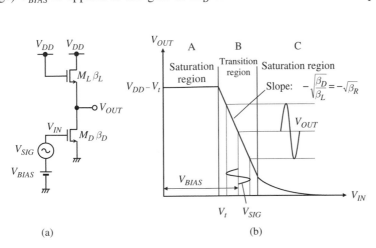

FIGURE 1.25. Inverter circuit used as an amplifier; schematic diagram (a), and transfer curve (b).

The operating point is thereby moved to the center of the transition region. If a small input signal V_{SIG} is superimposed on the bias voltage, the signal is amplified by the slope of the transition region $(-\sqrt{\beta_R})$ and appears at the output terminal, as shown later. It is convenient to use an equivalent circuit for the analysis of an analog circuit [38]. In this section, equivalent circuits and the analysis of basic analog circuits using them are described.

1.5.2. Equivalent Circuits

Large-signal and small-signal equivalent circuits are used for the analysis of analog circuits. The former is a model for large analog signals and is mainly used for DC and transient characteristics. These equations are complex and often incomprehensible, because the model must deal with wide-range voltage/current conditions. On the other hand, the small-signal equivalent circuit deals with only small signals, allowing the equations to be simple (linear) and more comprehensible. This is used for the analysis of AC characteristics, such as loop stability and PSRR (Power Supply Rejection Ratio). The analysis of basic circuits using both equivalent circuits is shown here. The analysis of more complex circuits will be described in Section 7.2.

Large-Signal Equivalent Circuit

The large-signal model of an nMOST shown in Fig. 1.26 is composed of a voltage-controlled current source, resistors, capacitors and diodes. Figure 1.26(a) is a complete model including parasitic capacitances and resistances. C_{GS}, C_{GD}, C_{GB}, C_{BD}, and C_{BS} are gate-source, gate-drain, gate-substrate, substrate-drain, and substrate-source capacitances, respectively. Note that C_{BD} and C_{BS} are added to the MOS-capacitor model shown in Fig. 1.8. Moreover, r_d and r_s are the drain electrode resistance and the source electrode resistance, respectively. Two diodes represent junctions between the drain diffusion layer and the substrate, and between the source diffusion layer and the substrate. This

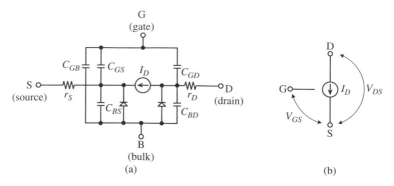

FIGURE 1.26. Large-signal model of nMOST; complete model (a), and simplified model (b).

model is mainly used for transient analysis. On the other hand, Fig. 1.26(b) shows a simplified model omitting the above-described parasitic elements. This model is used for DC analysis. The value I_D of the voltage-controlled current source is expressed by Eq. (1.1) or (1.3), as described in Section 1.2.

Let us replace the MOSTs of the circuit in Fig. 1.25 with the large signal model, and derive the relation between V_{SIG} and V_{OUT} in region B. Since all MOSTs operate in the saturation region, we have

$$V_{OUT} = (V_{DD} - V_t) - \sqrt{\beta_R}\,(V_{BIAS} - V_t) - \sqrt{\beta_R}V_{SIG}. \tag{1.18}$$

Thus, it is found that V_{OUT} is $-\sqrt{\beta_R}$ times V_{SIG}.

Small-Signal Equivalent Circuit

Figure 1.27 shows the small-signal model of an nMOST. Figure 1.27(a) is a complete model including parasitic elements, and Fig. 1.27(b) shows a simplified model omitting them. Here, g_{bd} and g_{bs} are the equivalent values of the conductance for the substrate-to-drain and substrate-to-source junctions, respectively. Since these junctions are normally reverse biased, the values are very small. The former is used for frequency-domain analysis, while the latter is used for DC or low-frequency analysis. The model includes only the small incremental portions of voltages or currents, which are obtained by differentiating the equations for the large-signal equivalent circuit at the operating point. Differentiating Eq. (1.3) for the saturation region of a MOST results in the following equation:

$$\frac{dI_D}{dV_{GS}} = \beta\,(V_{GS} - V_t). \tag{1.19}$$

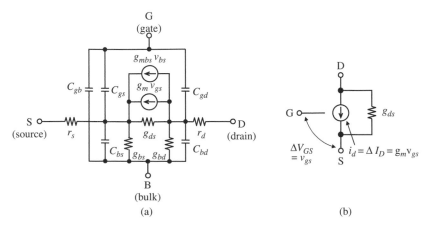

FIGURE 1.27. Small-signal model of MOST; complete model (a), and simplified model (b).

Here, substituting the equation obtained from Eq. (1.3), $V_{GS} - V_t = \sqrt{2I_D\beta}$, into Eq. (1.19) results in the following equation:

$$\frac{dI_D}{dV_{GS}} = \sqrt{2I_D\beta} \equiv g_m. \tag{1.20}$$

It means that the increment of the drain current, i_d, is proportional to the increment of gate-source voltage, v_{gs}. The proportionality constant g_m is called transconductance and is an important parameter for small-signal analysis. The drain current is expressed by the voltage-controlled current source $g_m v_{gs}$ in the model. There is another voltage-controlled current source $g_{mbs}v_{bs}$ in this figure. This is the change of the drain current by the change of the substrate-source voltage. This is introduced by the change of V_t in Eq. (1.5). However, this is often disregarded because it is smaller than $g_m v_{gs}$ in usual usage. Moreover, there is a conductance element g_{ds} in parallel with the voltage controlled current source. This is due to the channel-length modulation parameter λ, which is explained in Fig. 1.28. The drain current of a MOST increases with the drain-source voltage even if the gate-sources voltage is constant as described in Section 1.2 (Fig. 1.6 and Eq. (1.10)). This effect can be expressed by inserting a conductance g_{ds} between drain and source in the small signal model. This is derived as follows. Differentiating Eq. (1.10) by V_{DS} results in the following equation:

$$\frac{dI_D}{dV_{DS}} = \frac{\beta}{2}(V_{GS} - V_t)^2 \lambda = \frac{I_D\lambda}{1 + \lambda V_{DS}} \equiv g_{ds} \approx I_D\lambda. \tag{1.21}$$

This is called a drain conductance. It is found from the above equation that g_{ds} is a nonlinear conductance dependent on I_D. However, g_{ds} can be assumed as linear because I_D is constant as long as only small signals are considered. The value of g_{ds} is generally about $10^{-5} - 10^{-4}$ S (Siemens) at $L = 1\,\mu m$ and $I_D = 1$ mA. This value is very small compared with $10^{-2} - 10^{-3}$ S of g_m. However, it cannot be ignored in the analysis of high gain amplifier circuits, as described later.

Let us analyze the circuit in Fig. 1.25 using the small-signal model of MOSTs. Redrawing the circuit in Fig. 1.25 using this model yields Fig. 1.29. Note that

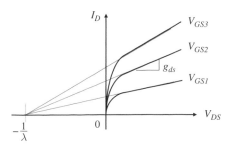

FIGURE 1.28. Drain conductance and coefficient of channel length modulation effect λ.

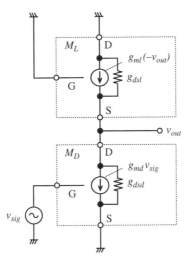

FIGURE 1.29. Small-signal equivalent circuit of inverter.

V_{DD} line is changed to a ground line. This is because a constant voltage source is equivalent to zero volts as long as only the incremental portions are considered. Deriving the relation between v_{SIG} and v_{OUT} from this equivalent circuit ignoring the drain conductance results in the following equations:

$$g_{mL}\left(-v_{OUT}\right) = g_{mD} v_{SIG} \tag{1.22}$$

$$\therefore \frac{v_{OUT}}{v_{SIG}} = -\frac{g_{mD}}{g_{mL}} = -\sqrt{\frac{\beta_D}{\beta_L}} = -\sqrt{\beta_R}. \tag{1.23}$$

This is the same as the result from Eq. (1.18). Thus, it turns out that the small signal equivalent circuit simplifies the analysis, as far as the voltage gain is concerned.

1.5.3. Basic Analog Circuits

The following basic analog circuits are extensively used in various voltage converters discussed in Chapters 6 and 7. This subsection describes the basics characteristics using the MOS models.

(1) Current Mirror Circuit

The current mirror circuit in Fig. 1.30 generates a constant current I_{OUT} proportional to the reference current I_{REF}. This circuit exploits the fact that the drain current of a MOST in the saturation region is almost constant independent of the drain-source voltage. The currents I_{REF} and I_{OUT} are given as

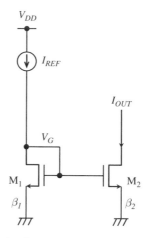

FIGURE 1.30. Current-mirror circuit.

$$I_{REF} = \frac{1}{2}\beta_1 \left(V_G - V_t\right)^2 , \qquad (1.24)$$

$$I_{OUT} = \frac{1}{2}\beta_2 \left(V_G - V_t\right)^2 , \qquad (1.25)$$

where V_G is the voltage of the common gates of M_1 and M_2. From these equations

$$I_{OUT} = \frac{\beta_2}{\beta_1} I_{REF} \qquad (1.26)$$

is obtained. Thus, I_{OUT} is the product of the β ratio of M_1 and M_2 and I_{REF}.

(2) Common-source Amplifier

Figure 1.31(a) shows a common-source amplifier composed of a driver nMOST and a constant-current load. Redrawing this circuit using the small-signal model in Fig. 1.27(b) results in Fig. 1.31(b). It should be noted that the current source I_B is not included in the equivalent circuit. This is because the impedance of a current source is infinity and no signal current flows through it. The relation between the input and output voltages of this circuit is obtained from the equivalent circuit as follows:

$$g_m v_{in} + v_{out} g_{ds} = 0, \qquad (1.27)$$

$$\therefore v_{out} = -\frac{g_m}{g_{ds}} v_{in}. \qquad (1.28)$$

Thus, the voltage gain of this circuit is g_m / g_{ds}.

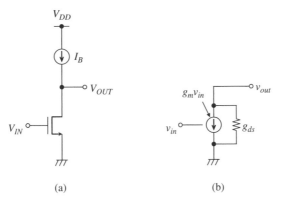

FIGURE 1.31. Common-source amplifier; circuit diagram (a), and small-signal equivalent circuit (b).

(3) Differential Amplifier

The differential amplifier in Fig. 1.32(a) receives two input signals V_{IN1} and V_{IN2} and amplifies their difference. The circuit is composed of a pair of input nMOSTs M_1 and M_2, a current-mirror load composed of M_3 and M_4, and a current source I_{SS}. The total current that flows through M_1 and M_2 is kept constant by the current source. Thus, if the current of one MOST increases, the current of the other decreases. If V_{IN1} rises and V_{IN2} is constant, the current of M_1 increases and the current of M_2 decreases. On the other hand, the current of M_4 increases because M_3 and M_4 compose a current mirror. Therefore, the voltage of the output node is raised due to a current difference between M_2 and M_4. Let us investigate the characteristics described above in more detail using the equivalent circuit in Fig. 1.32(b). First, equations based on KCL (Kirchhoff's current law) are set up at the nodes of v_1, v_{out}, and v_C as follows.

$$-g_{md}\left(v_{in1}-v_c\right)-g_{dsd}\left(v_1-v_c\right)-\left(g_{ml}v_1+g_{dsl}v_1\right)=0, \tag{1.29}$$

$$-g_{md}\left(v_{in2}-v_c\right)-g_{dsd}\left(v_{out}-v_c\right)-\left(g_{ml}v_1+g_{dsl}v_2\right)=0, \tag{1.30}$$

$$g_{md}\left(v_{in1}-v_c\right)+g_{md}\left(v_{in2}-v_c\right)+g_{dsd}\left(v_1-v_c\right)+g_{dsd}\left(v_{out}-v_c\right)=0. \tag{1.31}$$

Solving these equations for v_{out} results in the following equations:

$$v_{out}=\frac{g_{md}g_{ml}}{D}\left[2\left(g_{dsd}+g_{md}\right)\left(v_{in1}-v_{in2}\right)\right], \tag{1.32}$$

$$D=\left(g_{dsd}+g_{md}\right)\left\{g_{dsd}g_{dsl}+2g_{ml}\left(g_{dsd}+g_{dsl}\right)\right\}. \tag{1.33}$$

Here, $g_{dsl} << g_{ml}$ is assumed. From Eq. (1.32) the differential gain of this amplifier is given as

$$A=\frac{v_{out}}{v_{in1}-v_{in2}}=\frac{2g_{md}g_{ml}\left(g_{dsd}+g_{md}\right)}{D}. \tag{1.34}$$

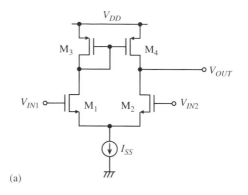

(a)

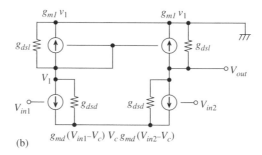

(b)

FIGURE 1.32. Differential amplifier; circuit diagram (a), and small-signal equivalent circuit (b), where g_{md} is the transconductance of M_1 and M_2, g_{ml} is that of M_3 and M_4, g_{dsd} is the drain conductance of M_1 and M_2, and g_{dsl} is that of M_3 and M_4.

Assuming g_{md}, $g_{ml} \gg g_{dsd}$, g_{dsl}, this is approximated by

$$A = \frac{g_{md}}{g_{dsd} + g_{dsl}}.$$ (1.35)

1.6. Basics of Memory LSIs

Figure 1.33 shows trends in memory cell size for various memories [1] which have been presented at major conferences. Both DRAM and SRAM cells have been miniaturized at a pace of about one-fiftieth per 10 years. Recently, however, a saturation has occurred due to the ever-more difficult process for device miniaturization. Flash memory has caught up with DRAM. Figure 1.34 shows trends in the memory capacity of VLSI memories at the research and development level [1]. DRAM has quadrupled its memory capacity every two and a half-years, although it has quadrupled every three years at the production level. As a result, standard (commodity or stand-alone) DRAMs have reached 4 Gb [15],

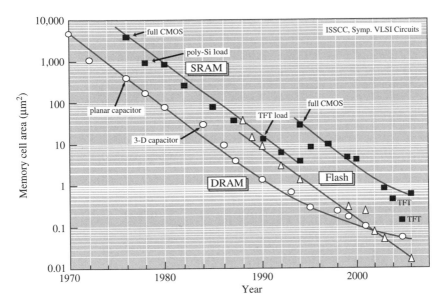

FIGURE 1.33. Trends in the memory-cell size of VLSI memories.

and 512 Mb to 2 Gb at the volume-production level. In the early days, the development of SRAMs was focused on low-power applications, especially with very low standby and data-retention power, while increasing memory capacity with high-density technology. After that, however, more emphasis had been

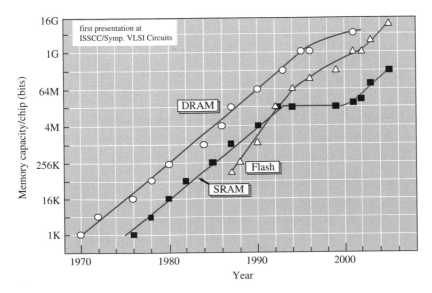

FIGURE 1.34. Trends in the memory capacity of VLSI memories.

placed on high speed rather than large memory capacity, primarily led by cache applications in high-speed MPUs. Most recently, the memory capacity started to increase again, reaching a 200-Mb cache [16]. Even a 256-Mb stand-alone SRAM [17] using thin film transistors has been presented. The memory capacity of flash memory has reached the 8-Gb level [18, 19] by using the two bits per cell scheme. The market for flash memories has expanded now with 4-Gb chips. They have become the key device for portable mass data storage applications such as digital still cameras, cellular phones, handheld devices, USB memories, and portable audio and video players, because of their exceptional feature of non-volatility, low bit cost with inherently small cell structures, large memory capacity, and high data throughput.

This section describes basics of memories in terms of the memory chip architecture and memory cells, exemplified by DRAMs, SRAMs, and flash memories.

1.6.1. Memory Chip Architectures

Figure 1.35 shows a typical memory chip architecture [1]. The chip is comprised of a memory cell array and peripheral circuits. Peripheral circuits include decoders and drivers for rows and columns, data-control circuits including amplifiers, I/O circuits, control logic circuits to control all blocks, and on-chip voltage converters. Here, the converters bridge the supply-voltage gap between the memory cell array and peripheral circuits and realize a single power-supply scheme. Due to its matrix architecture, the chip features iterative circuit blocks, which govern the chip size, such as the cell array itself and the decoder blocks and relevant driver blocks. Memory chip performance, as represented by access and cycle times for RAMs or access, programming, and erase times for flash

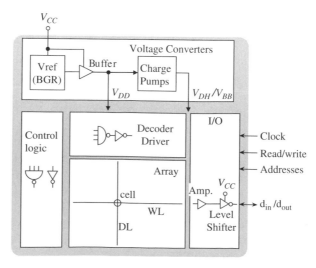

FIGURE 1.35. Architecture of a memory chip [1, 3].

memories, is mostly determined by the memory array and its relevant circuits. For example, low signal-to-noise-ratio (S/N) memory cells, the complicated timing sequences necessitated by memory-cell operations, and the small current-driving capabilities of the memory array circuits all degrade chip performances. The memory cell array is especially problematic in terms of performance and stable operations of the whole chip.

The memory array is usually large. For example, for stand-alone SRAMs, the array occupies 60–70% of the total area of the chip, for high-end embedded SRAMs it occupies about 50%, and for stand-alone DRAMs it is usually 55–70%, but for high-performance DRAM and protocol-intensive DRAMs such as Rambus it is lower. Thus, the word and data lines are heavily capacitive. In addition, they are resistive because those in high-density memory cells are formed using refractory metals such as poly-Si, which are suitable for self-aligned processing and fine patterning despite being resistive materials. Thus, the memory array inevitably suffers from large RC line delays. To reduce the delays, the multi-division of a memory array with full usage of multilevel metal wiring, and small-signal transmissions on the lines are often used [1]. Large matrix-structured memory arrays also generate various kinds of capacitive coupling noises [1], destabilizing the operations of memory cells. This calls for a high-S/N memory cell design. The overwhelming number of memory cells and/or decoders and drivers of the memory array also determines the subthreshold current of the whole chip, as discussed in Chapter 4.

1.6.2. Memory Cells

Memories can be categorized as volatile memories such as DRAMs and SRAMs, and non-volatile memories such as flash memories. Figure 1.36 compares the cell circuits and operating voltages with an assumption of a single external supply voltage of 1.8 V. Obviously, DRAM and flash memory cells need many internal voltages. The one-transistor one-capacitor (1-T) DRAM cell (a) consisting of a MOST (M_0) that works as a switch, and a capacitor for storing charge. For example, the absence of charge (electrons for nMOST) at the capacitor corresponds to "1", while the existence of charge corresponds to "0". In other words, for the voltage expression, a high stored voltage corresponds to "1" while a low stored voltage corresponds to "0". The write operation is performed by turning on the switch and applying a voltage corresponding to the write data from the data line (DL) to the capacitor. Here, the switch is turned on by applying a high enough voltage to the word line (WL) to eliminate the V_t drop. The read operation is performed by turning on the switch. A resultant signal voltage developed on the floating data line, depending on the stored data at the capacitor, is detected by a sense amplifier on the data line. In principle, the cell holds the data without power consumption. Actually, however, leakage currents in the storage node degrade an initial high stored voltage, finally causing the loss of information. The loss can be avoided by a "refresh" operation: The cell is read before the stored voltage has become excessively decayed, and then it

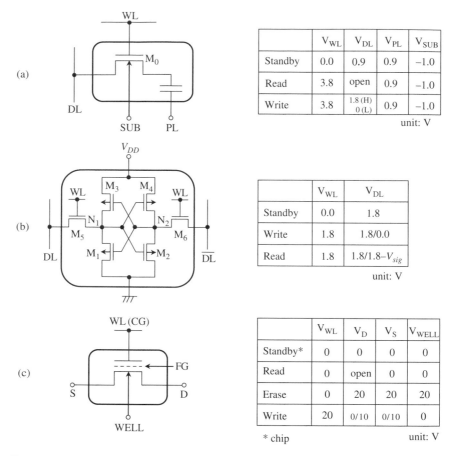

FIGURE 1.36. Memory cell operations of DRAM cell (a), SRAM cell (b), and NAND flash memory cell (c). A single external supply of 1.8 V is assumed.

is rewritten by utilizing the resultant read information, so that the voltage is restored to its initial value. A succession of the read-rewrite operation at a given time interval retains the data. The interval, which is determined by the leakage current, is about 2–64 ms. The name DRAM is derived from the fact that data is dynamically retained by refresh operations, which differs from SRAM.

The six-transistor (6-T) SRAM cell consists of a flip-flop circuit (M_1-M_4) that is constructed between a power supply voltage (V_{DD}) and the ground (V_{SS}), and two switching MOSTs (M_5, M_6). Data is communicated between a pair of data lines, DL and $\overline{DL}$, and the flip-flop by turning on M_5 and M_6 with an activating word line (WL). The write operation is performed by applying a differential voltage between a high voltage (H) and a low voltage (L) to a pair of data lines and thus to the storage nodes, N_1 and N_2. For example, "1" is written for a polarity of H at DL (N_1) and L at $\overline{DL}$ (N_2) while "0" is written for an opposite polarity of L at DL and H at $\overline{DL}$. The read operation is performed by detecting the polarity of a differential signal voltage developed on the data lines. No refresh operation is needed, because leakage currents at N_1 and N_2, if any, are compensated for by a static current from the power supply as long as V_{DD} is supplied. Thus it allows ease of use, although need for many MOSTs in a cell increases the memory cell size to more than four times that of DRAM.

A typical flash memory cell consists of one transistor. The operation is based on the floating-gate (FG) concept, in which the threshold voltage of the transistor can be changed repetitively from a high to a low state, corresponding to the two states of the memory cell, i.e., the binary values ("1" and "0") of the stored bit. Unlike RAMs, a flash cell needs an additional erase operation, because it must be initialized to a state of non-existence of electrons by extracting electrons from the floating gate (i.e. the storage node). After that, cells can be "written" into either state "1" or "0" by programming method, which is achieved by either injecting electrons to the floating gate or not doing so. For example, "0" is written when electrons are injected, while "1" is written when electrons are not injected (i.e., the erased state), resulting in the two threshold voltages. For NAND cell, the read operation is performed by turning off the selected word line, as explained later. For "0", the transistor is kept off because electrons at the floating gate prevent the MOST from turning on. For "1", however, the MOST turns on. Thus, a sense amplifier on the data line can differentiate the currents to discriminate the information. Note that, in principle, the electrons injected never discharge because they are stored at the floating gate and surrounded by pure insulators. Data retention is ensured even when the power supply is off, thus realizing a non-volatile cell.

Memories can also be categorized as destructive read-out (DRO) memories such as DRAMs, and the non-destructive read-out (NDRO) memories such as SRAMs and flash memories. The DRO memory cell needs successive read and rewrite (or restoring) operations to retain the data, causing a slow cycle time. Since such operations are needed for each cell along the selected word line, a detector (i.e., sense amplifier) and rewrite circuit must be placed on each pair of data lines. Thus, the circuits must be small enough to meet a small data-line pitch. Fortunately, for DRAMs, a simple cross-coupled CMOS circuit enables such functions with negligible power dissipation. If coupled with a half-V_{DD} sensing, no dummy cell is required to generate the reference level [1]. In contrast, in the NDRO memory, only one detector for the whole array is sufficient, accepting a larger detector. In general, the NDRO favors current sensing, while the DRO favors voltage sensing. The current sensing needs more area and more difficult circuit techniques than voltage sensing. Generation of an accurate reference level is also more difficult.

1.7. Basics of DRAMs

Figure 1.37 shows a conceptual 1-T cell array of n rows by m columns. Plural memory cells, a precharge circuit and equalizer (PC) and a cross-coupled CMOS sense amplifier (SA) are connected to each pair of data lines (DLs) which

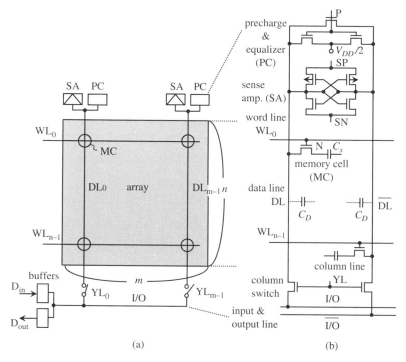

FIGURE 1.37. A conceptual DRAM array (a), and an actual data-line configuration (b). Reproduced from [1] with permission of Springer.

communicates with a pair of common data input/output lines (I/O and $\overline{\text{I/O}}$) through a column switch (YL). The 1-T cell operation consists of read, write and refresh operations, as described previously. All operations entail common sub-operations: precharging (i.e. initializing) all pairs of data lines to a floating voltage of a half-V_{DD} by turning off the precharge circuit and equalizer, and then activating a selected word line. A memory cell has a capacitance C_S, and each data line (or bit line) has a parasitic capacitance C_D.

1.7.1. Read Operation

A stored data voltage, V_{DD} ("1") or 0 V ("0"), at the cell node (N) of each cell along the word line is read out on the corresponding data line, as shown in Fig. 1.38. The available signal voltage component in the cell for "1" and "0" is $V_{DD}/2$ for the reference voltage of $V_{DD}/2$. As a result of charge sharing, the signal voltage ($\pm v_S$) developed on the floating data line (for example, DL) is expressed by

$$v_S \left(= v_{S\,\text{max}}\right) = \frac{C_S}{C_D + C_S} \cdot \frac{V_{DD}}{2} = \frac{Q_S}{C_D + C_S}. \tag{1.36}$$

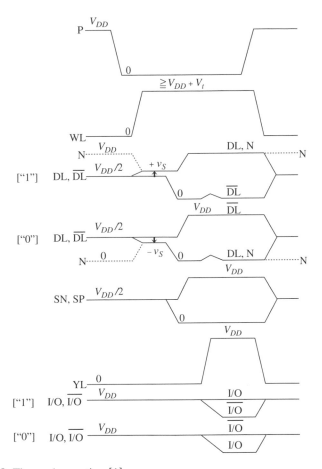

FIGURE 1.38. The read operation [1].

Here the amplitude of the word-line voltage for the read operation is sufficiently boosted, which is usually the same as that for the write operation. Otherwise the high signal voltage $(+v_S)$ is degraded because the signal voltage component $(V_{DD}/2)$ at the cell node is not completely transferred to the data line. We call it a full read operation, which is carried out by "word-boosting or word-bootstrapping" (word-line voltage $V_W \geq V_{DD} + V_t$) so that the V_t drop at the cell is eliminated. Here, V_t denotes the lowest necessary V_t at the source-follower mode, as explained previously (Fig. 1.5(a) and Eq. (1.7)) or in Chapter 2 (V_{tFW} in Fig. 2.10). Unfortunately, v_S is inherently small (100–200 mV) because C_D is much larger than C_S. A small C_S and a large C_D come from the needs for a small cell size and for connecting a large number of cells to a data line, respectively. Hence the original large signal component ($V_{DD}/2$, usually s0.6–2.5 V) at the storage-node collapses to v_S. Thus, the 1-T DRAM cell features the destructive readout (DRO) characteristics, necessitating a successive amplification and restoration for each of the cells along the word line, as discussed previously. This is performed by a cross-coupled (latch-type) differential CMOS sense amplifier on each data line, with the other data line ($\overline{DL}$) as a reference. Then, one of the amplified signals is outputted as a differential voltage to the I/O lines by activating a selected column line, YL.

1.7.2. Write Operation

The write operation (Fig. 1.39) is always accompanied by a preceding read operation. After almost completing the above amplification, a set of differential data-in voltages of V_{DD} and $0\,V$ is inputted from the I/O lines to the selected pair of data lines. Hence, the old cell data is replaced by the new data. Note that the above amplification/restoration operation is done simultaneously for each of the remaining cells on the selected word line to avoid loss of information.

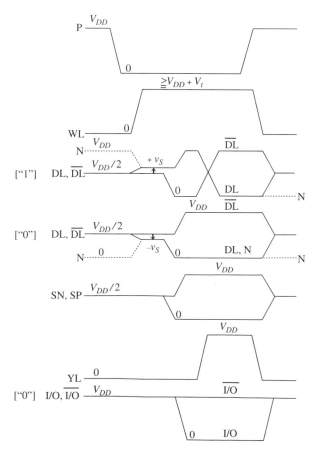

FIGURE 1.39. The write operation for low-voltage ("0") binary information [1].

1.7.3. Refresh Operation

The initial high stored voltage, V_{DD}, decays during a data retention period due to charge losses by leakage currents, or α -particle or cosmic-ray neutron irradiations. The resultant degraded voltage must be restored by a refresh operation that is almost the same as the read operation, except that all YLs are kept inactive. This is done by reading the data of cells on the word line and restoring them for each word line so that all of the cells retain the data for at least t_{REFmax}. Here t_{REFmax} is the maximum refresh time for the cell, which is guaranteed in catalog specifications, and is exemplified by $t_{REFmax} = 64$ ms for a 64-Mb chip. Thus, each cell is periodically refreshed at intervals of t_{REFmax}. Obviously, although the read signal voltage is maximum (v_{smax}) just after the write operation, that is, v_S given by (1.36), it becomes minimum just before the refresh operation, which is expressed as

$$V_{S\,min} = v_{S\,max} - \frac{Q_L + Q_C}{C_D + C_S}. \tag{1.37}$$

For a successful refresh operation, v_{Smin} must be larger than the data-line noise, v_N, that is caused by capacitive couplings to the data line from other conductors, the electrical imbalance between a pair of data lines, and amplifier offset voltage, as discussed in Chapter 2. Thus, $v_{Smin} > v_N$. The formula can be changed with the charge expression as

$$Q_S > Q_L + Q_C + Q_N. \tag{1.38}$$

Here, Q_S is signal charge ($= C_S V_{DD}/2$), Q_L is leakage charge that is the product of cell leakage current, i_L, and t_{REFmax}, and Q_C is the maximum charge collected at the cell node by α -particle or cosmic-ray neutron irradiations. Q_N is data-line noise charge ($= v_N(C_D + C_S)$). To cope with cell miniaturization and V_{DD} reduction, v_{Smin} must be increased. Consequently, increasing v_{Smax} by maintaining Q_S and reducing C_D is essential. Reducing Q_L, Q_C and v_N are also crucial. In other words, design and technology must be developed so that the charge relationship in Eq. (1.38) is always maintained through high S/N techniques aiming at a larger Q_S and smaller effective noise-charges (Q_L, Q_C, Q_N).

There are two kinds of refreshing schemes, the distributed refresh and the lumped (or burst) refresh, when each cell in a memory array configuration with n rows by m columns shown in Fig. 1.37 is refreshed at the interval of t_{REFmax}. The distributed refresh, as shown in Fig. 1.40, is more popular, which distributes the n refresh operations uniformly within t_{REFmax}, enabling a refresh time interval of t_{REFmax}/n. Thus, for example, the interval is $16\,\mu$s for $n = 4096$ and $t_{REFmax} = 64$ ms, interrupting random operations with one refresh cycle every $16\,\mu$s. Hence,

$$t_{REFmax} = nt_{RC}/\eta,$$

$$\eta = nt_{RC}/t_{REFmax}, \tag{1.39}$$

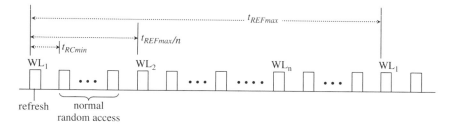

FIGURE 1.40. The refresh operation. t_{RCmin}, minimum cycle.

where n is the refresh cycle, t_{RC} is the memory cycle time, and η is the refresh busy rate that expresses the percentage of the time for which data is not accessible from outside the chip. The lumped refresh repeats the n-word-line selections at the fastest cycle (t_{RC}). During a resulting period of nt_{RC}, the chip cannot accept any random operation from outside. Instead, during the remaining period, $t_{REFmax} - nt_{RC}$, it can accept any operation without needing to be paused for a refresh operation.

1.8. Basics of SRAMs

Figure 1.41 shows an SRAM array using the 6-T cell that consists of transfer MOSTs, M_5 and M_6, driver MOSTs, M_1 and M_2, and load MOSTs, M_3 and M_4. The 6-T cell has offered a wide voltage margin and noise immunity and ease of fabrication with a process compatible with CMOS logic in MPUs (microprocessors). Although a cache SRAM does not need a density as high as that of a DRAM, the memory cell size is still a prime concern to reduce the fabrication cost, especially for MPUs that have a denser on-chip cache.

1.8.1. Read Operation

It starts with activation of a selected word line after equalizing all pairs of data lines to V_{DD}, as shown in Fig. 1.42. Each cell along the selected word line develops a small signal voltage, v_S, on one of data lines, depending on the stored cell information. For example, for "1", where cell node (N_1) is at a low voltage and cell node (N_2) is at a high voltage, a cell read current that results from turning on the transfer MOST (M_5) and driver MOST (M_1) discharges the data line (DL). If the data-line load (Z) is a static circuit (e.g., gate-drain connected nMOS circuit), the current develops a small static signal voltage on the data line DL, as a result of a ratio of M_5, M_1, and Z. If it is a dynamic circuit (e.g., pMOST circuit controlled by the precharge pulse P) that only precharges the data lines during standby periods, as in DRAMs (Fig. 1.37), the current continues to discharge the DL. In practice, the word pulse is turned off once the

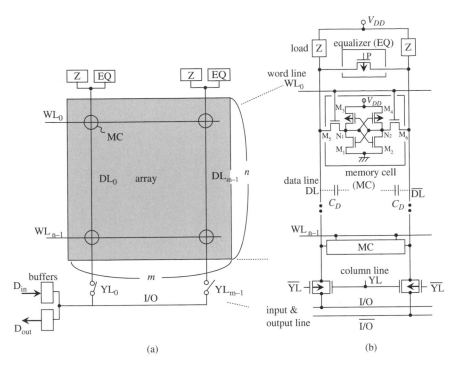

FIGURE 1.41. A conceptual SRAM array (a) and an actual data-line configuration (b). Reproduced from [1] with permission of Springer.

signal is detected by a highly sensitive amplifier on the I/O line, so excessive DL discharging is prohibited and thus a low DL charging power and a fast cycle time are achieved. Note that N_1 is raised due to ratioed operation while activating the WL. Another data line ($\overline{DL}$) remains at the equalized voltage with cutting off both M_2 and M_6, as long as the raised N_1 voltage is smaller than the V_t of M_2. A read data from the selected pair of data lines is transferred to a pair of I/O lines by turning on the column switch selected by a column line (YL) activation. The polarity of the differential signal voltage is detected and amplified by a sense amplifier on a pair of I/O lines, so as to be outputted as the data output (D_{out}) from the chip. For "0" stored information, the same operation is carried out, except that the opposite voltage polarity is applied. Here, to ensure NDRO characteristics the M_5 conductance is set to be about 1.5 times smaller than the M_1 conductance. Thus, the cell read current is almost equal to the constant saturation current of M_5. NDRO characteristics allow a sense amplifier on each pair of data lines to be eliminated. A possibly degraded voltage difference between two nodes (N_1, N_2) that is developed during the read operation, finally recovers to a full V_{DD} after the word-line voltage is turned off, with the help of the cell feedback loop.

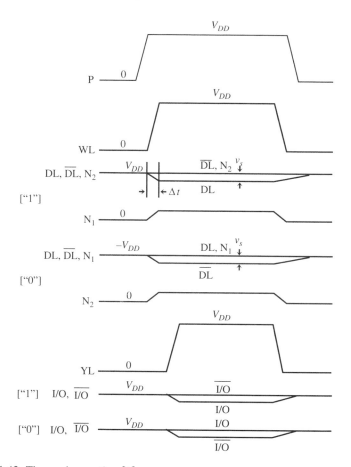

FIGURE 1.42. The read operation [1].

1.8.2. Write Operation

Write operation is carried out by applying a differential input, whose polarity corresponds to the write binary information, to the I/O lines through the data input buffer, as shown in Fig. 1.43. To quickly transfer the full differential voltage to the data line and then to the cell nodes (N_1, N_2), each column switch is composed of parallel-connected nMOST and pMOST. An nMOST can fully discharge the data line with its high gate voltage, while a pMOST can fully charge up the data line with its low gate voltage. The write operation is completed by turning off the word line so that the differential voltage is held as stored information. Note that each of the remaining cells on the selected word line continues to develop a signal voltage on the corresponding data line, in the same manner as in the read operation.

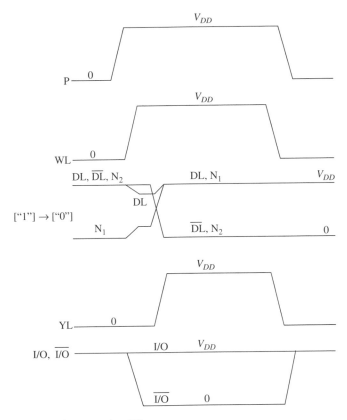

FIGURE 1.43. The write operation [1].

1.9. Basics of Flash Memories

1.9.1. The Basic Operation of Flash Memory Cells

The operations of flash memory cells [21] are based on the floating-gate concept, in which the threshold voltage of a transistor can be changed repetitively from a high to a low state, corresponding to the two states of the memory cell, i.e., the binary values ("1" and "0") of the stored bit. Cells can be "written" into either state "1" or "0" by either "programming" or "erasing" method. One of the two states is called "programmed," the other "erased". In some kinds of cells, the low-threshold state is called "programmed"; in others, it is called "erased". Although this may induce some confusion, the different terms are related to the different organizations of the memory array. Their read operation is performed by applying a gate voltage that is between the above two threshold voltages, and by sensing the current flowing through the transistor.

The simple model shown in Fig. 1.44 helps in understanding the electrical behavior of the flash memory cell. Consider the case when the charge, Q_{FG}, is

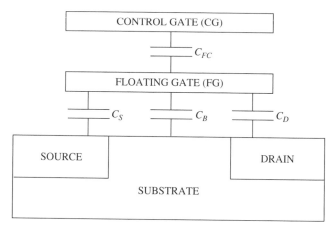

FIGURE 1.44. A schematic of the cross-section of a flash memory cell [1].

stored in the floating gate (FG). Then,

$$Q_{FG} = C_{FC}(V_{FG} - V_{CG}) + C_S(V_{FG} - V_S)$$
$$+ C_D(V_{FG} - V_D) + C_B(V_{FG} - V_B) \qquad (1.40)$$

where V_{FG} is the potential at the FG, V_{CG} is the potential at the control gate (CG), and V_S, V_D, and V_B are potentials at the source, drain, and bulk (substrate), respectively. Hence, the potential at the FG due to capacitive coupling is given by

$$V_{FG} = \frac{Q_{FG}}{C_T} + \frac{C_{FC}}{C_T}V_{CG} + \frac{C_S}{C_T}V_S + \frac{C_D}{C_T}V_D + \frac{C_B}{C_T}V_B, \qquad (1.41)$$
$$C_T = C_{FC} + C_S + C_D + C_B.$$

If the source and bulk are both grounded, (1.41) can be rearranged as

$$V_{FG} = \frac{Q_{FG}}{C_T} + \frac{C_{FC}}{C_T}V_{CG} + \frac{C_D}{C_T}V_{DS}. \qquad (1.42)$$

Here, we define V_t (FG) as the threshold voltage of the FG transistor, the gate of which is the FG. If V_{FG} is lower than V_t (FG) the FG transistor turns off, while if it is higher than V_t (FG) it starts to turn on. Consider another transistor (i.e. CG transistor) with the same source and drain, the gate of which is the CG. Obviously, the CG transistor also starts to turn on when $V_{FG} = V_t$ (FG). This point is just the threshold voltage, V_t (CG), of the CG transistor. Thus, V_t (CG) is derived from (1.42) with $V_{FG} = V_t$ (FG) and $V_{CG} = V_t$ (CG), as follows:

$$V_t(CG) = \frac{C_T}{C_{FC}}V_t(FG) - \frac{Q_{FG}}{C_{FC}} - \frac{C_D}{C_{FC}}V_{DS}. \qquad (1.43)$$

Therefore, V_t (CG) which is the memory-cell threshold voltage, depends on Q_{FG}. The V_t (CG) shift, ΔV_t, is thus given by

$$\Delta V_t = -\frac{\Delta Q_{FG}}{C_{FC}}, \qquad (1.44)$$

where ΔQ_{FG} is the change in Q_{FG}. This equation shows that the role of injected charge is to shift the I-V curves of the cell (i.e. the CG transistor). If the reading biases are fixed (usually at $V_{CG} \sim 5\,\mathrm{V}$ and $V_{DS} \sim 1\,\mathrm{V}$), the presence of charge greatly affects the current level used to sense the cell state. Figure 1.45 shows two curves: curve A represents the "0" state, while curve B is for the "1" state with a V_t(CG) shift. The current is approximately $100\,\mu\mathrm{A}$ for "0" while it is 0 for "1", when V_{CG} is chosen to be an appropriate voltage between the two threshold voltages. It is indispensable for the normally-off transistor to start to turn on only by the application of V_{CG}; otherwise, the cells connected to the same data line may turn on even without the application of a word pulse. Thus, to ensure a successful read operation the following relationships must be satisfied:

$$V_R > \text{low } V_t\ (CG) > 0,$$

$$V_R < \text{high } V_t\ (CG).$$

There are two typical mechanisms to transfer electric charges from and into the FG. They are the hot-electron injection (HEI) mechanism and the Fowler-Nordheim (FN) tunneling mechanism [22].

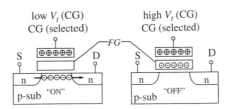

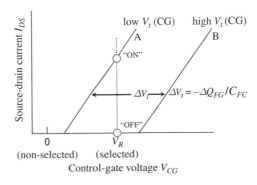

FIGURE 1.45. The principle of the read operation of a Flash memory cell [1].

Hot Electron Injection. Electrons traveling from the source to the drain are "heated" by a high lateral electric field (between source and drain), causing avalanche breakdown phenomena in the vicinity of the drain. The resultant impact ionization generates hole-electron pairs in the drain region. The generated electrons are injected to the floating gate through the oxide by a transversal electric field (between the channel and the control gate), while the generated holes flow to the substrate as the substrate current. Figure 1.46 shows the relationship between the gate current (I_G) and the gate voltage (V_{GS}) of an FG transistor [1, 43]. The gate current starts to flow in accordance with the start of the channel (source-drain) current flow when V_{GS} is increased. It increases with V_{GS} because of the increase in the channel current. An excessive V_{GS} for a fixed V_{DS}, however, decreases I_G because of a reduced drain-gate potential difference. Thus, if the bias condition for the drain and floating gate is set so that the gate current is maximized, this realizes the fastest injection speed. This injection always entails a large channel current, causing a high-power dissipation. For example, the current is as high as about 0.5 mA at a control-gate voltage of 12 V.

Fowler-Nordheim Tunneling. The current density (J) of the tunneling current is obtained in the simplest form as

$$J = AE^2 \exp(-B/E) \tag{1.45}$$

where A and B are almost constant, and E is the electric field. Figure 1.47 shows log J versus E. There is a large variation, of about seven orders of magnitude, in the tunnel current when the field is changed from $7\,\mathrm{MV\,cm^{-1}}$ to $10\,\mathrm{MV\,cm^{-1}}$. Since the field is roughly the applied voltage divided by the oxide thickness, a

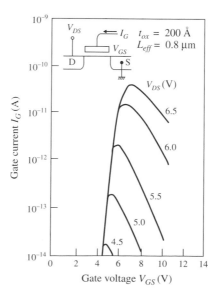

FIGURE 1.46. The gate current caused by channel hot-election injection [1, 43].

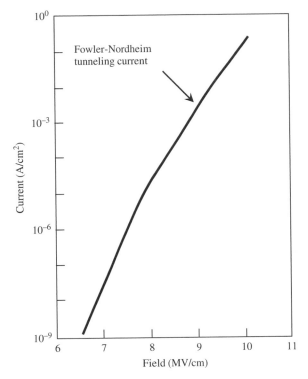

FIGURE 1.47. The Fowler-Nordheim tunneling current, as a function of the electric field [1, 21].

reduction in the oxide thickness for a fixed applied voltage produces a rapid increase in the tunneling current. An optimum thickness, of about 10 nm with $10\,\mathrm{MV\,cm^{-1}}$, however, is chosen in present-day devices as the result of the trade-off between performance constraints and reliability concerns. Thin oxides reduce the erasing speed, but also increase power consumption and degrade the reliability of the oxide. The above exponential dependence of the tunnel current on the oxide-electric field calls for very good process control. Otherwise, some critical problems are caused. For example, a very small variation in the oxide thickness among the cells in a memory array produces a great difference in the erasing or programming currents, thus spreading the threshold-voltage distribution in both logical states. Moreover, the tunneling currents may become important for device reliability at low fields, either in the case of poor-quality tunnel oxides or when thin oxides are stressed many times at high voltages. In fact, poor-quality oxides are rich in interface and bulk traps, and trap-assisted tunneling is made possible since the equivalent barrier height seen by electrons is reduced. Thus, tunneling requires a much lower oxide field than $10\,\mathrm{MV\,cm^{-1}}$. The oxide defects, whose density increases with decreasing oxide thickness, must be avoided to control the erasing and programming characteristics and to ensure good reliability.

Many types of flash cells have used the above-described mechanisms. Some of them differ in their charge-transfer mechanisms, their cell structures, and the logical connections of their cells. Even multilevel schemes using these cells have been proposed. Of the cells, the NOR cell and the NAND cell are explained below in detail because they are popular.

1.9.2. The NOR Cell

The name of the cell is derived from the logical connection of the cells. Figures 1.48 and 1.49 show the industry-standard one-transistor stacked-gate flash memory cell [22] and its operation [1, 46]. The tunnel oxide under the floating gate is about 10 nm thick. To have a junction that can sustain the high applied voltages without breaking down, the source junction is carefully designed to achieve a lighter and deeper junction. Therefore, the source diffusion is realized differently from the drain diffusion, which does not undergo such high bias conditions. Oxide/nitride/oxide (ONO) interpoly dielectrics are used to realize a high C_{FC}. The cell uses the FN tunneling current for erasing and the hot-electron injection for programming (write).

The erase operation is performed by a combination of a 0-V control-gate voltage and a high source-voltage, so that any electrons at the floating gate are ejected to the source by the tunneling effect of the thin oxide. All cells are erased simultaneously, since a high voltage is applied concurrently to the sources of all cell transistors. Here, the drain is made open to prevent a turn-on current in the transistor, caused by capacitive coupling from the source to the floating gate. Note that different initial values of the cell threshold voltage and different gate oxide thicknesses of the FG transistors may cause a variation in the threshold voltage at the end of the erase operation. A higher threshold voltage after erasing results in a slower read operation. Thus, in practice, before applying the erase pulse, all of the cells in the array/block are programmed so that all of the thresholds start at approximately at the same value. After that, an erase pulse that has a controlled width is applied. After an erase pulse, however, there may be typical bits and fast erasing bits due to a gate-oxide thickness variation.

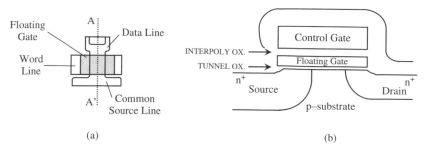

(a)   (b)

FIGURE 1.48. The NOR cell [1, 21]. (a) The layout of a typical double-polysilicon stacked gate cell; (b) a schematic cross-section along line A-A'.

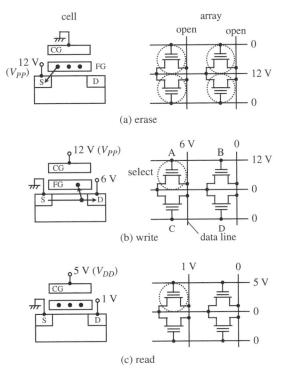

FIGURE 1.49. The basic operation of the NOR cell [1]. (a) Erase; (b) write; (c) read.

Therefore, subsequently, the whole array/block is read to check whether or not all the cells have been erased. If not, another erase pulse is applied and another read operation follows. This algorithm is applied until all of the cells have threshold voltages that are lower than the erase verify level. A similar algorithm is also useful to avoid "over-erase" that may cause normally-on transistors. Typical erasing times range from 100 ms to 1 s.

As for the write operation (i. e. programming), a high-voltage pulse is applied to the control gate so that the capacitive coupling makes the transistor turn on, with a floating-gate voltage raised from a voltage that was set by the preceding erase operation. If the write data corresponds to a high voltage (for example, "1") applied to the drain (i.e. the data line), a high source-drain current of about 0.6 mA flows. As a result, hot electrons generated in the vicinity of the drain are injected into the floating gate. Consequently, after turning off the word pulse, the FG transistor is deeply cut off with a negative gate voltage. If the write data is for a drain voltage of 0 V ("0"), the floating gate remains at the previous erasing state of having no electrons. Thus, after programming, the cell-threshold voltage, V_t (CG), becomes high or low. The shift in the threshold voltage, ΔV_t, depends upon the width of the programming pulse. To have $\Delta V_t = 3 - 3.5$ V, a pulse width with typical values in the $1-10\,\mu s$ range must be applied. The read operation is done by the application of 5 V ($V_R = V_{DD}$) to the control gate and

1 V to the drain, as described before. An access time of 50–100 ns is obtained. Note that a higher drain-voltage may cause "soft-write" for the cell during the read operation, although it offers faster sensing with a larger read current.

The reliability issue regarding the programming/erasing cycles and retention characteristics is highlighted for flash memories. The following is an example of programming disturbs [46]: there are two major disturbs when programming the selected cell (e.g. cell A) in Fig. 1.49(b). One is due to the high voltage (12 V) applied to the control gates of the non-selected cells (e.g. cell B) on the same word line. The other is due to the medium high voltage (6 V) applied to the drains of the non-selected cells (e.g. cell C) on the same data line. For cell B there might be tunneling of electrons from the FG to the control gate through the interpoly oxide if the FG is filled with electrons. This induces charge loss, reducing the margin for the high threshold voltage. There might also be tunneling of electrons from the substrate to the FG if the FG is "empty". This induces a charge gain, reducing the margin for the low threshold voltage. For cell C electrons tunnel from the FG through the gate oxide to the drain, reducing the margin for the high threshold voltage. Disturbances similar to the above are present even during read operations. That is why a drain voltage as low as 1 V is applied, to avoid the soft-write that can occur during read operations. The influence of disturbances becomes more and more prominent when increasing the number of reading-programming or programming-erasing cycles.

Flash memory needs multiple power supplies internally, calling for on-chip voltage converters. A voltage down-converter, which converts the high external supply voltage to a low internal supply voltage, enables driving an internal load with a large output current. On the other hand, a voltage up-converter, by means of charge pumps, suffers from a small output current, as discussed in Chapter 8. Thus, to achieve a single external supply, the current flowing into the drain terminal at a voltage of 6 V that is generated by a pump circuit is limited to a small avalanche-breakdown current by reducing the number of cells (for example, 16 cells) that are written simultaneously. Thus, the drain voltage must be generated from the external V_{PP} (12 V) through an on-chip voltage down-converter. An on-chip voltage up-conversion from the external V_{DD} (5 V) to 6 V by using a charge pump fails to supply a high enough current. A read voltage of 1 V is also generated from V_{DD}, using another voltage down-converter. Thus, the cell needs two external power supplies, V_{PP} and V_{DD}.

A negative word-voltage erasing scheme [23, 43] allows even the NOR cell to operate at a single external V_{DD}, as shown in Fig. 1.50. Here, there are minor changes, caused by an advanced device, in the values of V_{PP} and currents, from those described so far. The scheme features a negative control-gate voltage and a V_{DD} source voltage in the erase operation. The setting gives almost the same tunneling current as that for the conventional setting shown in Fig. 1.49. The V_{DD} at the source can supply a large tunneling current of 5 mA, which is necessary for a block erase. An on-chip negative voltage (−10 V) generator from V_{DD}, which uses a charge pump, manages to drive its load despite the charge pump

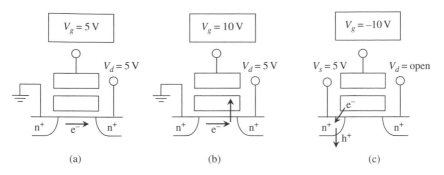

FIGURE 1.50. The negative word-line voltage scheme [1, 23, 43]. (a) Read, 60μA/cell; (b) program, 0.5mA/cell (8 mA for 16I/Os); (c) erase, 10nA/cell (5 mA for 64Kb block).

because of the purely capacitive CG load. Even for write and read operations, single V_{DD} operation is realized: The external V_{DD} supplies a large enough hot-electron current of 8 mA at the V_{DD} drains and generates a high CG voltage of 10 V through a voltage up-converter. Note that reduction of the source voltage simplifies the source structure.

1.9.3. The NAND Cell

In this type of cell the elementary unit is not composed of the single three-terminal cell, but of more FG transistors connected in a series (8–32 transistors), which constitutes a chain connected to the data line and to ground through two selection transistors, as shown in Figs. 1.51 and 1.52 [20, 24, 47, 48]. This organization eliminates all contacts between word lines, which can be separated by their minimum design rule, thus reducing the occupied area to 44%. Figure 1.52 shows the cross section of an 8-bit elementary block for a NAND array with peripheral circuits. The FN tunneling current is used for both erasing and programming. The erase voltages are 20 V to the n-substrate, the p-well2, and the drain (DL) and source of the chain, and 0 V to all the control gates of the chain when the selection transistors (M_{S1}, M_{S2}) are turned on. This biasing induces electron tunneling toward p-well2 from the FGs, resulting in a low threshold voltage for all of the cells. Note that a low threshold voltage is set to be negative (i.e. normally on), which differs from the NOR cell. There is no voltage difference between the drain of each cell transistor and p-well, so that there is no breakdown of the junction. The programming voltages are 20 V to the control gate of the selected cell and 10 V to all of the remaining control gates of non-selected cells, with p-well2 grounded. The selection transistors are biased to connect the chain to the data line and isolate it from the ground. If a "0" is to be stored, the data line is grounded. Hence, the sources, drains and channels of the cell transistors are grounded, and only the selected cell-transistor has such an electric field in the oxide to induce electron injection from the channel into the FG, causing a high threshold voltage(i.e., normally off). This is because the FG voltage of

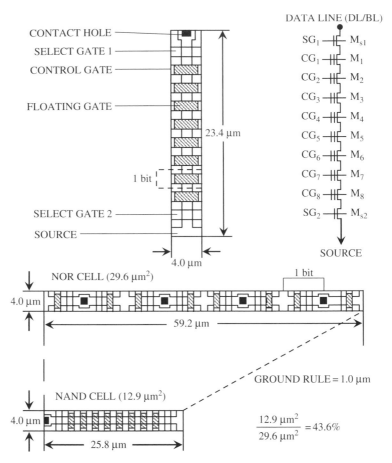

FIGURE 1.51. NAND architecture. The dimensions of a NAND array are compared to those of a NOR array [1, 47, 48]. Reproduced from [47, 48] with permission; © 2006 IEEE.

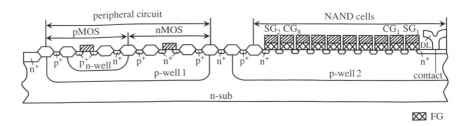

FIGURE 1.52. A cross-section of NAND cells and the peripheral circuit [1, 47, 48]. Reproduced from [47, 48] with permission; © 2006 IEEE.

the selected cell is raised to a sufficiently high tunneling voltage of about 10 V, while the other FG voltages are raised to a sufficiently low voltage, of about 5 V. This is justified by an exponential relationship between the tunneling current and the electric field, as expressed in Eq. (1.45). If a "1" is to be stored, the data line is biased at 10 V. Here, there is no tunneling for the selected cell because of a small voltage difference between the FG and the substrate, which keeps the threshold voltage low (i.e, negative/normally on). Obviously, to transfer the data-line voltage to any selected cell in the chain, the non-selected cell transistors must always be conductive, independent of their stored information. Thus, the FG voltages of the non-selected cells must exceed the high threshold voltage sufficiently. The read operation is performed by applying 5 V (V_{PASS}) to all of the control gates except the selected one, which is grounded. The selection transistors are turned on to connect the chain to the ground and data line. Thus, the data line, which has been precharged to a high voltage (≈ 1 V), is discharged if the stored data is "1" (normally on). It holds the precharged voltage if the stored data is "0" (a threshold voltage high enough to cut off the transistor). Figure 1.53 shows the voltage relationships for the selected and non-selected cells. A read operation succeeds when the following conditions for the selected cell and for the non-selected cells are satisfied, respectively:

$$\text{low } V_t(\text{CG}) < 0 < \text{high } V_t(\text{CG}),$$

$$V_{PASS} > \text{high } V_t(\text{CG}).$$

The raised supply voltages can be internally generated by on-chip charge-pumps from a single external power supply, due to small tunneling currents. The small currents also allow an increase in the number of parallel programmed cells without increasing power consumption. Electron tunneling from whole channel region is uniform through the oxide. This makes erasing fast, exemplified by 6 ms

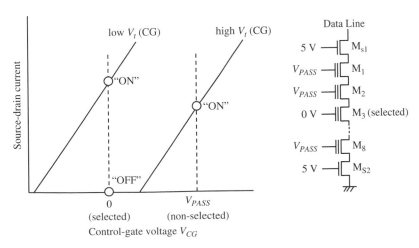

FIGURE 1.53. The principle of the read operation of the NAND cell [1].

per block and around 100 ms per chip. However, the access time is slow because of a small cell current of less than 1 μA. The chain structure is responsible for the reduced drive current.

Nano-scale NAND flash memories necessitate noise reduction techniques to cope with the data(bit)-line to data-line (DL-DL) and word-line to word-line capacitive couplings [19, 25], which are common to all high-density memory cell arrays in nano-scale technologies. The shielded DL sensing method [25] is thus used to suppress the interference of neighbor DLs during read operation, because 90% of DL capacitance is from adjacent DLs. In addition, they need various on-chip voltage conversion techniques [19, 26] for stable memory cell operations, as discussed above, and for realizing a single external supply voltage compatible with the previous-generation voltage despite scaled devices. For example, NAND flash needs four kinds of charge-pump circuits, which are a 5-V(V_{PASS}) pump for non-selected word lines in read operation, a 10-V pump for non-selected word lines in program operation and two 20-V pumps for the p-well during the erase operation and for the selected word line during the program operation. The pumped voltages are not scalable even if devices are scaled down. For example, the read voltage, which is 5 V, is hardly lowered due to wide erased-V_t distribution and the need for high-speed operation. The FN tunneling current also requires an almost constant program/erase voltage V_{PP}, as explained previously. The pumped voltages are all generated from one internal supply voltage (e.g., 2.5 V) that is generated by an on-chip voltage down-converter from a single external supply voltage (e.g., 3.3 V) [19, 26]. Since the external supply is gradually lowered, the voltage ratio of the pumped (boosted) voltage to the external supply voltage increases with device scaling. Here, the Dickson V_{PP} charge pump and its switch [27, 28] are widely used because it takes advantage of low power and smaller circuit area over the capacitor-switched booster.

1.10. Soft Errors

The soft error issue has been important to high-density RAM designs since the mid-1970s. There are two kinds of soft errors (i.e. non-permanent failures) [1, 29]: alpha-particle induced soft errors and cosmic-ray neutron-induced soft errors. The importance of the alpha-particle induced soft errors has long been recognized, and the studies are thus quite advanced. Recently, cosmic-ray-induced soft errors have been a serious problem even at sea level. A high-energy neutron penetrates into LSI devices and has a nuclear reaction with a silicon atom in the substrate, giving off secondary ions. As shown in Fig. 1.54 [30, 44], the secondary ions generate ten times as many free charges as an alpha particle, so a single cosmic-ray incidence often induces soft errors in multiple cells. A multicell error, however, cannot usually be corrected by an on-chip error checking and correction (ECC) circuit, as discussed later. The ECC is only capable of correcting one error at each address. That is, a multicell error where

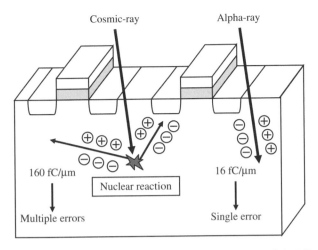

FIGURE 1.54. The schematic of charge generation by a cosmic ray [30, 44].

all of the affected cells belong to the same address is not corrected by the ECC circuit. However, the circuit does correct a multicell error where the affected cells all belong to different addresses. To increase the rate of multicell error correction by ECC circuits, we need to arrange the cell addresses so that simultaneous cosmic-ray-induced multiple errors are most likely to occur at different physical addresses. Thus, the mechanism responsible for multicell errors must be clarified. The details are discussed in Chapter 3.

There is a critical charge (Q_C) for failure that is the amount of charge required to upset the information of the node, and thus equal to the signal charge (Q_S) of the node. Q_C (i.e., Q_S) decreases with down-scaling of devices and V_{DD}. In the past, the issue has been more serious for the RAM cells, especially for SRAM cells having a small Q_S, than for logic circuits. This is because Q_S of each RAM cell is enormously small due to tiny storage node, and the number of such RAM cells is huge. Recently, however, the issue is becoming increasingly important even for logic circuits because the Q_C of logic circuit nodes has drastically reduced, and ECC is not effective the way it is in memories. Thus, in addition to adding an extra capacitance to the logic circuit node to increase the critical charge, soft-error hardened latch circuits and level keepers have been proposed [31]. FD-SOI devices are also effective to reduce the soft error rate due to the thin SOI layer (see Fig. 2.27(c)).

1.11. Redundancy Techniques

Redundancy techniques have been widely used as effective methods of enhancing production yield and reducing cost per bit of DRAMs since the 64–256-Kb generations [1]. The techniques replace defective memory elements (usually word lines and/or data lines) by on-chip spare elements. Figure 1.55 shows a

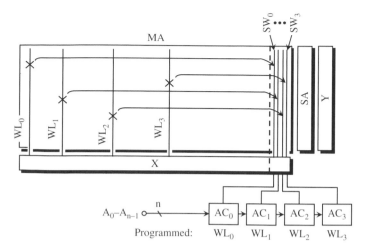

FIGURE 1.55. A conventional redundancy technique applied to a DRAM [1]. X, row decoder; Y, column decoder; SA, sense amplifier; AC, address comparator.

well-known redundancy technique [1] applied to a DRAM without memory-array division. Redundant data lines are omitted here for simplicity. The memory has L (here, $L = 4$) spare word lines $SW_0 - SW_3$ and the same number of address comparators $AC_0 - AC_3$. If the addresses of the defective word lines ($WL_0 - WL_3$) have been programmed into the address comparators during wafer testing, the address comparators allow one of the spare word lines ($SW_0 - SW_3$) to be activated whenever a set of input address signals ($A_0 - A_{n-1}$) during actual operations matches one of the defective addresses. In modern DRAMs, as many as over one hundred defective elements could be replaced by spare elements in the early stage of production, with an additional chip area of less than 5%. The program elements are usually poly-Si fuses, which are blown by means of a laser beam or a pulsed current, although they do accept memory-cell capacitors [1]. Laser programming occupies a smaller chip area and does not normally affect circuit performance, but it does require special test equipment and increased wafer handling and testing time. Also, the laser spot size and beam-positioning requirements will become more stringent for ever finer line widths. On the other hand, electrical programming by a pulsed current is carried out using standard test equipment. Usually, a hole is cut in the passivation glass over such fuses to reduce the amount of programming current needed. The possibility of mobile-ion contamination of active circuit areas can be eliminated by using guard-ring structures surrounding the fuse area, or other techniques [1]. The area and performance penalties of electrical programming can be minimized by careful circuit design. Electrical programming is used when the number of fuses required is not large enough to offset the negative aspects of laser programming. In any event, laser programming has been widely accepted due to the small area and performance penalties, simplicity, assurance of cutting, and the ease of laying out fuses.

1.12. Error Checking and Correcting (ECC) Circuit

On-chip error checking and correcting (ECC) is effective for correcting both soft errors and hard errors (defects) of RAMs [39–41]. The principle of ECC is shown in Fig 1.56. During write operation, the encoding circuit (EN) generates check bits from the input data bits. The generation rule is determined by the error-correcting code used. Both the data bits and the check bits are stored in the memory array so that the stored data have a certain amount of redundancy. If there is a defect in the memory array or a soft error occurs before reading, the read data contains one or more errors. However, the error(s) can be detected and corrected by the decoding circuit (DE) if the number of errors does not exceed the correction capability of the error-correcting code. The operation of DE is as follows. First, check bits are generated from the read data bits just like the EN and are compared with the read check bits. If no error exists, both are the same. If they are not the same, the position(s) of erroneous bit(s) are detected by analyzing the comparison results and the errors are corrected before outputting. The corrected data are usually written back to the memory array to prevent the accumulation of soft errors. In DRAMs, checking and correcting are performed during every refresh operation as well as during read operation.

The key to designing an on-chip ECC circuit is the selection of a suitable error-correcting code. Error-correcting codes are classified into single-error correction (SEC) codes, single-error correction and double-error detection (SEC-DED) codes, double-error correction (DEC) codes, etc., according to the error-detection/correction capability. Table 1.1 shows the minimum number of check bits, ΔN, required for N_0 data bits. Generally, higher error-detection/correction capability requires more check bits. It should be also noted that the ratio $\Delta N/N_0$

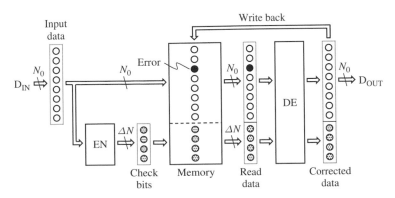

FIGURE 1.56. Principle of on-chip ECC for RAMs.

TABLE 1.1. Minimum number of required check bits, ΔN, for single-error correction (SEC), single-error correction and double-error detection (SEC-DEC) and double-error correction (DEC) codes.

Number of data bits N_0	8	16	32	64	128	256
SEC	4	5	6	7	8	9
SEC-DED	5	6	7	8	9	10
DEC	8	9	10	12	14	16

decreases as N_0 increases. Although a large N_0 reduces the memory-area penalty, it results in both large circuit-area and access-time penalties. This is because the number of gates n the encoding and decoding circuits is approximately proportional to $N_0 \cdot \log N_0$ and the number of stages of the circuits is proportional to $\log N_0$. Therefore, the design of an ECC circuit requires a compromise among the memory area, circuit area and access time. The error correcting codes that have been applied to RAMs until now include Hamming codes [41], modified Hamming codes [40], and bidirectional parity codes [39]. The former one is an SEC code, while the latter two are SEC-DED codes. A DEC code is not realistic for on-chip ECC because of considerable enlargement of the encoding/decoding circuits.

1.13. Scaling Laws

The performance of the above-described memory and logic LSIs has been improved by scaling down of design parameters such as the physical size, impurity, and V_t of MOSTs, the physical size of interconnects, and the operating voltage V_{DD}, as shown in Fig. 1.57 [1, 46]. Scaling under a constant electric field, as shown in Table 1.2 [46], is well-known as ideal scaling. There have been two other scaling approaches that are modifications of ideal scaling; constant operation-voltage scaling and the combination of the above two kinds of scaling methods.

1.13.1. Constant Electric-Field Scaling

In this scaling method, the physical size and threshold voltage V_t of MOST, the interconnect in Fig. 1.57, and the power supply V_{DD} are scaled down by a factor $k(> 1)$. The substrate doping concentration N of the MOST is increased by the same factor. This is done to suppress the resulting short-channel effects, such as drain-source punch-through, by decreasing the depletion length, which is proportional to $\sqrt{V_{DD}/N}$, to $1/k$. As a result, the scaling factor of the MOST current I_{DS}, which is proportional to $(W/L)(V_{DD} - V_T)^2/t_{ox}$, is also $1/k$. This scaling enables not only high density, but also high performance, while maintaining

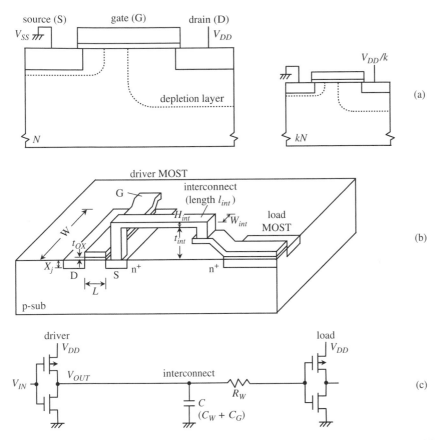

FIGURE 1.57. Scaling in MOS LSIs [1, 46]. (a) The scaling of a MOST ($k = 2$); (b) the conceptual structure of a MOS LSI; (c) the basic circuit. Reproduced from [1] with permission of Springer.

device reliability under constant electric field conditions: The delay, power dissipation, and power-delay product of a circuit, ignoring interconnect capacitances, are improved by $1/k$, $1/k^2$, and $1/k^3$, respectively. The charging power of the load is also reduced by $1/k^3$, with a reduction in the load capacitance of $1/k$. However, the drawbacks are as follows. The interconnect resistance is increased by a factor of k, which makes the interconnect delay more prominent as compared to the circuit delay. The voltage drop along a power-supply line cannot be scaled down, which makes the voltage margin of the circuit narrow. The reliability of the metal interconnect is degraded because of electron migration caused by the increased current density.

In practice, there are some parameters that depart from ideal scaling. Because of the issue of V_{DD} standardization, the external supply V_{DD} cannot necessarily be scaled down, as discussed later. To partly solve the issues regarding resistance and the current density of the interconnect, the thickness and width are not

TABLE 1.2. Various scaling approaches [46].

Approaches	constant electric field $\overset{V}{\underset{\text{core}}{\bigcirc}}$ (scaled) V	constant external voltage $\overset{V}{\bigcirc}$ (fixed) V	constant external voltage $\overset{V}{\bigcirc}$ (fixed) V'(scaled)
Parameters			
Transistor			
dimensions (L, W, t_{ox}, X_j)	$1/k$	$1/k$	$1/k$
impurity (N)	k	k	k
voltage (V_{DD}, V_{tn}, V_{tp})	$1/k$	1	$1/k(V')$
electric field	1	k	1
current (I_{DS})	$1/k$	k	$1/k$
on resistance ($R_{ON} \propto V_{DD}/I_{DS}$)	1	$1/k$	1
delay ($\tau_c = R_{ON}C_G$)	$1/k$	$1/k^2$	$1/k$
power ($P = IV$)	$1/k^2$	k	$1/k$
power delay product ($P\tau_c$)	$1/k^3$	$1/k$	$1/k^2$
area (A)	$1/k^2$	$1/k^2$	$1/k^2$
power density (P/A)	1	k^3	k
Load			
gate capacitance ($C_G \propto WL/t_{OX}$)	$1/k$	$1/k$	$1/k$
interconnect capacitance ($C_W \propto W_{int}l_{int}/t_{int}$)	$1/k$	$1/k$	$1/k$
interconnect resistance ($R_W \propto l_{int}/W_{int}H_{int}$)	k	k	k
interconnect delay ($\tau_W = R_W C_W$)	1	1	1
time constant ratio (τ_W/τ_C)	k	k^2	k
current density ($I_{DS}/W_{int}H_{int}$)	k	k^3	k
voltage drop ratio ($R_W I_{DS}/V_{DD}$)	k	k^3	k
charging power ($\propto CV_{DD^2}$)	$1/k^3$	$1/k$	$1/k^3$

necessarily scaled down in accordance with the scaling law. Other non-scalable parameters, such as parasitic source and drain resistances, contact (through-hole) resistances, short-channel effects, V_t, and device-parameter fluctuations, can also degrade the MOST driving current. In particular, excessive scaling down of V_t is not allowed because of an unacceptably large subthreshold current. In addition, the ever-increasing V_t variation with device scaling causes large variations in speed and subthreshold currents. The details are described in Chapters 4 and 5.

1.13.2. Constant Operation-Voltage Scaling

In the past, DRAM has maintained its power-supply voltage V_{DD} at the same level for as long as possible to solve the V_{DD} standardization issue. For example, a 5-V V_{DD} was used for four generations, 64 Kb to 4 Mb, despite the successive scaling down of internal devices. Even small devices for the 4-Mb chip were tailored to withstand 5-V V_{DD} operation with the help of stress-voltage-immune MOST structures such as LDD (Lightly-Doped Drain). In this scaling method, the MOST delay is improved by a factor of $1/k^2$. However, the electric field and power dissipation of the MOST, and the current density and voltage drop along the interconnect are degraded by k and k^3, respectively. The resulting serious problems are velocity saturation, conductance degradation, degraded reliability due to hot carriers, and the gate-insulator and pn-junction breakdown of the MOST. In addition, electromigration in metal interconnect, CMOS latch-up, and noise are also problems.

1.13.3. Combined Scaling

This scaling necessitates an on-chip voltage-down converter. The converter adjusts the internal supply voltage V_{INT} (i.e., V' in the figure, and V_{DL} discussed later) to match the lowering of the breakdown voltage of the scaled devices, so that the electric field of each device is held constant without changing V_{DD}. This scaling solves almost all of the problems involved in constant-electric-field scaling. Thus, this scaling has been widely used in modern DRAMs, flash memories, and low-end MCUs. However, a power loss by $1/k$, as compared with the constant electric-field scaling, is involved.

1.14. Power Supply Schemes

In principle, LSIs consist of a core and an I/O circuit. The supply voltage to the core must be low to achieve a high density with scaled devices that have low breakdown voltages, while the supply voltage to the I/O circuit must be high to meet the high-voltage requirement of the I/O interface specification, inevitably requiring large devices with fairly high breakdown voltages. Hence, LSIs operate at the two supply voltages. To make them easier to use, however, some LSIs are designed with a single external supply. Though it entails a power loss at the converter, the chip internally generates a low supply voltage for the core with an on-chip voltage down-converter. This can be done only if the core consumes relatively low average and spike currents, so the converter can manage the current with an acceptable power loss and area penalty. This is the preferred design of stand-alone (i.e., general purpose) memories and low-end MPUs/MCUs, which consume a relatively low current and whose design prioritizes ease of use and thus a single external power supply. In contrast, high-end MPUs and embedded (e)-RAMs, which always consume a large current and whose design prioritizes high speed, generally have a dual external power supply. The resulting two external power supply schemes are shown in Fig. 1.58 [1]. Here, ARRAY, PERI, LOGIC, and I/O are memory array, peripheral circuits, logic block, and I/O circuit, respectively.

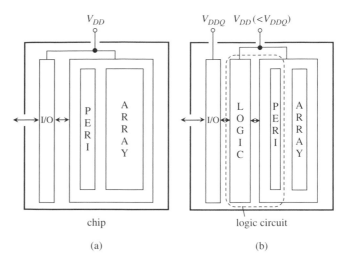

FIGURE 1.58. Examples of applications of a low-voltage memory circuit [1]. (a) Single power supply; (b) dual power supply.

The voltage-down converter also plays an important role in the power-supply standardization [1] in the nano-meter era. It bridges a voltage gap between the high external supply (V_{DD}) and the low internal supply (V_{DL}) voltage tailored below the break-down voltage. Thus, it maintains the external supply voltage, which is dictated by systems, at the same level for as long as possible, even if the breakdown voltage of devices in the core is rapidly lowered below the external supply voltage with device scaling. In addition, the converter approach eventually reduces the power dissipation to $1/k$ (k: device scaling factor, see Table 1.2) despite a power loss at the converter by a voltage difference between the two, and reduces cost because of accepting scaled devices. However, to further reduce the power with reduced power loss, the external supply voltage must be switched to a lower standard voltage when the devices are scaled down to the extent that the power loss is intolerable due to an excessive voltage difference. That is why the external supply has been reduced step by step from 5 V (see Fig. 1.61).

Both power supply schemes need not only the voltage-down converter, but also various on-chip voltage converters. They also need level shifters to accommodate the voltage differences between the core and the I/O. In addition, two kinds of MOSTs with different gate-oxide thicknesses (t_{OX}s) and threshold voltages (V_ts) (i.e., thin-t_{OX} and low-V_t MOSTs for the core, and thick-t_{OX} and high-V_t MOSTs for the I/O), or single thin-t_{OX} MOSTs coupled with a stress-voltage tolerant circuit, mentioned in Chapter 9, are needed to ensure high speed in the core and device reliability at the I/O.

Figure 1.59 [1, 36] shows an example of a single power supply scheme for a stand-alone 16-Mb DRAM. The lowered supply voltage (V_{DL}) is generated from the single external V_{DD} by voltage down-converters using an on-chip reference

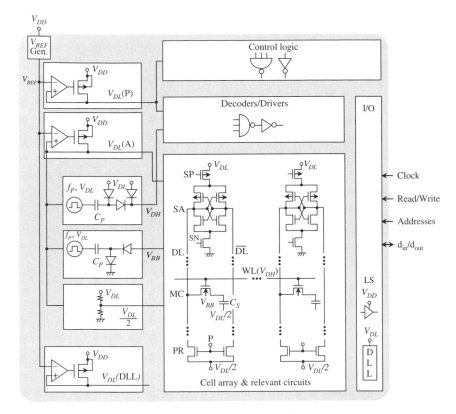

FIGURE 1.59. Schematic architecture of DRAM chip [1, 36]. SA, sense amp; MC, memory cell; DL/$\overline{\text{DL}}$, data lines; WL, word line; PR, precharger; SP/SN, SA activation signals; LS, level shifter.

voltage (V_{REF}). The raised dc voltage (V_{DH} or V_{PP}) enables word-bootstrapping to perform a full-V_{DL} read and write of the high-V_t cells by eliminating the V_t drop in the cell. The negative voltage (V_{BB}) ensures the cells' retention characteristics by applying it to the substrate of an nMOS cell array to protect it from forward biasing of the pn-junction at the cell-storage node, and/or to the word lines during non-selected periods when low-V_t cells are used (i.e., negative word-line scheme). Here, the V_{BB} and V_{DH} circuits consist of a charge-pumping capacitor and MOS diodes that function as rectifiers. A half-V_{DL} power supply realizes a quiet and low-power array by precharging the data lines to the half-V_{DL} level, and double the cell capacitance for a given stress voltage by supplying it to the cell-capacitor plate. At the I/O, in addition to a level shifter, even analog delay-locked loop (DLL) circuits [32] is necessary to solve the high-speed clocking problem when the output data timing of synchronous DRAMs (SDRAMs) is aligned with the input clock. A typical voltage down-converter (VDC) for a 16-Mb DRAM [1, 36] is shown in Fig. 1.60. It consists of a current-mirror differential amplifier and a pMOS-output, so the V_{DL} is well regulated

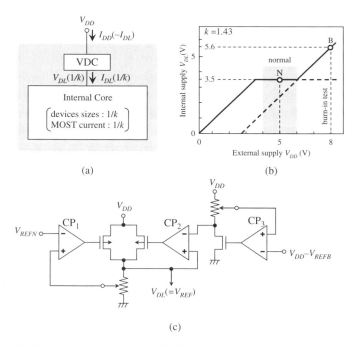

(a)   (b)

(c)

FIGURE 1.60. Voltage down-converter (VDC) capable of burn-in test [1, 36]. Concept (a), internal supply (b), and circuits (c).

to a reference voltage of 3.5 V (i.e., V_{REFN}) at the normal voltage of 5-V V_{DD}. The burn-in test at a high stress voltage of 8 V is made possible by raising the V_{DL} with another parallel-connected converter operating at another reference voltage, V_{REFB}. The W/L of the out-put pMOST was about 6,000 for a 16-Mb load of $R_L = 5$ ohm and $C_L = 650$ pF. By using a pole-zero phase compensation of $Rc = 7$ ohm and $Cc = 650$ pF, a phase margin of about 53 degrees, and a well-regulated V_{DL} of less than 5% despite a sharp spike current of 100 mA with a pulse width of 20 ns were realized with an area penalty of about 5% of the total chip. The more detail is discussed in Chapter 7.

1.15. Trends in Power Supply Voltages

Figure 1.61 shows trends in the supply voltage of RAMs [3]. For stand-alone DRAMs, the external supply voltage has been reduced from 1.8 V to 1.5 V, and further to 1.2 V, along with reductions of the internal supply voltage V_{INT} (to be more exact, for the data line and some peripheral circuits) using voltage down-converters. This is also the case for embedded (e-) DRAMs. For stand-alone SRAM products, the single external supply voltage, but without the down-converter, has been reduced with almost the same pace as that for the stand-alone DRAMs. For e-SRAMs, however, the external supply voltage has been reduced

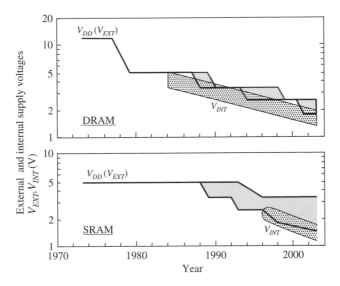

FIGURE 1.61. Trends in the power supply voltage of RAMs. Reproduced from [3] with permission of IBM.

even more than that for DRAMs under the influence of MPUs, now reaching below 1.2 V. For high-end MPUs, the supply voltage has been almost the same as that of e-SRAMs which are usually used as on-chip cache memories. As for low-end MCUs (Microcontroller Units), however, the single external supply has dominated, allowing various supply voltages to be coexistent, reflecting wide applications from high-end to low-end systems, as shown in Fig. 1.62. Note that single 5-V MCUs for car-engine controls exist even in the nanometer era with the help of a voltage down-converter. Stand-alone NAND flash memories have used the same single external supply voltage as stand-alone RAMs and low-end MPUs/MCUs, although various voltages are used internally, as discussed

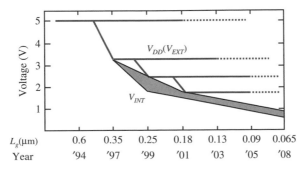

FIGURE 1.62. Trends in the power supply voltage of microcontrollers. Reproduced from [45] with permission; © 2006 IEEE.

previously. Embedded-flash memories also need various internal supply voltages for stable memory cell operations.

1.16. Power Management for Future Memories

Many technical issues must be resolved to design ultra-low voltage nano-scale memories. Designers must: ensure high signal-to-noise-ratio (S/N) in the memory cells; reduce leakage currents, and variations in the voltage margin of memory cells and the speed and leakage current of peripheral circuits and; manage internal power-supply. A high S/N cell design is necessary because signal charge and signal voltage of memory cells decrease and noise increases as device size and voltage decrease. Reduction in leakage currents is also vital because, for example, the subthreshold current increases exponentially as V_t decreases, eventually dominating even the active current of a chip. Examples of this are the data retention current of a 1-Mb SRAM array (Fig. 1.63(a)) and the active current of peripheral circuits in DRAMs (Fig 1.63(b)) [1, 3, 37]. Variations in the voltage margin of memory cells and the speed and leakage of peripheral circuits must be reduced because they increase rapidly as device size and voltage decrease. Management of internal supply voltages is the key to this reduction.

Figure 1.64 is a schematic of a power management design that reduces the subthreshold current (leakage) of the core using a low-actual V_t MOST. The basic design concept proposed to date (and illustrated in the figure) [1] is an effectively high-V_t MOST with various dynamic back-biasing schemes. One such scheme is a high-speed, low-leakage LSI in which the V_t is controlled so that it

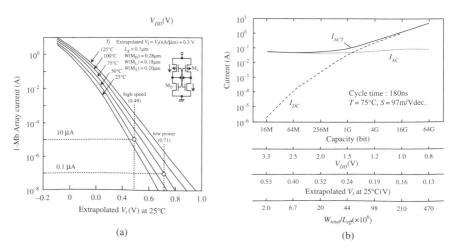

(a)

(b)

FIGURE 1.63. V_t of cross-couple MOSTs versus subthreshold current of 1-Mb SRAM (a), and active current of peripheral circuits in DRAMs (b) [1, 3, 37].

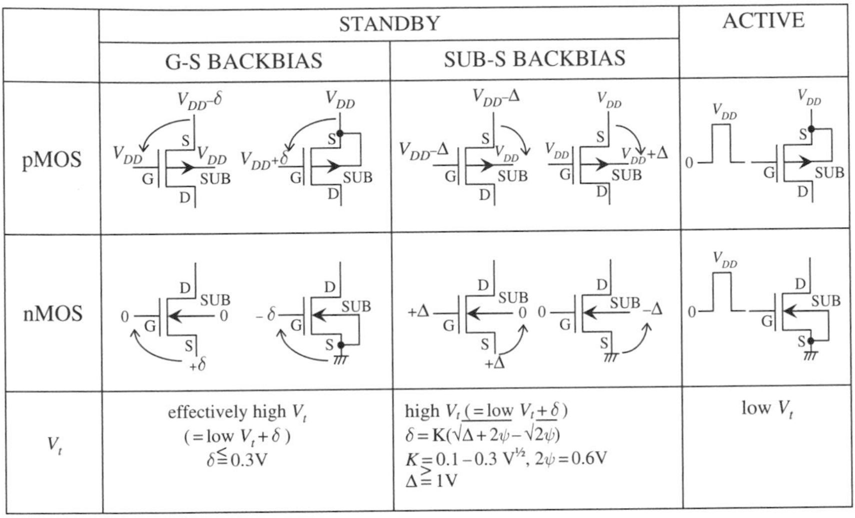

	STANDBY		ACTIVE
	G-S BACKBIAS	SUB-S BACKBIAS	
pMOS	$V_{DD}-\delta$ V_{DD}	$V_{DD}-\Delta$ V_{DD}	V_{DD} V_{DD}
nMOS			V_{DD}
V_t	effectively high V_t $(=\text{low } V_t + \delta)$ $\delta \lesssim 0.3\text{V}$	high $V_t (=\text{low } V_t + \delta)$ $\delta = K(\sqrt{\Delta + 2\psi} - \sqrt{2\psi})$ $K = 0.1 - 0.3 \text{ V}^{1/2},\ 2\psi = 0.6\text{V}$ $\Delta \gtrsim 1\text{V}$	low V_t

FIGURE 1.64. The concepts behind the variable-V_t approaches for reducing the standby subthreshold current of low- V_t MOSTs. Reproduced from [1] with permission of Springer.

is low enough in the active mode to operate at high speed but high enough in the standby (or non-selected) mode for a low leakage. Obviously, the necessary low V_t in the active mode favors no back-biasing scheme, that is, directly connecting the substrate and source of pMOSTs and nMOSTs during active periods, even though they have a large subthreshold current. Gate-source (G-S) back biasing and substrate SUB or well-source (SUB-S) back biasing reduces the subthreshold current during the standby period. The G-S back bias provides a sufficiently high V_t by changing the source voltage from V_{DD} to $V_{DD} - \delta$ for a pMOST, and from $0\,\text{V}$ to δ for an nMOST or by changing the gate voltage from V_{DD} to $V_{DD} + \delta$ for a pMOST and from $0\,\text{V}$ to $-\delta$ for an nMOST. In practice, δ can be less than $0.3\,\text{V}$, because even a small δ greatly reduces the subthreshold current. SUB-S back biasing provides a high V_t due to the body effect, in which the source voltage is changed by Δ with a fixed substrate voltage, or the substrate voltage is changed by Δ with a fixed source voltage. However, to realize the same high V_t (i.e. the same δ) as in the G-S back bias, Δ must be quite large, as shown by the following equation

$$\delta = K \left(\sqrt{\Delta + 2\psi} - \sqrt{2\psi} \right)$$

For example, a δ of $0.2\,\text{V}$ that makes possible a two-order reduction in the subthreshold-current for a MOST with $S = 0.1\,\text{V/dec.}$ requires a Δ as large as $2.5\,\text{V}$ for $K = 0.2\,\text{V}^{1/2}$. The pn-junction leakage of the resulting large Δ may generate excessive stand-by power. Note that the G-S back biasing caused by changing the source voltage by δ also raises the V_t. However, the effect is negligible because of a small δ.

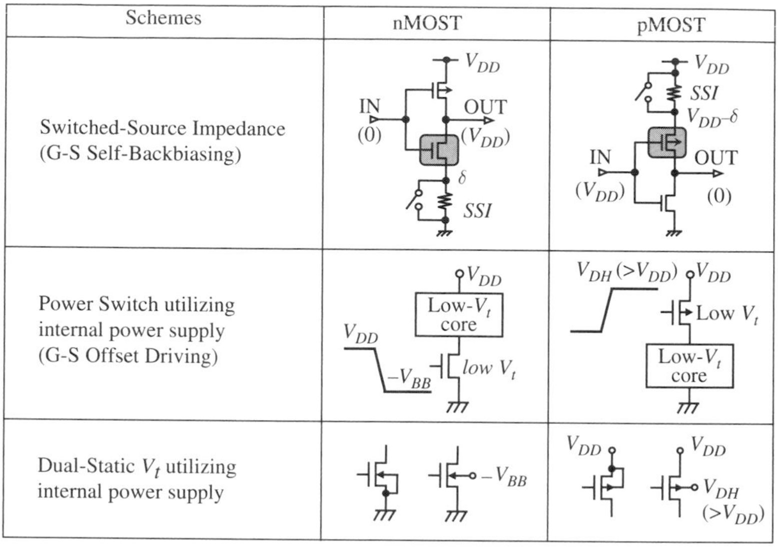

Schemes	nMOST	pMOST
Switched-Source Impedance (G-S Self-Backbiasing)	IN (0) → OUT (V_{DD}), V_{DD}, δ, SSI	V_{DD}, SSI, $V_{DD-\delta}$, IN (V_{DD}) → OUT (0)
Power Switch utilizing internal power supply (G-S Offset Driving)	V_{DD}, Low-V_t core, $-V_{BB}$ low V_t	$V_{DH} (>V_{DD})$ V_{DD}, Low V_t, Low-V_t core
Dual-Static V_t utilizing internal power supply	$-V_{BB}$	V_{DD} V_{DD}, V_{DH} ($>V_{DD}$)

FIGURE 1.65. Practical circuits to reduce the subthreshold current, applicable even to active mode.

Figure 1.65 shows schematics of three practical examples for memories [3] applicable even to the active mode. These are the switched-source impedance (SSI), power switch, and dual-static V_t circuits. An SSI is stacked at the source of the MOST that generates a subthreshold current. Thus, it is located at the nMOS source for low input but at the pMOS source for high input. When the switch is turned off, a voltage δ is developed by a leakage, so the leakage is automatically reduced, resulting in gate-source self-back biasing. The power switch shuts off the power supply to the low-V_t internal core, eliminating the leakage. This circuit uses a negative supply V_{BB} or a raised supply V_{DH} (or V_{PP}) to cut off a low-V_t nMOST or pMOST switch, resulting in G-S offset driving. For high-speed random cycle switching, however, the gate voltage of the power-switch MOST must be well regulated, particularly when it is generated by the on-chip charge pump explained in Chapter 4. Dual-static V_t also uses a V_{BB} or V_{DH} to increase the V_t of the nMOST or pMOST with a substrate biasing, enabling a dual-static V_t.

Managing power is surprisingly difficult. Indeed, in the past, DRAM designers have encountered many problems even in static or quasi-static controls of V_{BB} and V_{DD}. In the following, the power-management designs [1, 3, 46] proposed to date are briefly described.

1.16.1. Static Control of Internal Supply Voltages

It is well known that DRAM has been the only large-volume production LSI using a substrate bias that is supplied from an on-chip V_{BB} generator. The on-chip

V_{BB} concept was firstly proposed in 1976 to suppress the V_t-lowering developed at a shorter channel [33, 34]. However, the 16-Kb nMOS DRAM, which was intensively developed using a channel length of $5\,\mu m$ as the most-advanced DRAM at that time, did not use the on-chip V_{BB} approach. Instead, it used a triple-power-supply of $V_{DD} = 12\,V$ for the internal core, $V_{CC} = +5\,V$ for chip I/O interface, and $V_{BB} = -5V$ for stable operations of the single p-type substrate. Despite the triple power supply, the 16-Kb DRAM was successful. It was in the 64-Kb DRAM generation that the on-chip V_{BB} approach was first industrialized to realize a single 5-V nMOS DRAM [1], where the generator supplied a quasi-static V_{BB} of about $-3\,V$ to the p-type substrate of the whole chip (i.e., both array and periphery) with $V_{DD} = V_{CC} = +5\,V$. Furthermore, a dual V_t of 0.5 V and 0.35 V, while a single V_t of 1 V in the preceding 16-Kb generation, was used to maintain the circuit speed despite a drastic voltage transition from 12 V to 5 V. The low V_t was for circuits necessitating a higher speed. In this scheme, however, the values of V_ts at a steady state of $V_{BB} = -0.3\,V$ and even V_{BB} were carefully chosen [1], considering the current drivability of the generator and substrate currents at the load, refresh-time characteristics of cells shortened by minority-carrier injection to cells, and possible instabilities in unusual conditions, such as a surge current at power-on due to a dynamically-changing V_t during a shallow V_{BB} on the way to the steady state V_{BB}, and degraded V_{BB}-level during burn-in high-voltage stress tests. Even so, the DRAM designers were fortunate because both the static bias setting of a deep V_{BB} of about $-2\,V$ to $-3\,V$, a sufficiently high V_t of about 0.5 V and 0.35 V, and slow cycle time of the DRAM allowed stable chip operations with small changes in V_t even with quite large quasi-static V_{BB} variations and V_{BB} noise [1]. In the CMOS era, the substrate bias has been removed from peripheral circuits to mainly eliminate the above-described instabilities involved in the generator, and it has only been supplied to the memory-cell array to ensure stable operations.

The on-chip V_{BB} approach has revived for LSIs in the nanometer era since the mid-1990's to reduce the subthreshold leakage with raised V_t, and to compensate for variations in leakage and speed caused by parameter variations in V_t, voltage, and temperature. When V_{BB} is statically or quasi-statically controlled based on parameter variations, inter-die leakage and speed variations of logic circuits can be suppressed although intra-die speed variations remain unimproved. Controlling forward V_{BB} is more effective in reducing speed variations because the $V_t - V_{BB}$ characteristics are more sensitive to V_{BB}. For example, it improved speed of operations by 10%. However, additional current consumption, in the form of bipolar current induced by the forward V_{BB}, is a matter that must be considered, as discussed previously. When compared with the above-described DRAMs, the smaller V_t, shallow V_{BB}, and faster cycle time necessary for the nano-scale LSIs increase instabilities although the lower V_{DD} decreases instabilities. Thus, the requirements to suppress noises at the source and substrate lines become more stringent, calling for a uniform distribution of the forward V_{BB} throughout the chip.

1.16.2. Dynamic Control of Internal Supply Voltages

Although the dynamic control of V_{DD} and/or V_{BB} reduces power dissipation and/or subthreshold currents, it may cause instability problems. Such dynamic controls are not welcome in terms of stable operations of LSIs, considering the voltage bump test which was carried out in the mid-1970's to guarantee the stable operation of the 16-Kb DRAM against noises on the memory board. At that time even a bump as small as $\pm 10\%$ V_{DD} made dynamic circuits unstable. The instability was due to charge being trapped at floating nodes when voltage bumps were applied, causing malfunctions at the next cycle. Unfortunately, almost all peripheral circuits and DRAM cells had been dynamic in the nMOS era until the early 1980's. Thus, small diode-connected nMOSTs (i.e. level keeper) were connected to floating nodes in peripheral circuits to allow trapped charges to escape. However, bumps degraded the voltage margin of nMOS cells, calling for grounded-plate cell capacitors [1] as a partial solution. Even in the CMOS era, memory cells, sensing relevant circuits (such as data-line precharge circuits and sense amplifiers), and row decoders/drivers have still been dynamic even if other peripheral circuits have been static. Half-V_{DD} (i.e., mid-point) sensing [1], coupled with a half-V_{DD} cell plate using the same half-V_{DD} generator and coupled with a boosted word line, has been a circuitry solution to maintain the wide margins of DRAM cells and sensing relevant circuits despite voltage bumps. A CMOS feedback level keeper that is familiar to logic designers has widely been used for other dynamic circuits.

Such instabilities might extensively occur if supply voltages of low-V_t circuits are dynamically and widely changed. Nevertheless, a few approaches have been proposed: Dynamic voltage scaling (DVS) [35], in which the clock frequency and V_{DD} dynamically vary in response to the computational load, is a good example, although it is only applied to the logic blocks. It provides reduced energy consumption per process during periods when few computations are made, while still providing peak performance when it is required. Applying DVS, however, would make dynamic circuits (e.g. e-DRAMs) unstable unless level keepers are used, although resultant instabilities depend on the changing rate of V_{DD} and clock frequency. Unfortunately, RAM cells and their relevant circuits are incompatible with such dynamic controls, and thus at least they should be "quiet". Moreover, they must operate at a higher V_{DD}. Their inherently small voltage margins are responsible for the requirements. Thus, as long as the controls never give detrimental effects to RAM cells and their related circuits, some of them could be applied to parts of peripheral logic circuits (e.g. static circuits) in RAMs. SRAMs using the 6-T cell, however, may accept dynamic voltage controls due to the cll's wide voltage margins as long as V_{DD} is quite high, although care should be taken if dynamic sensing schemes are adopted. Note that the highest V_{DD} (V_{max}) and lowest V_{DD} (V_{min}) that DVS can accept are eventually determined by the breakdown voltage of MOSTs, and the lowest necessary V_{DD} in terms of the stability of RAM cells and speed variation of logic circuits, as discussed in Chapters 2 to 5. Since V_{max} is lowered while V_{min} is usually not lowered, with device scaling, the range across which it is possible to vary V_{DD} becomes narrower. This implies that this approach becomes less effective with device scaling. In addition, successful operations over a wide range of V_{DD} require the accurate tracking of all circuit delays.

1.17. Roles of On-Chip Voltage Converters

On-chip voltage converters are becoming increasingly important for ultra-low voltage nano-scale memories to accomplish the above-described challenges. They include the reference generator, the voltage down-converters, the voltage up-converter and negative voltage generator with charge pump circuits, and level shifters. Key design issues in the converters are the efficiency of voltage conversion, the degree of voltage setting accuracy and stability in the output voltage, load-current delivering capability, the power of the converter itself (especially during the standby period), speed and recovery time, cost of implementation, and reliability. The roles of the converters are summarized in the following. More details of circuit configurations are described in Chapters 6 to 9.

The reference voltage generator must create a well-regulated supply voltage for any variations of voltage, process, and temperature, because other converters operate based on the reference voltage. The voltage down-converter allows a single external power supply while standardizing the power supply, as explained previously. Thus, the converter has made it possible to quadruple memory capacity with the same V_{DD} despite the ever-lower device breakdown voltage. Moreover, it has realized successive chip-shrinking with scaled devices under a fixed memory capacity and V_{DD}, enabling a reduction in bit-cost. In addition, the converter provides the many advantages of a well-fixed internal voltage regardless of the unregulated supply-voltages from various batteries, protection of internal core-circuits against high voltages within a wide range of voltage variations, and an adjustable internal voltage (V_{DL}) in accordance with variations in supply voltage (V_{DD}), temperature, and device parameters to compensate for speed and leakage variations.

The voltage up-converter consisting of charge pump circuits has been widely used to eliminate the V_t-drop in DRAM circuits and to program and erase a flash memory cell, as explained previously. The requirements for the converter, however, are different for both memories. The raised-voltage (V_{PP} or V_{DH}) necessary for DRAMs is much lower than that for flash memory. Instead, the cycle time (t_{cyc}) of the V_{PP}-pulse for DRAMs is much shorter: In modern memory design with $V_{DD} = 3.3\,V$ or $5\,V$, the DRAM-V_{PP} ranges from $4\,V$ to $8\,V$ with a boost ratio (V_{PP}/V_{DD}) of 1.3–1.5 while the flash memory-V_{PP} is $10\,V$ to $20\,V$ with a boost ratio of 2–6. The cycle time is about $100\,ns$ for DRAMs while for flash memories it is $1\,\mu s$ at most even for V_t-verify operation. Thus, the V_{PP}-generator for DRAMs must provide more charges to compensate for the charge loss at the load every cycle, if the same load capacitance and conversion efficiency of V_{PP}–generator are assumed. The negative-voltage generator also consisting of charge pump circuits has been widely used to generate a negative substrate bias (V_{BB}) for DRAMs to prevent nMOSTs from being forward-basing, especially for cell-transfer MOSTs in a noisy cell array. Coupled with a triple well substrate, it protects the DRAM cell array from minority carrier injection from peripheral circuits [1].

Even for ultra-low voltage nano-scale memories, both the up-converter and the negative-voltage generator are indispensable to precisely control the V_t of MOSTs and power switches, and to compensate for the leakage and speed variations. Unfortunately, however, the designs are more and more difficult because of the ever-larger pumping current due to a poor pumping efficiency at a low V_{DD}. In addition, a deep understanding of the voltage generation scheme and the load characteristics is required for successful designs. In particular, in low-V_t LSIs, special attention must be paid to instabilities involved in the floating nodes. Indeed, floating nodes, especially heavily-capacitive floating node are sources of malfunctions. For example, in unusual conditions [1], such as during power on/off, and burn-in test with high stress voltage and high temperature, MOSTs may fall into extremely low V_t or even into a depleted state, despite a normal low-V_t in the normal operation, unexpectedly causing a large rush current [1, 3].

Level shifters are also necessary to bridge resultant voltage differences between internal blocks, and the internal core and I/O circuits. In addition to the voltage up-converter, they call for stress voltage-immune devices and/or circuits to ensure the reliability of devices.

References

[1] K. Itoh, *VLSI Memory Chip Design*, Springer-Verlag, NY, 2001.

[2] S. Rusu, S. Tam, H. Muljono, D. Ayers, and J. Chang, "A dual-core multi-threaded Xeon processor with 16 MB L3 cache," ISSCC Dig. Tech. Papers, pp. 102–103, Feb. 2006.

[3] Y. Nakagome, M. Horiguchi, T. Kawahara, K. Itoh, "Review and prospects of low-voltage RAM circuits," IBM J. R & D, vol. 47, no. 5/6, pp. 525–552, Sep./Nov. 2003.

[4] M.I.Elmasry, Ed., *Digital MOS Integrated Circuits 2* (IEEE Press, New York 1992).

[5] E. Hamdy and A. Mohsen, "Characterization and modeling of transient latch-up in CMOS technology," IEDM Dig. Tech. Papers, pp.172–173, 1983.

[6] J. Kedzierski, E. Nowak, T. Kanarsky, Y. Zhang, et al., "Metal-gate FinFET and fully-depleted SOI devices using total gate silicidation," IEDM Dig. Tech. Papers, pp. 247–250, Dec. 2004.

[7] International Technology Roadmap for Semiconductors, 2004 Update, Emerging Research Devices, pp. 9–14.

[8] T.C. Chen, "Where CMOS is going: Trendy Hype vs. Real Technology," ISSCC Dig. Tech. Papers, pp. 22–28, Feb. 2006.

[9] K. Nii, Y. Tenoh, T. Yoshizawa, S. Imaoka, Y. Tsukamoto, Y. Yamagami, T. Suzuki, A. Shibayama, H. Makino, and S. Iwade, "A 90-nm low-power 32K-Byte embedded SRAM with gate leakage suppression circuit for mobile applications," Symp. VLSI Circuits Dig., pp. 247–150, June 2003.

[10] T. Inukai et al., "Suppression of stand-by tunnel current in ultra-thin gate oxide MOSFETs by dual oxide thickness MTCMOS," Int. Conf. on Solid-State Dev. and Mat. Ext. Abst., pp. 264–265, Aug. 1999.

[11] K. Osada, Y. Saitoh, E. Ibe, and K. Ishibashi,"16.7fA/cell tunnel-leakage-suppressed 16-Mbit SRAM based on electric-field-relaxed scheme and alternate ECC for handling cosmic-ray-induced multi-errors," ISSCC Dig. Tech Papers, pp. 302–303, Feb. 2003.

[12] D. J. Frank, "Power-constrained CMOS scaling limits," IBM J. Res. & Dev., vol. 46, pp. 235–244, Mar. 2002.

[13] G. Ono, M. Miyazaki, H. Tanaka, N. Ohkubo, and T. Kawahara, "Temperature referenced supply voltage and forward-body-bias controled (TSFC) architecture for minimum power consumption," ESSCIRC Dig. Tech. Papers, pp. 391–394, 2004.

[14] S. Heo and K. Asanovic, "Leakage-biased domino circuits for dynamic fine-grain leakage reduction," Symp. VLSI Circuits Dig. Tech. Papers, pp. 316–319, June 2002.

[15] H. Yoon, J. Y. Sim, H. S. Lee, K. Nam Lim et al., "A 4Gb DDR SDRAM with gain-controlled pre-sensing and reference bitline calibration schemes in the twisted open bitline architecture," ISSCC Dig. Tech. Papers, pp. 378–379, Feb. 2001.

[16] S. Naffziger, B. Stackhouse, and T. Grutkowski, "The implementation of a 2-core multi-threaded Itanium®-family processor," ISSCC Dig. Tech. Papers, pp. 182–183, Feb. 2005.

[17] Y. H. Suh, H. Y. Nam, S. B. Kang, B. G. Choi, H. S. Mo, G. H. Han, H. K. Shin, W. R. Jung, H. Lim, C. K. Kwak, H. G. Byun, "A 256 Mb synchronous-burst DDR SRAM with hierarchical bit-line architecture for mobile applications," ISSCC Dig. Tech. Papers, pp. 476–477, Feb. 2005.

[18] D-S Byeon, S-S Lee, Y-H Lim, J-S Park et al., "An 8Gb multi-level NAND Flash memory with 63nm STI CMOS process technology," ISSCC Dig. Tech. Papers, pp. 46–47, Feb. 2005.

[19] K. Takeuchi, Y. Kameda, S. Fujimura, H. Otake et al., "A 56nm CMOS 99mm^2 8Gb multi-level NAND Flash memory with 10 MB/s program throughput," ISSCC Dig. Tech. Papers, pp. 144–145, Feb. 2006.

[20] F. Masuoka, M. Asano, H. Iwahashi, T. Komuro, and S. Tanaka, "A new Flash EEPROM cell using triple polysilicon technology," IEDM Dig. Tech. Papers, pp. 464–467, Dec. 1984.

[21] P. Pavin, R. Bez, P. Olivo, and E. Zanoni, "Flash memory cells-an overview," IEEE Proc. Vol. 85, No.8, 1248–1271, 1997.

[22] V. N. Kynett, A. Baker, M. Fandrich, G. Hoekstra, O. Jungroth, J. Kreifels, and S. Wells, "An in-system reprogrammable 256K CMOS Flash memory," ISSCC Dig. Tech. Papers, pp. 132–133, Feb. 1988.

[23] S. Haddad, C. Chang, A. Wang, J. Bustillio, J. Lien, T. Montalvo, and M.V. Buskirk, "An investigation of erase-mode dependent hole trapping in Flash EEPROM memory cell," Electron Device Letters, vol. 11, no.11, pp. 514–516, Nov. 1990.

[24] F. Masuoka, M. Momodori, Y. Iwata, and R. Shirota, "New ultra high density EPROM and Flash EEPROM cell with NAND structure cell," IEDM Dig. Tech. Papers, pp. 552–555, 1987.

[25] T. Hara, K. Fukuda, K. Kanazawa, N. Shibata, et al., "A 146-mm^2 8-Gb multi-level NAND Falsh memory with 70-nm CMOS technology," IEEE J. Solid-Sate Circuits, Vol. 41, No. 1, pp.161–169, Jan. 2006.

[26] K. Imamiya, Y. Sugiura, H. Nakamura, T. Himeno, K. Takeuchi, T, Ikehashi, K. Kanda, K. Hosono, R. Shirota, S. Aritomo, K. Shimizu, K. Hatakeyama,

and K. Sakui, "A 130-mm², 256-Mb NAND Flash with shallow trench isolation technology," IEEE J. Solid-State Circuits, Vol.34, No.11, pp. 1536–1543, Nov. 1999.

[27] T. Tanzawa and S. Atsumi, "Optimization of word-line booster circuits for low-voltage Flash memories," IEEE J. Solid-State Circuits, Vol.34, No.8, pp. 1091–1098, Aug. 1999.

[28] T. Tanzawa, T. Tanaka, K. Takeuchi, and H. Nakamura, "Circuit techniques for a 1.8-V-only NAND Flash memory," IEEE J. Solid-State Circuits, Vol.37, No.1, pp. 84–89, Jan. 2002.

[29] J.F. Ziegler, H. W. Curtis, H. P. Muhlfeld, C. J. Montrose, et al., "IBM experiments in soft fails in computer electronics (1978–1994)," IBM J. Res. Develop., vol.40, no.1. pp. 3–18, Jan. 1996.

[30] K. Osada, K. Yamaguchi, Y. Saitoh, and T. Kawahara, "SRAM immunity to cosmic-ray-induced multierrors based on analysis of an induced parasitic bipolar effect," IEEE J. Solid-State Circuits, vol. 39, No.5, pp. 827–833, May 2004.

[31] Y. Komatsu, Y. Arima, T. Fujimoto, T. Yamashita, and K. Ishibashi, "A soft-error hardened latch sceme for SoC in a 90 nm technology and beyond," CICC Dig. Tech. Papers, pp. 329–332, Oct. 2004.

[32] T. Saeki, Y. Nakaoka, M. Fujita, A. Tanaka, et al., "A 2.5-ns clock access, 250-MHz, 256-Mb SDRAM with synchronous mirror delay," IEEE J. Solid-State Circuits, vol. 31, no.11, pp. 1656–1668, Nov. 1996.

[33] M. Kubo, R. Hori, O. Minato and K. Sato, "A threshold voltage controlling circuit for short channel MOS integrated circuits," ISSCC Dig. Tech. Papers, pp. 54–55, Feb. 1976.

[34] E. M. Blaser, W. M. Chu, and G. Sonoda, "Substrate and load gate voltage compensation," ISSCC Dig. Tech. Papers, pp. 56–57, Feb. 1976.

[35] T. Burd, T. Pering, A. Stratakos and R. Brodersen, "A dynamic voltage scaled microprocessor system," ISSCC Dig. Tech. Papers, pp. 294–295, Feb. 2000.

[36] K. Itoh, "Analog circuit techniques for RAMs-present and future-," Analog VLSI Workshop, 2005 IEEJ, Dig. Tech. Papers, pp. 1–6, Bordeaux, Oct. 2005.

[37] K. Itoh, K. Osada, and T. Kawahara, "Reviews and prospects of low-voltage embedded RAMs," CICC Dig. Tech. Papers, pp. 339–344, Oct. 2004.

[38] Phillip E. Allen and Douglas R. Holberg: "CMOS analog circuit design," New York: Holt, Rinehart and Winston, INC.1987.

[39] T. Mano, J. Yamada, J. Inoue and S. Nakajima, "Circuit techniques for a VLSI memory," IEEE J. Solid-State Circuits, vol. SC-18, pp. 463–469, Oct. 1983.

[40] H. L. Kalter, C. H. Stapper, J. E. Barth Jr., J. DiLorenzo, C. E. Drake, J. A. Fifield, G. A. Kelley Jr., S. C. Lewis, W. B. van der Hoeven and J. A. Yankosky, "A 50-ns 16-Mb DRAM with a 10-ns data rate and on-chip ECC," IEEE J. Solid-State Circuits, vol. 25, pp. 1118–1128, Oct. 1990.

[41] K. Arimoto, K. Fujishima, Y. Matsuda, M. Tsukude, T. Oishi, W. Wakamiya, S. Satoh, M. Yamada and T. Nakano, "A 60-ns 3.3-V-only 16-Mbit DRAM with multipurpose register," IEEE J. Solid-State Circuits, vol. 24, pp. 1184–1190, Oct. 1989.

[42] F. Morishita, K. Suma, M. Hirose, T. Tsurude, Y. Yamaguchi, T. Eimori, T. Oashi, K. Arimoto, Y. Inoue, and T. Nishimura, "Leakage mechanism due to floating body and countermeasure on dynamic retention mode of SOI DRAM," Symp. VLSI Tech. Dig. Tech. Papers, pp. 141–142, 1995.

[43] *Low-power High-speed LSI Circuits & Technology*, REALIZE INC., 1998 (in Japanese).

[44] S. Satoh, Y. Tosaka, and T. Itakura, "Scaling law for secondary cosmic-ray neutron-induced soft-errors in DRAMs," Ext. Abstract, Int'l Conf. Solid-State Devices and Materials, pp. 40–41,1998.

[45] M. Hiraki, K. Fukui and T. Ito, "A low-power microcontroller having a $0.5 - \mu A$ standby current on-chip regulator with dual-reference scheme," IEEE J. Solid-State Circuits, vol. 39, pp. 661–666, Apr. 2004.

[46] K. Itoh, *VLSI Memory Design*, Baifukan, Tokyo, 1994 (in Japanese).

[47] M. Momodori, Y. Itoh, R. Shirota, Y. Iwata, R. Nakayama, R. Kirisawa, T. Tanaka, S. Aritome, T. Endoh, K. Ohuchi, and F. Masuoka, "An experimental 4-Mbit CMOS EEPROM with a NAND-structured Cell," IEEE J. Solid-State Circuits, vol. 24, no. 5, pp. 1238–1243, Oct. 1989.

[48] M. Momodori, T. Tanaka, Y. Iwata, Y. Tanaka, H. Oodaira, Y. Itoh, R. Shirota, K. Ohuchi, and F. Masuoka, "A 4-Mb NAND EEPROM with tight programmed V_t distribution," IEEE J. Solid-State Circuits, vol. 26, no. 4, pp. 492–496, Apr. 1991.

[49] H. Gotou, Y. Arimoto, M. Ozeki, and K. Imaoka, "Soft error rate of SOI-DRAM," IEDM Dig. Tech. Papers, pp. 870–871, 1987.

[50] K. Ishibashi, T. Yamashita, Y. Arima, I. Minematsu and T. Fujimoto, "A 9mW 50MHz 32b adder using a self-adjusted forward body bias in SoCs," ISSCC Dig. Tech. Papers, pp. 116–117, Feb. 2003.

[51] Y. Komatsu, K. Ishibashi, M. Yamamoto, T. Tsukada, K. Shimazaki, M. Fukazawa, and M. Nagata, "Substrate-noise and random-fluctuations reduction with self-adjusted forward body bias," CICC Dig. Tech. Papers, pp. 35–38, 2005.

[52] K. Ishibashi, T. Fujimoto, T. Yamashita, H. Okada, Y. Arima, Y. Hashimoto, K. Sakata, I. Minematsu, Y. Itoh, H. Toda, M. Ichihashi, Y. Komatsu, M. Hagiwara, and T. Tsukada, "Low-voltage and low-power logic, memory, and analog circuit techniques for SoCs using 90nm technology and beyond (invited)," IEICE Tran. on Electronics, Vol.E89-C, No.3, pp. 250–262, 2006.

[53] *Leakage in Nanometer CMOS Technologies*, Edited by S. G. Narendra and A. Chandrakasan, Springer, 2006.

2
Ultra-Low Voltage Nano-Scale DRAM Cells

2.1. Introduction

Ultra-low-voltage nano-scale DRAMs are becoming increasingly important [1–4] to meet the need for low-power and large memory capacity memories from the rapidly growing mobile market, while ensuring the reliability of nano-scale devices. They could replace embedded (e-) SRAMs in MPUs (Microprocessor Units) and MCUs (Microcontroller Units) because low-voltage nano-scale e-SRAMs are more and more difficult to realize with a small enough memory cell [5–7], as discussed in Chapter 3. Thus, sub-1-V DRAMs have been actively researched and developed, exemplified by a 0.6-V 16-Mb e-DRAM [8], and a 1.2- to 1-V 16-Mb e-DRAM [9]. To create such DRAMs, however, many challenges remain with the memory cell and cell related circuits [1–4]. First, the signal-to-noise-ratio (S/N) of the memory cell must be maintained for reliable sensing despite lower voltage operations. Thus, noise components and their generation mechanisms in the array must be intensively investigated, which is especially important for the well-known mid-point sensing approach (i.e., the half-V_{DD} data-line precharge) that always entails a slow speed. Second, the leakage current and its variation in the memory cell must be reduced by using a high enough threshold voltage (V_t) for the transfer MOST to ensure sufficiently long refresh time. Third, the memory cell must be simple and small enough, especially for low-voltage e-DRAMs, since the memory block dominates the chip size of many LSIs.

This chapter describes the state-of-the-art DRAM cells in terms of ultra-low-voltage operations. First, trends in the development of various DRAM cells including gain cells are reviewed with respect to S/N and cell size. Second, the cell type and the sensing scheme are summarized comprehensively, focusing on the one-transistor one-capacitor (1-T) DRAM cells. Third, the folded-data-line 1-T cell that has been the de-facto standard cell is discussed in detail, clarifying limiting factors of low-voltage operations. Fourth, the open-data-line 1-T cell that was popular in the 1970's is investigated in terms of noise, because it could revive due to the small cell size if noise is reduced sufficiently. Fifth, the two-transistor two-capacitor (2-T) cell suitable most for low-voltage operations due to minimized noise is discussed. Finally, low-voltage potential of double-gate fully-depleted SOI cells is investigated.

2.2. Trends in DRAM-Cell Developments

For large memory-capacity DRAMs, reducing the memory-cell size, while maintaining a high S/N high, is a prime concern [1]. Figure 2.1(a) compares cell sizes of various practical DRAM cells [2] which have been used in the products. With helps of a self-aligned contact (SAC), triple-poly silicon, and vertical capacitors process, the 1-T cell has reached its minimum size of $8F^2$ (F: feature size) for the folded-data-line arrangement [1–4]. The cell becomes larger when the contact is replaced by a non-self-aligned contact that a logic-compatible process needs. The three-transistor (3-T) cell and the four-transistor (4-T) cell, which were used for 1–4-kb DRAMs in the early 1970's, and the six-transistor (6-T) SRAM cell are also shown in the figure. In principle, the 3-T, 4-T, and 6-T cells do not require special capacitors and can be fabricated in a logic-compatible process with non-self-aligned contacts and single poly-silicon. Numerous other memory cells have been proposed to reduce the cell size. In fact, even a one-transistor gain cell has recently been proposed, as explained later.

2.2.1. The 1-T Cell and Related Cells

The 1-T cell has been dominating the DRAM market for the last 30 years due to its smallest cell size. The cell, however, has taken different directions for stand-alone DRAMs and e-DRAMs since the 1990's, in accordance with the different requirements. Stand-alone DRAMs have given the first priority to a small chip with the smallest cell possible. Thus, a self-aligned contact for memory cells is necessary despite the speed penalty inflicted due to the increased

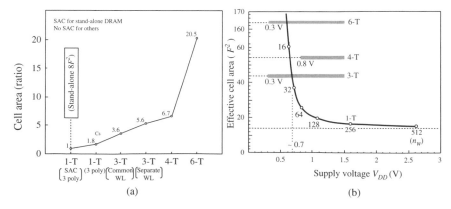

FIGURE 2.1. Cell area of RAM cells (a), and the effective cell area including overhead area coming from shared sense amplifier (b). SAC; self-aligned contact. n_W; the number of word lines connected to one pair of data lines to maintain a constant signal voltage of 200 mV. Reproduced from [2] with permission of IBM.

contact resistance. Moreover, a long data line is indispensable to minimize the area penalty caused by the data-line division, thus calling for a large C_S process and quite a high voltage for a given signal voltage, as discussed later. Leading developments in the cell size are a 4–6F^2 (F: feature size) trench-capacitor vertical-MOST cell [10, 11] and a 6F^2 stacked-capacitor open-DL cell [12]. In contrast, e-DRAMs have given the first priority to the fastest and simplest cell possible. Thus, a non-self-aligned contact (thus low contact resistance) cell and a small subarray are essential, as also discussed later. The small physical size of the subarray improves the array speed due to lower RC delay. In addition, the resultant small data-line capacitance (C_D) allows the use of a small cell capacitance (C_S) (e.g., 5 fF) and thus a logic-compatible simple-capacitor process (e.g., a MOS-planar capacitor) for a given signal voltage. Coupled with high-speed circuit techniques, such as multi-bank interleaving, pipelined operation, and direct sensing [1], the array speed is maximized, solving the speed problem in the row-cycle of DRAMs. If the resultant cell area is still significantly smaller than the 6-T full-CMOS SRAM cell, such e-DRAMs could replace SRAM cells due to a faster cycle time than SRAMs for a fixed memory capacity. A good example is the so-called 1-T SRAM™ [13] which incorporated a 1-T DRAM cell with a C_S smaller than 10 fF using a single poly-Si planar capacitor, and an extensive multi-bank scheme with 128 banks (32 Kb in each) that can operate simultaneously. Somasekhar et al. achieved a row-access frequency higher than 300-MHz for a 0.18-μm 1.8-V 2-Mb e-DRAM with a planar capacitor cell [14]. Another example is a 1.2-V 322-MHz random-cycle 16-Mb e-DRAM in 90-nm CMOS [9, 35], as shown in Fig. 2.2, in which a short data line connecting 32 cells (i.e., 64 cells per sense amplifier) and a small-RC delay cell (5-fF C_S, 10-Ω contact resistance) are used. A 5-fF C_S was realized using a 0.24-μm^2 cell (30 F^2) with a Ta$_2$O$_5$ metal-insulator-metal (MIM) stacked capacitor under the data line (DL). The refresh time was 32 ms, and data retention power was 60 μW at $V_{DD} = 1$ V and 85 °C.

	e-DRAM	conventional
Structure	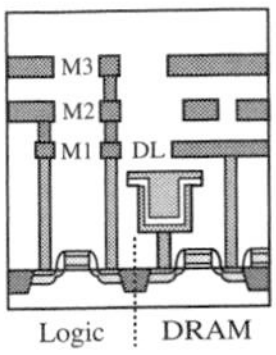	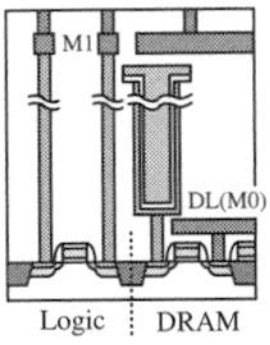
Cells/DL	32	128
C_s	5 fF (Ta$_2$O$_5$; MIM)	≥15 fF (MIS)
Additional wire	No	local wire (M0)
Thermal budget	no impact on logic	intolerable impact
Cell RC delay	W storage cont.	Non-metalized cell
	Co-salicided S/D	
Cell contact R	10 Ω	10 kΩ

FIGURE 2.2. e-DRAMs compared with the conventional DRAM [9, 35].

2.2.2. Gain Cells

3-T and 4-T Cells. Gain cells are suitable for low-voltage operations [2], despite a large cell size, because of a high enough signal voltage even at a low V_{DD}. In particular, 3-T and 4-T cells, which were industrialized in the early 1970's, may be candidates for e-DRAMs because they accept logic compatible processes and simple designs. Their advantages become more prominent at a lower V_{DD}. Figure 2.1(b) compares effective cell areas for V_{DD}. Here, the effective cell area is the sum of the actual cell area and overhead area involved in data-line (DL) divisions. Note that even a large-C_S 1-T cell requires more DL-divisions at a lower V_{DD} to maintain the necessary signal, causing a rapid increase in the effective cell area with decreasing V_{DD} [2]. The lack of gain in the 1-T cell is responsible for the increase. On the other hand, the 3-T, 4-T, and 6-T cells are all gain cells. Thus, in principle, they can develop a sufficient signal voltage without increasing the number of DL-divisions, even at a lower V_{DD}, and thus provide a fixed effective cell area that is almost independent of the V_{DD}. Actually, however, the V_{DD} has a lower limit for each cell. For the 3-T cell, it would be around 0.3 V, assuming a V_t for the storage MOST of around 0 V, a negative word-line scheme (see Fig. 2.11(b)) of $V_{WL} = -0.5$ V for both read/write lines, and a low V_t for the read/write MOSTs of $V_t(r) = 0$ and $V_t(w) = 0.3$ V. An initial stored voltage (V_{STORE}) of 0.3 V for the cell, and even a decayed V_{STORE} of 0.1 V, can be discriminated because of the gain if an improved sensing scheme is developed. The detection of and compensation for V_t-variations and an additional capacitor at the storage node would further improve stability and reliability. For the 4-T cell, the V_{DD} limit would be higher than 0.8 V, because the V_t of cross-coupled MOSTs must be higher than 0.8 V to ensure enough t_{REFmax}, and thus the V_{DD} must be higher than this voltage. The 6-T SRAM cell would be around 0.3 V if a raised supply voltage (V_{DH}) (e.g., 1 V) were supplied from an on-chip charge pump, as explained in Chapter 3. Consequently, the effective cell area of the 3-T cell would be smaller than other cells at a V_{DD} less than 0.7 V. In any event, in addition to the low junction temperature caused by the ultra-low V_{DD}, the wide voltage margin provided by gain cells would meet the specification of the maximum refresh time (t_{REFmax}). Adjusting the potential profile of the storage node to suppress the pn-leakage current further lengthens the t_{REFmax} and preserves the refresh busy rate, even in larger memory-capacity DRAMs [1], or it lowers the data retention current in the standby mode. Even if the t_{REFmax} is short, fast e-DRAMs, combined with a small subarray and new architectures, would accept a drastically-shortened t_{REFmax}, as discussed later. However, a challenge is to suppress increase in area or process complexity caused by adding a storage capacitor to address the soft error issue, and by a multi-divided data-line array necessary for high-speed operations. Otherwise, the advantage of the gain cells may be offset. In addition, the detection of and compensation for V_t-variations, and a high-speed sensing in such low-V_{DD} regions are real challenges.

1-T Gain Cells. Figure 2.3 shows the concept of the one-transistor gain cell proposed recently [15]. The cell named the floating-body-transistor cell (FBC)

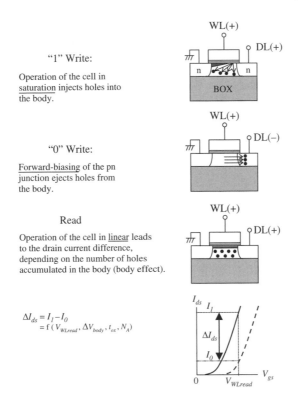

"1" Write:

Operation of the cell in
saturation injects holes into
the body.

"0" Write:

Forward-biasing of the pn
junction ejects holes from
the body.

Read

Operation of the cell in linear leads
to the drain current difference,
depending on the number of holes
accumulated in the body (body effect).

$$\Delta I_{ds} = I_1 - I_0$$
$$= f(V_{WLread}, \Delta V_{body}, t_{ox}, N_A)$$

FIGURE 2.3. Principle of the FBC. Reproduced from [15] with permission; © 2006 IEEE.

has the ability to achieve a $4F^2$ cell with self-aligned contact technologies. The nMOST forms on partially-depleted silicon-on-insulator (PD-SOI). The number of majority carriers (holes for nMOST) is changed in write operations, and different drain current is read and sensed due to the body effect that depends on the number. To write data "1", the nMOST is operated in saturation, leading to impact ionization which injects holes into the body, as discussed in Fig. 2.17. To write data "0", the pn-junction between the body and the drain is forward-biased, ejecting the stored holes from the body. Thus, the body (i.e., substrate) voltage and the V_t become different, depending on the data. To read the data, the nMOST is operated in the linear region to keep the data state "0" from being violated by impact ionization. This makes the drain current different due to the different V_t, enabling a successful read and sensing. The drain current difference ΔI_{ds}, which is the "1" current I_1 minus the "0" current I_0, is a function of the WL voltage during read, the body voltage difference between "1" and "0", ΔV_{body}, the gate oxide thickness, t_{OX}, and the acceptor concentration in the body N_A. Figure 2.4 shows the cell array and the cross sections [15]. An n-doped poly-silicon pillar is connected to an n-type diffusion layer spread under the BOX (buried oxide) which is biased at -1 V. The pillar forms a capacitor coupled to the body which accumulates holes. Thus, it serves as a stabilizing capacitor C_S which helps

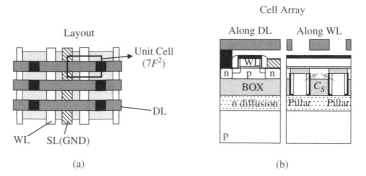

FIGURE 2.4. Layout of the cell array (a), and cross sections along a data line (DL) and word line(WL) (b). Reproduced from [15] with permission; © 2006 IEEE.

increase the signal and the retention time as well. In other words, it is added to increase the signal charge of the cell. The FBC, however, is volatile and effectively a destructive read-out (DRO) memory cell [16, 17]. It needs to be refreshed, because holes are generated in the body at "0" state due to the pn-junction reverse-bias leakage current between the body and the source/drain [18] (see Fig. 2.12). The resultant refresh time is around 100 ms at 85 °C. In addition, a few holes stored in the body are eliminated in each WL cycle due to charge pumping, calling for rewrite operations with a cross-coupled sense amplifier on each data line. This is because a few electrons are trapped at the Si-SiO$_2$ interface during inversion. After turning off the WL, a few holes recombine with trapped electrons during accumulation, as shown in Fig. 2.5 [17]. This implies that the cell current (I_1) for "1" depends on the pumping count. In fact, I_1 becomes different for reading continuously with one WL pulse, being held without reading, and being pumped, as clearly seen in Fig. 2.6 [17]. Thus, the cell is effectively a DRO cell, calling for almost the same refresh operations and circuit configurations as those for conventional DRAMs. In addition, the small and floating body is expected to increase the soft-error rate, as it did for PD-SOI DRAM [18]. Such instabilities may be involved in the access and storage MOSTs in two-transistor PD-SOI gain cell [19].

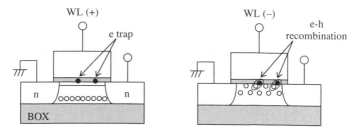

FIGURE 2.5. WL pumping mechanism [17].

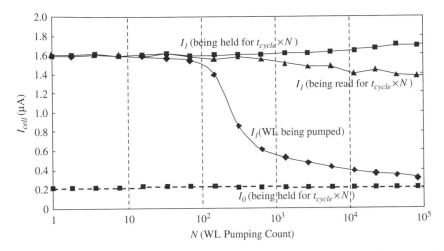

FIGURE 2.6. Measured cell read currents with and without WL pumping. Reproduced from [17] with permission; © 2006 IEEE.

2.3. 1-T-Based Cells

Due to a small floating signal voltage that results from the absence of gain in the 1-T cell, its operation is quite sensitive to various noise sources in the array, which are inherently large because a large voltage swing simultaneously occurs on a huge number of heavily-capacitive data lines in a high-density array. The amount of noise that affects sensing depends on the data-line arrangement, the cell structure, the data-line precharging scheme, and the array configuration and layout. There are two types of noise that must be reduced; differential-mode noise (v_{ND}) and common-mode noise (v_{NC}). The differential-mode noise degrades the sensing stability by effectively reducing the signal voltage. It comes from capacitive coupling to the data lines from many other conductors in the array. Moreover, it effectively comes from the offset voltage δV_t (i.e., V_t-mismatch between paired MOSTs) of the differential sense amplifiers (SAs). The common-mode noise slows down the sensing speed by effectively reducing the gate voltage (V_G) (i.e., gate-over-drive voltage) of the turned-on MOST in the SA. It comes from not only capacitive coupling but also from the V_t variation of the MOSTs in the SAs. The common-mode noise has been tolerated so far due to a high enough V_{DD}.

2.3.1. The Data-Line Arrangement

Differential sensing with differential sense amplifiers is always indispensable for reliable sensing through rejecting a large amount of common-mode noise. There are three types of sensing schemes, categorized in terms of the data-line (DL) arrangement, as shown in Fig. 2.7. They are two types of folded-DL arrangements (a) and (c), and the open-DL 1-T cell arrangement (b). Type (c) is

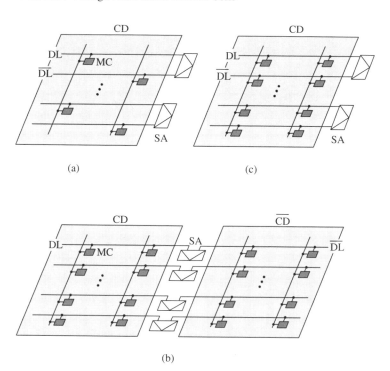

FIGURE 2.7. The folded-data-line 1-Tcell (a), the open-data-line 1-T cell (b), and the folded-data-line 2-T cell (c)[1].

the 2-T cell, which consists of twin 1-T cells and thus called the twin cell. For the folded-DL arrangement, each pair of DLs runs on the same conductor (CD), so any common-mode noise coupled to the data lines from the conductor via the conductor-DL capacitance is canceled by the sense amplifier. On the contrary, for the open-DL arrangement, each pair of DLs runs on the different conductors (CD and $\overline{\text{CD}}$). Since the voltage swing of the conductors can differ, the voltage coupled to each pair of data lines can be different. Thus, a differential noise is generated on the pair of data lines. In general, the open-DL cell (c) is smallest, but the largest noise, while the 2-T (twin) cell (b) is largest, but the smallest noise. The folded-DL cell (a) compromises the cell size with noise.

The folded-DL 1-T cell (a) has been the sole type of cell for the last 25 years. With the help of a self-aligned contact, triple-poly silicon, and vertical capacitor process, this type of cell has reached its minimum size of $8F^2$ (F: feature size) [1], as mentioned previously. The greater difficulty of device minia-turization and the increasing cost of fabrication in the nanometer era lead to a requirement for new memory cells which are smaller than $8F^2$. Thus, a $6\text{-}F^2$ trench-capacitor folded-DL cell has been proposed [10, 11]. However, it requires a vertical transistor along with an additional tight-pitch layer for its vertically folded-DL arrangement. Recently, the open-DL 1-T cell, widely used in the 4–16-kb generations in the 1970's, has revived due to its small ($6F^2$) and simple structure [12], although reducing noise is the key to stable operations. The 2-T cell is not only suitable for high-speed DRAMs [20, 21], but also most promising for ultra-low voltage DRAMs. Despite the large cell size, the 2-T cell DRAM may potentially replace the 6-T SRAM if its size remains much smaller than the 6-T SRAM cell, as discussed in Chapter 3.

2.3.2. The Data-Line Precharging Scheme

There are two types of data-line precharging schemes independent of the data-line arrangement; the full-V_{DD} DL-precharging scheme and the half-V_{DD} (midpoint) DL-precharging scheme [1]. Full-V_{DD} sensing was used in the nMOS era from the mid-1970's to the early 1980's. However, it suffered from high power consumption and large spike current, a large amount of noise in the array, and a slow cycle time [1]. An inaccurate setting of the reference level was also problematic, even if dummy cells and dummy word-lines were used [1]. Even so, it was tolerated at that time because the memory capacity and V_{DD} were still small and high (i.e., 12 V in 4–16-kb generations), respectively, and the signal voltage was thus relatively large. Mid-point sensing replaced full-V_{DD} sensing in the early 1980's, with the help of cross-coupled (latch-type) CMOS sense amplifiers, because it has unique features; that is, halving the data-line charging power that dominates the total chip power, a quiet array with differential driving of data lines, and automatic creation of an accurate reference level of a half-V_{DD} without any area and speed penalty [1]. Instead, it strictly limits low-voltage operations due to a low-voltage sensing based on $V_{DD}/2$.

2.4. Design of the Folded-Data-Line 1-T Cell

This section clarifies the limiting factors of low-voltage operation of the 1-T cell. The folded-DL 1-T cell combined with mid-point sensing is cited as an example because it has been the de-facto standard DRAM-cell circuit. Discussion here includes investigations of the lowest necessary V_t of the transfer MOST in the cell, the lowest necessary word voltage for a full read/write, the necessary signal charge and signal voltage, noise sources and noise reduction strategies, sensing speed and gate over-drive voltage of sense amplifiers (SAs), and the lowest necessary V_t of the SA-MOSTs. They are all related to the minimum V_{DD} of DRAMs for successful sensing.

2.4.1. The Lowest Necessary V_t and Word Line Voltage

Each non-selected cell must hold its stored data for the t_{REFmax} (i.e., the maximum refresh time) by reducing leakage currents at the storage node. If the gate oxide thickness of the transfer MOST is thick enough, there are two leakage currents at the node [1]: the subthreshold current (i_1) to the data line (DL), and the pn-junction leakage current (i_2) to the substrate, as we can see in Fig. 2.8. Thus, for

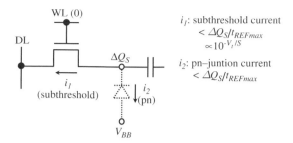

FIGURE 2.8. Two leakage-current components of non-selected DRAM cell. Reproduced from [1] with permission of Springer.

an acceptable signal-charge decay (ΔQ_S) of the cell, each leakage current must be reduced to satisfy the specification of t_{REFmax} with i_1 ($\propto 10^{-Vt/S}$) $< \Delta Q_S/t_{REFmax}$, and $i_2 < \Delta Q_S/t_{REFmax}$.

The subthreshold current is estimated by measuring the data retention time under the condition where 0 V is applied to the data line for as long as possible. In practice, the condition is achieved by a set of successive low-level ("L") data-line disturbances that is done with successive operations of other cells on the same data line, as shown in Fig. 2.9 [1]. The refresh time measured under the disturbance is called the dynamic refresh time. The higher the V_t, the lower the subthreshold current. For a fixed cycle time, the lowest necessary $V_t (= V_{tmin})$ for the transfer MOST must be gradually increased with memory capacity because the t_{REFmax} must be longer to preserve the refresh busy rate $\eta (= nt_{RC}/t_{REFmax})$ [1]. This is true in a logical array configuration with n rows by m columns, in which n is the refresh cycle, and t_{RC} is the memory cycle time (i.e., RAS cycle), as shown in Fig. 1.37. The V_{tmin} can be lower with a faster

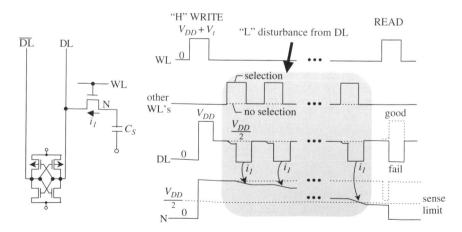

FIGURE 2.9. The loss of stored data due to the data-line 'L' disturbances. $Q_C = 0$ and $Q_N = 0$ are assumed. Reproduced from [1] with permission of Springer.

cycle time for a given η, as explained later. In contrast, the pn-current can be estimated by measuring the date retention time while applying a static V_{DD} to the data line and $0\,V$ to the word line so that the subthreshold current to the data line is eliminated (so-called static refresh time). It is becoming increasingly difficult to simultaneously reduce both subthreshold current and pn-junction current in the nanometer era. The need for a high enough V_t requires increased doping concentration for the ever-shorter channel-length MOST, and thus causes increased pn-leakage current. Thus, non-planar MOSTs (e.g., a recessed MOST and a FinFET [22]) and/or the negative word-line scheme are necessary to allow lower doping concentration.

Let us investigate the lowest necessary $V_t(=V_{tmin})$ in more detail, since it eventually determines the lowest necessary word voltage of the 1-T cell. The 1-T DRAM cell needs a full-V_{DD} write to store a full V_{DD} from the data line to the cell node without a V_t drop. Thus, the word-line voltage, V_W, must be higher than the sum of V_{DD} (i.e., the data input voltage from the data line), V_{tmin}, the V_t-increase due to the body effect developed by raising the source (i.e., storage node) by V_{DD}, and the V_t variation caused by narrow channel effects and process variations. This raised word voltage necessitates a stress voltage-immune transfer MOST. If the V_t variation is neglected, V_{tmin} is expressed as (see Fig. 1.5(a))

$$V_W > V_{DD} + V_{tFW}$$

$$V_{tFW} = V_{t0} + K\left(\sqrt{|V_{BB}| + V_{DD} + 2\psi} - \sqrt{2\psi}\right),$$

$$V_{t\min} = V_{t0} + K\left(\sqrt{|V_{BB}| + 2\psi} - \sqrt{2\psi}\right),$$

where V_{tFW} is the lowest necessary V_t for a full write, which is the sum of V_{tmin} and the V_t-increase by the body effect. Note that V_{tmin} is defined as the V_t at a source (i.e., the data line) voltage of $0\,V$, which differs from V_{tFW}. Here, V_{tmin} can be different in a low-cost design and a high-speed design. The low-cost DRAM needs a large number of memory cells connected to each DL-pair to achieve a small chip by reducing the overhead caused by each DL-division. The resultant long data line and associated large DL capacitance C_D call for a large C_S, which has been attained using sophisticated stacked and trench capacitors and high dielectric constant (high-k) thin-films [1]. The requirement for a large C_S can be relaxed by using a high V_{DD}. That is why a high V_{DD} is needed for low-cost stand-alone DRAMs. In addition, the slow cycle time due to the large C_D calls for a long refresh time t_{REFmax} for a given refresh busy rate, necessitating a high V_{tmin}, as explained previously. In contrast, the high-speed DRAM needs a short data-line with small C_D. The resultant fast cycle time accepts a shorter t_{REFmax} for a given refresh busy rate, allowing a low V_{tmin}. Accordingly, the lowest necessary word voltage can be different. Figure 2.10 [3] shows the calculated values of t_{REFmax}, V_{tmin}, V_{tFW}, and V_W for a hypothetical low-cost 64-Mb DRAM that operates at $t_{RC} = 100\,ns$ and $V_{DD} = 2\,V$ with a pair of data lines connecting 1,024 cells. They are 8 ms, 0.7 V, 1.3 V, and 3.3 V, respectively. However, they are 1.3 μs, 0.3 V, 0.7 V, and 1.7 V for a high-speed design that operates at

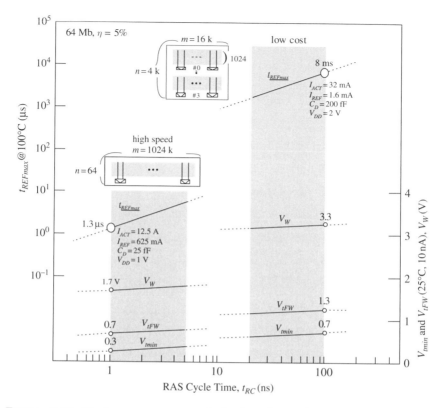

FIGURE 2.10. Maximum necessary refresh time (t_{REFmax}) and necessary cell V_t. Reproduced from [3] with permission; © 2006 IEEE.

$t_{RC} = 1$ ns and $V_{DD} = 1$ V with a pair of data lines connecting 64 cells. Here, $\eta = 5\%$, $C_S = 30$ fF, $k = 0.5$ V$^{1/2}$, $\Delta Q_S/Q_S = 0.1$ as an acceptable signal-charge decay, $S = 120$ mV/dec ($100\,°C$), and $V_t(25\,°C) = V_t(100\,°C) + 0.15$ V, and the V_t defined at a constant current of 10 nA per cell and $25\,°C$ are assumed.

In the past, the word-line voltage has not been scaled down at the same pace as the data-line voltage, due to a large V_{tmin} and V_{tFW}. On the contrary, the data-line voltage, V_{DD}, has been scaled down to reduce the power dissipation of the chip. Note that DRAM's power dissipation is almost governed by the data-line voltage rather than the word-line voltage, since only one selected word-line is activated while a huge number of heavily-capacitive data lines are simultaneously activated at V_{DD}. It should be noted that the boost ratio, $V_W/V_{DD} > (1 + V_{tFW}/V_{DD})$, must be increased as the data-line voltage (V_{DD}) is lowered, which charges the voltage up-converter with a larger boost ratio.

As for the word boosting scheme, three alternatives [1] are well known. Figure. 2.11(a) shows the conventional word boosting (or word bootstrapping) scheme for a high-actual V_t ($\geq V_{tmin}$) MOST. The word-line high level is usually generated by a boosted power supply, V_{DH} (i.e., V_{PP}). Figure 2.11(b) shows

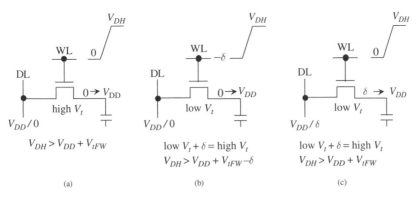

FIGURE 2.11. The conventional word boosting (a), the negative word-line scheme (NWL)(b), and the boosted sense-ground (BSG)[23, 24].

the negative word-line (NWL) scheme for a low-actual V_t MOST, in which the word-line low level is negative $(-\delta)$ to cut the leakage with an effective high $V_t (= V_{tmin} + \delta)$. Figure 2.11(c) depicts the boosted sense-ground (BSG) scheme for a low-actual V_t MOST [23, 24]. The NWL and BSG are the gate-source offset driving schemes, as discussed in Chapter 4. V_{DH} and δ are generated by charge-pump circuits, as discussed in Chapter 8, while the boosted ground level is generated with a similar circuit configuration as a voltage down-converter explained in Chapter 7. The BSG has not been so popular as the word boosting and NWL because the signal voltage is reduced due to the lower data-line voltage (by δ).

2.4.2. The Minimum V_{DD}

The minimum $V_{DD}(= V_{min})$ of DRAMs is usually determined by the S/N of cells or the gate-over-drive of sense amplifiers (SAs). For the mid-point sensing shown in Fig. 2.12(a), the differential signal voltage v_S with the positive or negative polarity is approximately expressed as $v_S = (V_{DD}/2)C_S/(C_D + C_S)$. The effective signal voltage v_{Seff}, taking the differential noise v_{ND} and the offset voltage δV_t (i.e., V_t mismatch between paired MOSTs) of SAs into consideration, is expressed as $v_{Seff} = v_S - v_{ND} - \delta V_t$. Thus, the minimum V_{DD} for a successful discrimination of the signal, $V_{min}(D)$, is given as

$$V_{min}(D) = 2(v_{ND} + \delta V_t)(C_D + C_S)/C_S. \qquad (2.1)$$

On the other hand, the gate-over-drive voltage (V_G) (i.e., effective gate voltage) of the turned-on MOST in the SAs depends on the signal polarity. For a positive signal the turned-on MOST is M_2, while for a negative signal it is M_1. The V_G is thus expressed as $V_G(M_2) = V_{GS} - V_t = V_{DD}/2 - (v_S + v_{NC}) - (V_{t0} + \Delta V_t)$ for the positive signal, and $V_G(M_1) = V_{DD}/2 - v_{NC} - (V_{t0} + \Delta V_t)$ for the negative signal. Here, v_{NC} is the common-mode noise, and V_{t0} and ΔV_t are the average

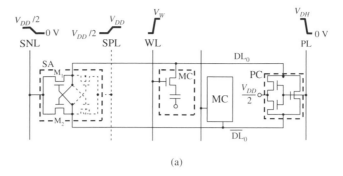

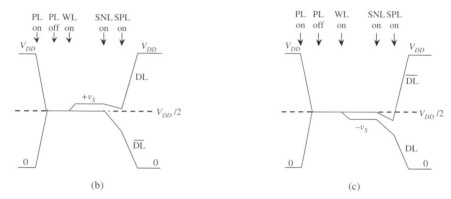

FIGURE 2.12. Data-line relevant circuits of the 1-T cell (a) and waveforms on a pair of data lines for a positive read signal (b) and a negative read signal (c), referred to a reference voltage (i.e., $V_{DD}/2$)[1].

threshold voltage and its variation in a chip, respectively. Hence, the minimum V_{DD} for a successful activation of SA, $V_{min}(C)$, is for the negative signal, which is expressed as

$$V_{min}(C) = 2\{v_{NC} + (V_{t0} + \Delta V_t)\}. \tag{2.2}$$

Note that the V_{min} of DRAMs is determined by the higher one of the two V_{min} values; $V_{min}(D)$ and $V_{min}(C)$. Thus, reducing δV_t, $(V_{t0} + \Delta V_t)$ and noise through investigating the S/N of the cell and noise generation mechanisms in the array is essential for low-voltage operations.

2.4.3. Signal Charge and Signal Voltage

The voltage at the storage node just after writing a full V_{DD} (i.e., 'H' write) from a data line decays due not only to cell leakage, i, but also to α-particle or cosmic-ray irradiation to the cell node, as shown in Fig. 2.13. Thus, the signal charge Q_S of DRAM cells [1] is time dependent. To be more exact, noise coupled

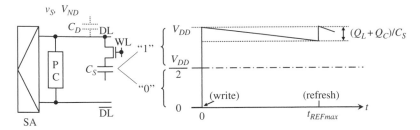

FIGURE 2.13. The operation of a DRAM cell. Reproduced from [1] with permission of Springer.

to the data line before and during sensing must be taken into consideration in the expression of Q_S, even if Q_S is for non-selected cells. This is because the necessary Q_S is determined as a result of a successful sensing. Eventually, the signal charge that is minimized at the maximum refresh time (t_{REFmax}) is given as

$$Q_S = Q_{SO} - Q_L - Q_C - Q_N,$$

where Q_{SO} is the initial signal charge ($= C_S V_{DD}/2$), Q_L is the leakage charge($= i t_{REFmax}$, $i = i_1 + i_2$), Q_C is the collected charges at the cell node by irradiation, and Q_N is the noise charge ($= v_{ND}(C_D + C_S)$). The signal voltage for a 'H' read, $v_S(H)$, is thus minimized at t_{REFmax} as

$$v_S(H) = (Q_{SO} - Q_L - Q_C - Q_N)/(C_D + C_S).$$

Thus, requirements for high S/N designs are summarized as follows. $v_S(H)$ must be at least positive and large enough for reliable sensing. This implies that Q_{SO} must be larger than the total effective noise charge, $Q_L + Q_C + Q_N$, although Q_{SO} usually decreases as V_{DD} and C_S are reduced with device scaling [1] (Fig. 2.14). Moreover, C_D must be minimized.

It should be noted that if a low voltage (i.e., 0 V) is written to the cell node (i.e., 'L' write), the voltage stays at almost the original value during non-selected periods. This is because a pn-junction leakage in the cell node, which makes the store voltage decay, is compensated for by a subthreshold current from the data line at a half-V_{DD}. Thus, the signal voltage for a 'L' read ($v_{SI}(L)$) that is larger than $v_S(H)$ is given as

$$v_S(L) = (V_{DD}/2)C_S/(C_D + C_S) = Q_{SO}/(C_D + C_S).$$

2.4.4. Noise Sources

The signal voltage developed on the data line is small and floating, and the subsequent sensing operation is thus sensitive to noises coupled to the data lines before and during sensing. In addition, some kinds of noise components slow down the sensing speed, especially for mid-point sensing. Since noises increase

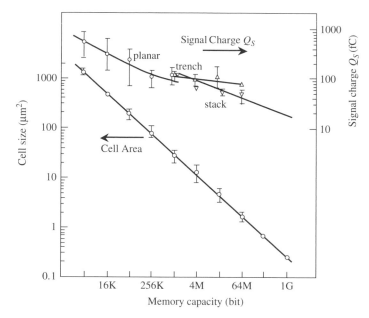

FIGURE 2.14. Trends in Q_S and the DRAM cell size. Reproduced from [1] with permission of Springer.

with device scaling, reducing noise is a key to low-voltage nano-scale DRAMs. In the past, many noise reduction techniques have been applied to the products. Of the techniques, the folded-DL cell, the half-V_{DD} DL precharge, and the triple-well structure [1, 36], as shown in Fig. 2.15, are extremely important. The folded-DL cell cancels various differential noise components on a pair of data lines. The half-V_{DD} DL precharge discriminates a positive-going or negative-going signal referred to the half-V_{DD} without dummy cells [1]. Coupled with the folded-DL cell, it enables low-noise low-power array by minimizing common-mode noises, although it suffers from a slow amplification because the sense amplifier must operate at the lowest voltage (i.e., the half V_{DD}) in a chip. A 1-V 1-Gb DRAM with $C_D = 100$ fF gives an example in which $64k$ to $128k$ pairs of data lines are simultaneously activated. For the V_{DD} DL-precharge, a total capacitance as large as 6 nF to 13 nF must be charged and discharged to a 1-V V_{DD} within 20 ns, while the signal voltage is as small as 100 mV. Thus, it couples large voltages to all non-selected word lines (WLs) in activated sub-arrays, through many DL-WL coupling capacitances (C_{WD}). It also couples a large voltage to the p-well substrate at a floating substrate-voltage (V_{BB}). These are common-mode noises. The coupled voltages to the word-lines and substrate re-couple differentially to data lines [1]. In contrast, the half-V_{DD} DL-precharge enables a quiet array, because the voltages coupled to the word lines and substrate are canceled due to an almost differential driving of pairs of data-lines. In addition, it halves the DL charging power, which dominates the total power of a chip, by halving the

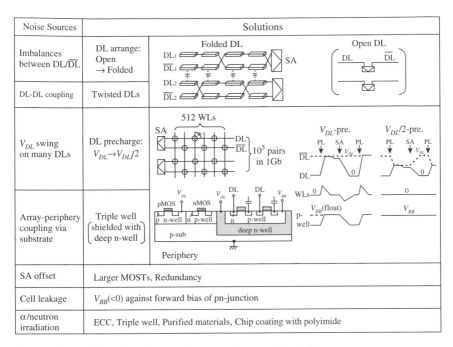

Noise Sources		Solutions	
Imbalances between DL/$\overline{DL}$	DL arrange: Open → Folded	Folded DL	Open DL
DL-DL coupling	Twisted DLs		
V_{DL} swing on many DLs	DL precharge: $V_{DL} \rightarrow V_{DL}/2$	512 WLs	V_{DL}-pre. $V_{DL}/2$-pre.
Array-periphery coupling via substrate	Triple well (shielded with deep n-well)		
SA offset	Larger MOSTs, Redundancy		
Cell leakage	$V_{BB}(<0)$ against forward bias of pn-junction		
α/neutron irradiation	ECC, Triple well, Purified materials, Chip coating with polyimide		

FIGURE 2.15. Noise-reduction techniques for DRAMs [1, 36].

voltage swing of data lines. Here, the deep n-well in the triple well structure eliminates the array-periphery coupling noise via the p-well sub, and works as a barrier against soft errors [1].

Noise that still remains despite such techniques gives detrimental effects to sensing operations. There are two kinds of noises, the differential (-mode) noise and the common (-mode) noise, as explained earlier. The differential noise coupled to the pair of data lines effectively reduces the signal voltage, causing unreliable sensing. The common noise couples the same voltage to the pair of data lines, and, although it is thus cancelled with a differential sense amplifier, it can make the subsequent sensing and amplification slow due to reduced gate-over drive, $V_G (= V_{GS} - V_t)$ in the sense amplifier. In the ultra-low voltage era, the common noise that has been neglected due to a high V_{DD} must also be taken into consideration, as mentioned previously. Let us investigate noise sources to the 1-T nMOST cell (MC) on the data line (DL_1) in an array shown in Fig. 2.16. Here, each pair of data lines has a precharger and equalizer (PC) and a CMOS sense amplifier (SA). After all data lines have been precharged to a half-V_{DD} by activating PL to a high level, they are left to a half-V_{DD} floating level by turning off PL. Then, a word line is activated by applying a high enough voltage. The resultant signal voltage is amplified by activating SNL of the nMOS SA from $V_{DD}/2$ to 0, followed by rewrite or restoring by activating SPL of the pMOS SA from $V_{DD}/2$ to V_{DD}. The voltages coupled to the pair of data lines from other conductors during the floating period prior to sensing works as a

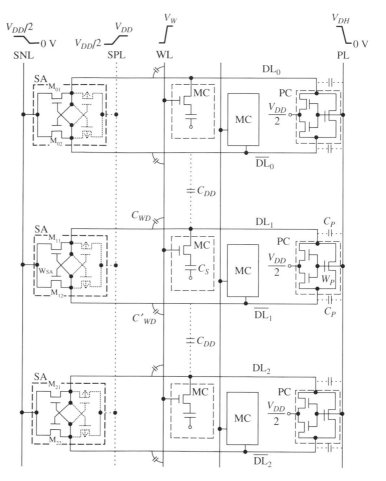

FIGURE 2.16. Memory-cell array and various coupling capacitances.

loss (subtractive) or a gain (additive) to the signal, depending on the polarity of the signal when referred to the $V_{DD}/2$ level. Here, a positive going signal is developed when a cell with a V_{DD}-stored voltage is read (i.e., 'H' read), while a negative going signal is developed when a cell with a 0-V-stored voltage is read (i.e., 'L' read). The following noise sources exist for the differential mode and the common mode.

Differential-Mode Noise Components. There are the following four kinds of differential noise components.

$\Delta(C'_{WD})$ and $\Delta(C'_{WD})$. These are positively-coupled voltages, via C_{WD} and C'_{WD}, from the word-line to a pair of data lines when the word-line is activated. $\Delta(C_{WD})$ is larger than $\Delta(C'_{WD})$ due to the existence of the gate capacitance of the transfer MOST. Thus, their voltage difference acts as a differential noise. Since a large gate capacitance starts to be developed when the word pulse reaches $V_{DD}/2 + V_{tM}$, they are given as $\Delta(C_{WD}) \cong (V_W - V_{DD}/2 - V_{tM})C_{GM}/(C_D + C_S + C_{GM})$, and $\Delta(C'_{WD}) = V_W C'_{WD}/(C_D + C'_{WD})$, where V_W, V_{tM}, and C_{GM} are the amplitude of the word pulse, and the V_t and gate capacitance of the transfer MOST, respectively. $\Delta(C'_{WD})$ is usually negligible. Assuming that $V_W = V_{DD} + V_{tM}$, the ratio to the signal voltage, $v_S = (V_{DD}/2)C_S/(C_D + C_S)$, is given as

$$\Delta(C_{WD})/v_S = (C_{GM}/C_S)(C_D + C_S)/(C_D + C_S + C_{GM}). \qquad (2.3)$$

For stand-alone DRAMs with a large enough C_S and C_D, the ratio becomes negligible. For e-DRAMs with a small C_S and C_D, however, it grows as Cs reaches C_{GM}.

$\Delta(C_{DD})$. This noise comes from the DL-DL capacitance C_{DD} whenever a voltage swing occurs on adjacent data lines. It randomly couples to each of the pair of data lines. Thus, it can work as a differential noise or even a common noise. In particular, it can be intolerably large when a large voltage swing occurs on the adjacent data lines, as discussed later.

$\Delta(C_D)$. This is an effective coupling voltage due to a capacitive-imbalance by C_S between a pair of DLs during sensing. It works as a loss for a 'L' read signal, but a gain for a 'H' read signal. This is because the additional C_S prevents the nMOS amplifier from amplifying the 'L' read signal, while it helps the 'H' read to stay at a high level. If the conductance of the transfer MOST in the cell is large enough, the noise is approximately given [1] as $\Delta(C_D) \cong (1/2)(C_D K/\beta)^{1/2} C_S/C_D$, where K is the slope with which the SNL voltage is decreased, and β is the nMOST's conductance. The noise becomes larger with a larger K; that is, faster activation of the amplifier.

δV_t. The offset voltage of the sense amplifier works as an effective random noise during sensing, due to its random variation. It should be noted that δV_t increases with device scaling.

Common-Mode Noise Components. There are the following three kinds of common noise components.

$\Delta(V_{DD}/2)$. This noise is generated by inaccurate setting of the $V_{DD}/2$-level, which is caused by output level fluctuations of the $V_{DD}/2$-generator, as discussed in Chapter 7. It can take a positive or negative value that is equally coupled to a pair of data lines.

$\Delta(C_P)$. This noise is a negatively-coupled common-mode voltage to the DLs from PC when the precharge pulse PL goes low. Since a large gate capacitance starts to be developed when the pulse reaches $V_{DD}/2 + V_{tP}$, it is given as $\Delta(C_P) \cong (V_P - V_{DD}/2 - V_{tP})C_{GP}/(C_D + C_{GP})$, where V_P, V_{tP} and C_{GP} are the amplitude of the pulse, the threshold voltage and gate capacitance of the precharge nMOST, respectively. Assuming that $V_P = V_{DD} + V_{tP}$, the ratio to the signal voltage is given as

$$\Delta(C_P)/v_S = (C_{GP}/C_S)(C_D + C_S)/(C_D + C_{GP}). \qquad (2.4)$$

For stand-alone DRAMs, the ratio becomes negligible, while for e-DRAMs it grows as Cs reaches C_{GP}.

ΔV_t. The V_t variation (ΔV_t) in a chip effectively works as a noise that prevents high-speed operations of sense amplifiers. Here, $V_t = V_{t0} + \Delta V_t$, and V_{t0} is the average threshold voltage in a chip.

2.4.5. The Effective Signal Voltage and the Gate-Over-Drive of SAs

The effective signal voltage and the gate-over-drive of the sense amplifier (SA) on the pair of data lines (DL_1 and $\overline{DL}_1$) are analyzed using the memory-cell array shown in Fig. 2.16. Here, the effective signal voltage, v_{Seff}, stands for the available signal voltage which results from subtracting the noise and the offset of the sense-amp from the original signal voltage, as mentioned earlier. To simplify the analysis, the two components $\Delta(V_{DD}/2)$ and ΔV_t are neglected because $\Delta(V_{DD}/2)$ can be reduced with a laser trimming of resistors in the generator [9], and ΔV_t can be compensated for by controlling the substrate-bias (V_{BB}) with an on-chip charge pump, or reduced by using the largest MOSTs possible. In the following formula, the absolute value is used for each parameter.

A. The 'H' Read

Voltages of the pair of data lines (DL_1 and $\overline{DL}_1$) just before sensing are given as

$$V(DL_1) = V_{DD}/2 - \Delta(C_P) + \Delta(C_{WD}) + v_S(H),$$
$$V(\overline{DL_1}) = V_{DD}/2 - \Delta(C_P) + \Delta(C'_{WD}) \pm \Delta(C_{DD}).$$

Note that the turned-on nMOS is M_{12} in this case, and only the signal voltage on one adjacent data line (DL_2) couples a $\Delta(C_{DD})$ to $\overline{DL}_1$ because another adjacent data line ($\overline{DL}_0$) is quiet. The "H" read signal $v_{Seff}(H)$ on DL_1 becomes minimum, if the memory cell is about to be refreshed, M_{12} has the maximum positive δV_t with $V_t = V_{t0} + \delta V_t$, and the signal polarity on DL_2 is positive. Here, the signal on DL_2 works as a loss for the differential signal of the pair of data lines. The gate-over-drive $V_G(H)$ of M_{12} also becomes minimum for the above V_t. Thus, they are approximately expressed as

$$v_{Seff}(H)_{min} = V(DL_1) - V(\overline{DL_1}) + \Delta(C_D) - \delta V_t = v_S(H) + v_{ND}, \tag{2.5}$$
$$V_G(H)_{min} = V(DL_1) - V_t(M_{12}) = V_{DD}/2 + v_S(H) + v_{NC}, \tag{2.6}$$
$$v_S(H) = (Q_{SO} - Q_L - Q_C)/(C_D + C_S),$$
$$v_{ND} = \Delta(C_{WD}) - \Delta(C'_{WD}) - \Delta(C_{DD}) + \Delta(C_D) - \delta V_t,$$
$$v_{NC} = -\Delta(C_P) + \Delta(C_{WD}) - V_t(M_{12}),$$
$$V_t(M_{12}) = V_{t0} + \delta V_t, \quad V_t(M_{11}) = V_{t0}.$$

Here, $\Delta(C_D)$ is simply added to $v_{Seff}(H)$ since it is additive to the signal.

On the other hand, they become maximum for the memory cell just after writing a "H" data, if M_{12} has a negative δV_t, and the signal polarity on DL_2 is negative. Thus, they are given as

$$v_{Seff}(H)_{max} = V(DL) - V(\overline{DL}) + \Delta(C_D) - \delta V_t = v_S(H) + v_{ND}, \qquad (2.7)$$

$$V_G(H)_{max} = V(DL) - V_t(M_{12}) = V_{DD}/2 + v_S(H) + v_{NC}, \qquad (2.8)$$

$$v_S(H) = Q_{SO}/(C_D + C_S),$$

$$v_{ND} = \Delta(C_{WD}) - \Delta(C'_{WD}) + \Delta(C_{DD}) + \Delta(C_D) + \delta V_t.$$

$$v_{NC} = -\Delta(C_P) + \Delta(C_{WD}) - V_t(M_{12}),$$

$$V_t(M_{12}) = V_{t0} - \delta V_t, \ V_t(M_{11}) = V_{t0}.$$

B. The 'L' Read

The voltages of the pair of data lines are given as

$$V(DL) = V_{DD}/2 - \Delta(C_P) + \Delta(C_{WD}) - v_S(L),$$

$$V(\overline{DL_1}) = V_{DD}/2 - \Delta(C_P) + \Delta(C'_{WD}) \pm \Delta(C_{DD}),$$

$$v_S(L) = Q_{SO}/(C_D + C_S).$$

In this case, the turned-on nMOST changes from M_{12} to M_{11}. The $v_{Seff}(L)$ and $V_G(L)$ become minimum, if M_{11} has a positive δV_t, and the signal polarity on DL_2 is negative. Thus, they are approximately expressed as

$$v_{Seff}(L)_{min} = V(DL) - V(\overline{DL}) + \Delta(C_D) + \delta V_t = -v_S(L) + v_{ND}, \qquad (2.9)$$

$$V_G(L)_{min} = V(\overline{DL}) - V_t(M_{11}) = V_{DD}/2 + v_{NC}, \qquad (2.10)$$

$$v_{ND} = \Delta(C_{WD}) - \Delta(C'_{WD}) + \Delta(C_{DD}) + \Delta(C_D) + \delta V_t, \qquad (2.11)$$

$$v_{NC} = -\Delta(C_P) + \Delta(C'_{WD}) - \Delta(C_{DD}) - V_t(M_{11}), \qquad (2.12)$$

$$V_t(M_{11}) = V_{t0} + \delta V_t, \ V_t(M_{12}) = V_{t0}.$$

On the other hand, they become maximum, if M_{11} has a negative δV_t, and the signal polarity on DL_2 is positive. They are given as

$$v_{Seff}(L)_{max} = V(DL) - V(\overline{DL}) + \Delta(C_D) + \delta V_t = -v_S(L) + v_{ND},$$

$$V_G(L)_{max} = V(\overline{DL}) - V_t(M_{11}) = V_{DD}/2 + v_{NC},$$

$$v_{ND} = \Delta(C_{WD}) - \Delta(C'_{WD}) - \Delta(C_{DD}) + \Delta(C_D) + \delta V_t,$$

$$v_{NC} = -\Delta(C_P) + \Delta(C'_{WD}) + \Delta(C_{DD}) - V_t(M_{11}),$$

$$V_t(M_{11}) = V_{t0} - \delta V_t, \ V_t(M_{12}) = V_{t0}.$$

It should be noted that $v_{Seff}(L)$ is always smaller than $v_{Seff}(H)$, and $V_G(L)$ is also smaller than $V_G(H)$. Thus, $v_{Seff}(L)_{min}$ and $V_G(L)_{min}$ must be increased by reducing their noises v_{ND} and v_{NC} for high S/N and fast sensing. This is done by reducing each noise component of v_{ND} and v_{NC}, as discussed later.

The V_G differences ($= V_G(\text{H}) - V_G(\text{L})$) causes a sensing problem: In practice, the sensing starts with a successive driving of the common source line SNL from $V_{DD}/2$ to 0 V and then SPL from $V_{DD}/2$ to V_{DD} by activating large driver MOSTs connected to SNL and SPL lines, each of which is shared with, for example, 8 pairs of data lines, as shown in Fig.2.17. The sharing is necessary to accommodate a sense amplifier (SA) within a small data-line pitch. However, it varies the starting time of activation for each SA. When sensing, the nMOST with $V_G(\text{H})$ starts to amplify the signal earlier than that with $V_G(\text{L})$ due to the V_G differences, even if the impedance consisting of the wiring and the driver MOST(M_{SNL}) is neglected. In practice, the 'H' read current flowing into the impedance raises the SNL voltage, making the amplification for a 'L' read signal further delayed due to the resultant reduced V_G. The delay depends on the above-described noise and the read data pattern along the selected word line. It is maximized when all turned-on MOSTs in SAs(SA_0-SA_6) have the $V_G(\text{'H'})_{max}$ given by Eq.(2.8), except the one turned-on MOST(SA_7) with $V_G(\text{'L'})_{min}$

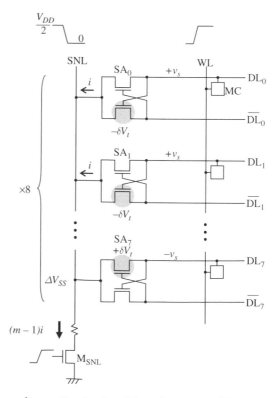

FIGURE 2.17. Dependence of activation delay of sense amplifiers on read data pattern along the selected word line.

given by Eq.(2.10). The delay has been reported to further increase $\Delta(C_{DD})$
noise [1, 25, 26]. Rapidly-discharging voltages on adjacent data lines that result
from earlier activation couple large noises to the pair of data lines that is still at
a small and floating voltage. The delay problem caused by the ever-increasing
noise and V_t-variations may be fatal in the ultra-low voltage nano-scale era.

2.4.6. Noise Reduction

A. Differential Mode

The differential noise v_{ND} in (2.11) is reduced as follows. $\Delta(C'_{WD})$ is usually
negligible. $\Delta(C_{DD})$ becomes negligible by twisting DLs [1] without an area
penalty [9], or by shielding data lines by means of devices and circuits. $\Delta(C_D)$
can be reduced to some extent with a slow activation of the sense amplifier [1].
Reduction of δV_t is difficult as long as bulk MOSTs are used. However, it can be
reduced in various ways: lithographically symmetric layout for sense amplifiers,
as exemplified by a circle-gate layout shown in Fig. 2.18 [33], is essential to
reduce the extrinsic δV_t caused by process variations and mask-misalignments.
In addition to a stringent control of the channel length and width, using the
largest MOSTs possible is a straightforward solution to reduce the extrinsic
and intrinsic δV_t. Fortunately, enlarging MOSTs (e.g., to 10–20 F^2, F; feature
size) is possible for DRAM sense amplifiers, as discussed in Chapter 3 (see
Fig. 3.23). Even so, δV_t increases with device scaling, as shown in Fig. 2.19,
because the intrinsic V_t variation increases. Even the typical offset for the 100-nm
process generation is reportedly as large as 30–40 mV [27]. Thus, new circuits
and/or new MOSTs to cope with the problem are necessary. However, many
δV_t-compensation circuits proposed since the end of the 1970's have suffered

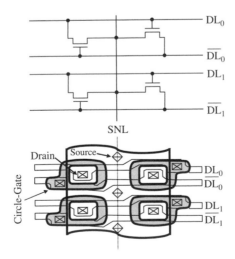

FIGURE 2.18. Circle-gate sense amplifier. Reproduced from [33] with permission;
© 2006 IEEE.

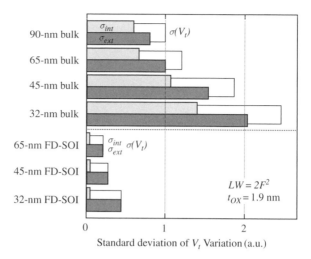

FIGURE 2.19. Standard deviations of V_t variation, $\sigma(V_t)$, and intrinsic (σ_{int}) and extrinsic (σ_{ext}) V_t variations. Reproduced from [29] with permission; © 2006 IEEE.

from area and speed penalties. Redundancy and/or on-chip ECC circuits [28] are thus essential to compensate for δV_t, although they are almost invalid for an excessive δV_t. A fully-depleted double-gate SOI [29] is reportedly a strong candidate despite expensive wafers, because the ultra-thin and lightly-doped channel of the SOI structure suppresses the intrinsic ΔV_t and thus δV_t (see Fig. 2.19). The details are described in Chapter 5.

B. Common Mode

$\Delta(C_P)$ can be reduced to some extent by differentially driving the precharger/equalizer (PC) [9]. $\Delta(C_{DD})$ can be negligible as in the differential mode. $\Delta(C'_{WD})$ is also negligible. The remaining problem is to reduce the V_t of the turned-on nMOST in sense amplifiers. In fact, there is the lowest necessary V_t which is determined by considering the different functions of sense amplifiers; sensing and holding of the amplified signal. A low V_t is better for fast sensing. An excessively low V_t, however, may collapse a floating signal voltage before activating the pMOS SA for further amplification. Figure 2.20 [1] shows waveforms on the data lines with solid lines for the high enough V_t, and with the broken lines for the too low V_t. For the high V_t, the turned-on MOST M_1 starts discharging DL while another MOST M_2 also gradually discharges $\overline{DL}$. When the resultant DL-voltage reaches the V_t (point A), however, M_2 stops discharging (point A') and the discharged voltage is held on $\overline{DL}$, and M_1 thus continues discharging with the held voltage. After DL has completely discharged, M_2 (with $V_{GS} = 0$) does not have significant subthreshold current because of the high enough V_t. For the low V_t, however, both data lines are rapidly discharged. Even when the DL has been discharged completely to point B, $\overline{DL}$ at point

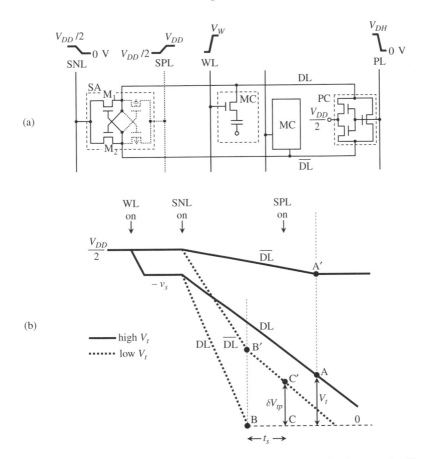

FIGURE 2.20. Dependence of data-line waveforms during amplification on the V_t of nMOSTs in the sense amplifier.

B′ continues to be discharged by M_2's constant subthreshold current until $\overline{DL}$ is completely discharged. This implies that the signal is lost, and the read has failed. The pMOS amplifier should be activated while the signal component is still larger than the offset of the pMOS amp (δV_{tp}), that is, before point C′.

The lowest necessary V_t at the sensing stage can be estimated as follows. If the signal voltage component after point B′ can be degraded by 50 mV $(= \delta V_{tp})$ during 5 ns $(= t_S)$ on the data line with a 100-fF C_D at 120 °C, the SA's nMOST with a channel width of 1 μm can have a subthreshold current of 1 μA at $V_{GS} = 0$. The V_t defined at a 1-μA constant current is thus around 0.2 V at 25 °C, with an assumption of $\Delta V_t / \Delta T_j = 2$ mV/°C. It corresponds to around −0.1 V at the commonly used current density of 1 nA/μm for $S = 100$ mV/decade, which is around 0.2 V by the extrapolated V_t definition. In practice, the drain-induced barrier lowering effect must be considered for the estimation because the sensing is carried out at a small V_{DS} region. Although the details remain unclear, the

V_t was reportedly lowered to 0.2 V, coupled with power switches, to achieve a 0.6-V-V_{DD} sensing for a 16-Mb e-DRAM with a 40-fF C_S [8].

Such an excessively low V_t, however, causes large subthreshold currents from the many CMOS SAs during the data holding period after amplification, thus calling for a higher V_t only for the period. The V_t is estimated as 0.5 V at 25 °C, assuming that each of the SA's nMOSTs can have only 1-nA subthreshold current to confine the accumulated leakage from 128-k SAs in a 1-Gb DRAM to 0.1 mA. The current difference by three orders (i.e., 1 μA and 1 nA) of magnitude makes a V_t-difference of 0.3 V for $S = 100$ mV/decade. Thus, it is ideal if a low V_t is used at the initial stage of sensing, while a high V_t limits leakage during data holding periods. Thus, a dynamic-V_t sensing with a partially-depleted (PD) SOI [1, 30, 31] and a sense amplifier controlled by a power switch [8] (see Fig. 4.22) are needed.

2.4.7. The Minimum V_{DD}

This section investigates the minimum V_{DD} ($= V_{min}$) of DRAMs, assuming that $\Delta(C_{DD})$, $\Delta(V_{DD}/2)$, ΔV_t, and $\Delta(C'_{WD})$ are negligible with noise reduction techniques. For stand-alone DRAMs with a large enough C_S and C_D, both $\Delta(C_P)$ and $\Delta(C_{WD})$ are negligible, as mentioned earlier. Thus, V_{min}s for the differential mode and the common mode are given by using $v_{Seff}(L)_{min}$ and $V_G(L)_{min}$, expressed as Eqs.(2.9) and (2.10), as

$$V_{min}(D) = 2\{\Delta(C_D) + \delta V_t\}(C_S + C_D)/C_S, \qquad (2.13)$$

$$V_{min}(C) = 2(V_{t0} + \delta V_t). \qquad (2.14)$$

For e-DRAMs, both $\Delta(C_P)$ and $\Delta(C_{WD})$ become significant. Thus,

$$V_{min}(D) = 2\{\Delta(C_{WD}) + \Delta(C_D) + \delta V_t\}(C_S + C_D)/C_S, \qquad (2.15)$$

$$V_{min}(C) = 2\{\Delta(C_P) + V_{t0} + \delta V_t\}. \qquad (2.16)$$

The V_{min} of DRAMs is determined by the higher one of the two V_{min} values expressed by Eqs.(2.13) and (2.14) for stand-alone DRAMs, and Eqs.(2.15) and (2.16) for e-DRAMs. For example, for $\Delta(C_{WD}) = 10$ mV [34], $\Delta(C_P) = 10$ mV [34], $\Delta(C_D) = 5$ mV [34], $\delta V_t = 50$ mV (see Fig. 5.6), $(C_S + C_D)/C_S = 5$, and $V_{t0} = 200$ mV, the V_{min} is equal to $V_{min}(D)$ and is 0.55 V for stand-alone DRAMs. It is also $V_{min}(D)$ and is 0.65 V for embedded(e-) DRAMs. This implies that reduction of δV_t is a key to low-voltage operations. This is the case for bulk CMOS.

2.5. Design of the Open-Data-Line 1-T Cell

The open data-line cell is attractive due to a small cell size of $6F^2$, as discussed previously. Unfortunately, it generates a large noise in the array, which is unpredictable and quite sensitive to the cell structure and the layout of the array.

The noise is always larger than that of the folded-data-line cell, because each of a pair of data-lines is laid out in two physically-separated subarrays. The noise cancellation by using such two subarrays is quite difficult, while the folded-data-line cell can cancel any noise at the two cross points in each cell in one array. The best way to reduce the noise is to suppress the voltage difference between the two arrays, which is realized by dividing an array into multiple copies of the smallest subarray possible and bridging between the two relevant subarrays with low impedance interconnects [12]. Even so, there is a residual noise that eventually may prevent low-voltage mid-point sensing. The noise generation mechanism of the cell array with the stacked-capacitor 1-T cell and the noise reduction are briefly described in the following.

2.5.1. Noise-Generation Mechanism

Open data-line cells with stacked-capacitors are shown in Fig. 2.21. The cell size is $6F^2$ at the word-line (WL) pitch of $2F$ and the data-line (DL) pitch of $3F$. The cell-plate (PL) on the storage node (SN) overlays the entire array, and the p-well (PW) and deep n-well (NW) on the p-substrate (SUB) are common to all memory cells of the array. The tilted active area allows a wider transfer gate and smaller data-line contacts (DCT). The storage capacitance (C_S) is 25 fF. The capacitances related to the data-line are the DL-WL capacitance ($C_{DW} = C_{WD}$), DL-storage node (SN) capacitance (C_{DS}), DL-p-well (PW) capacitance (C_{DP}),

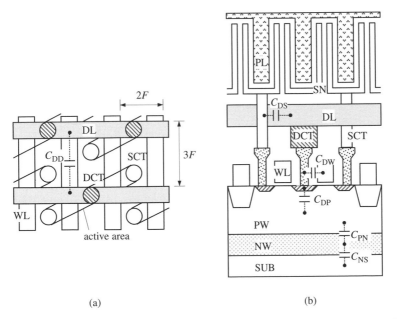

(a)   (b)

FIGURE 2.21. $6F^2$ open-DL memory cell. (a) Top view. (b) Cross-sectional view. Reproduced from [12] with permission; © 2006 IEEE.

TABLE 2.1. Data-line related capacitances of the open data-line array in 0.13-μm technology. Reproduced from [12] with permission; © 2006 IEEE.

	(aF/cell)	Simulated (measured)
C_{DW}	DL to WL	59 (56)
C_{DS}	DL to SN	71 (73)
C_{DP}	DL to PW	47 (58)
C_{DD}	DL to DL	1 (2)
C_{PN}	PW (-0.7 V) to NW (3.1 V)	43
C_{NS}	NW to SUB (0 V)	12
Cell size		0.109 μm^2
C_S		25 fF
V_t of cell transistor		1.1 V @ $V_{BB} = -0.7$ V

and DL-DL capacitance (C_{DD}). Table 2.1 gives, for comparison, the values as simulated by using a 3D capacitance-extraction tool along with values measured with the test chip. In the open data-line cell, a wider data-line pitch and the shielding effect of denser storage-node contacts (SCT) than in the folded data-line cell lead to a reduced C_{DD}.

Figure 2.22 shows an open-DL-cell array where each of a pair of data lines is arranged in two adjacent arrays (MA$_0$, MA$_1$) that are separated by sense-amplifiers (SA). Four kinds of conductors work as noise sources in each array; the p-well, the plate, the group of non-selected word lines, and adjacent data-lines. The above-described capacitors that are related to the conductors generate noise

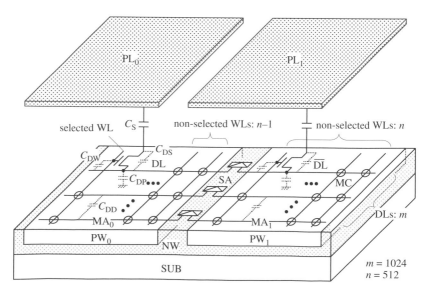

FIGURE 2.22. Open-DL cell array with stacked-capacitor cell. Reproduced from [12] with permission; © 2006 IEEE.

against the cell-signal voltage on the pair of data lines whenever the voltages on the conductors are changed. C_{DD} generates the well-known inter-data-line coupling noise [1]. The other capacitances generate noise due to the differential voltage swing on each of the m pairs of DLs when the mid-point sensing scheme is applied. The generation of noise can be conceptually explained by a simplified array (Fig. 2.23), in which two separate noise sources (NS_0, NS_1) represent p-wells, plates and groups of non-selected word-lines, R_A is the resistance of each noise source and R_B is the connection resistance, C represents C_{DP}, C_{DS}, and C_{DW} and C_0 and C_{D0} are the capacitances of the noise source and data line, respectively. The simultaneous development of a large voltage swing (V) on the background data-lines couples a voltage δV to the noise source (NS_0). The magnitude of δV depends on the data-pattern of the cells along the selected WL. The value of δV is maximized when $(m-1)$ DLs in an array are driven with the same polarity. The maximum voltage (δN) is then re-coupled to the target data-line in the same array through C. Since the two groups of $(m-1)$ DLs and $(m-1)$ $\overline{DL}$s are differentially driven, the maximum differential voltage $2\delta N$ is generated across the target pair of data-lines. The $2\delta N$ acts as differential-mode noise when its polarity is opposite to the cell-signal polarity on the target data-line. When both polarities are the same, it is added to the signal. The above-described noise is generated whenever a large voltage swing is involved. Thus, it is generated after precharging data-lines and when cell-signals are sensed and amplified.

2.5.2. Concepts for Noise Reduction

One guiding principle for reducing the level of noise is to reduce the impedance R_A (Fig. 2.23). This enables suppression of the noise peak and quicker recovery to the original voltage level because of the lower time constants of the conductors. Bridging of noise sources NS_0 and NS_1 with a low-impedance (R_B) conductor

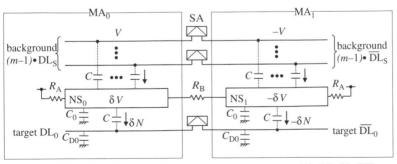

NS: WL, PL, PW
C : C_{DP}, C_{DS}, C_{DW}
C_0 : NS capacitance
C_{D0} : DL capacitance

FIGURE 2.23. Noise-generating mechanisms in an open-DL array. Reproduced from [12] with permission; © 2006 IEEE.

provides a further reduction in the voltage difference, even though the respective voltages remain coupled to each noise source. The low-impedance array shown in Table 2.2 was thus proposed [12]. Here, the original array featuring simple V_{BB}-supply and plate layouts was cited as a reference, which has been accepted for the folded data-line cell because of its inherently low-noise features. A row of V_{BB}-supply contacts is placed at one end of the p-well surrounded by the SA's n-well and the deep n-well. Both plates are composed of TiN, and they are only connected, with Al wiring, at the ends of the arrays. In the proposed array, however, a row of V_{BB}-supply contacts is placed at each end of each p-well region with an area penalty of only 0.6%. Moreover, the impurity concentration of the p-well is increased to reduce the p-well's sheet resistance from $2k\Omega/\square$ to $500\Omega/\square$. This increase is applied only to the middle part of the p-well to avoid detrimental effects on the memory-cell MOSTs and to keep the p-well to deep n-well breakdown voltage constant. The plate layer itself provides the bridge over the SA region between the pair of adjacent plates without area penalty. The tungsten (W), which is used for DLs, is also utilized to reduce the sheet resistance from $20\Omega/\square$ to $2\Omega/\square$ by placing it on the TiN. The WL's sheet resistance is reduced from $10\Omega/\square$ to $2\Omega/\square$ by using a W and poly-Si dual layer instead of the WSi and poly-Si dual layer of the original array. Noise of the above-described low-impedance array was evaluated by means of circuit simulations. Figure 2.24 shows dependence of the noise level on array size. Here, all conditions except the DL's length are kept the

TABLE 2.2. Proposed open-data-line array and the original array. Reproduced from [12] with permission; © 2006 IEEE.

	original array	proposed array
p-well	p-well $(2\,k\Omega/\square)$ $R_A = 620\,\Omega$, $R_B = 70\,\Omega$	p-well $(500\,\Omega/\square)$ dual contact (64 contacts/row) $R_A = 70\,\Omega$, $R_B = 60\,\Omega$
cell plate	TiN-cell plate $(20\,\Omega/\square)$ $R_A = 16\,\Omega$, $R_B = 5\,\Omega$	W on TiN-cell plate $(2\,\Omega/\square)$ bridging (128 lines/array) $R_A = 0.3\,\Omega$, $R_B = 0.4\,\Omega$
word-line	WSi / poly-Si WL $(10\,\Omega/\square)$ $R_A(WL, WD) = 16\,\Omega$ $R_B(V_{SS}) = 1\,\Omega$	W / poly-Si WL $(2\,\Omega/\square)$ $R_A(WL, WD) = 8\,\Omega$ $R_B(V_{SS}) = 1\,\Omega$

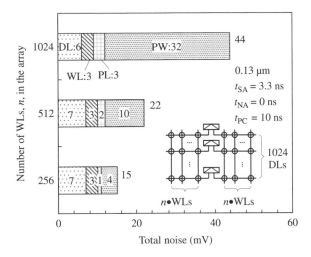

FIGURE 2.24. Dependence of the noise level on array size. Reproduced from [12] with permission; © 2006 IEEE.

same as in the basic proposed array. Obviously, the noise becomes smaller with a smaller array size, because a smaller array can be considered as, in a sense, a reduced impedance array. Furthermore, the array noise was compared with that of the $8F^2$ folded-data-line cell, as shown in Fig. 2.25 and Table 2.3. Consequently,

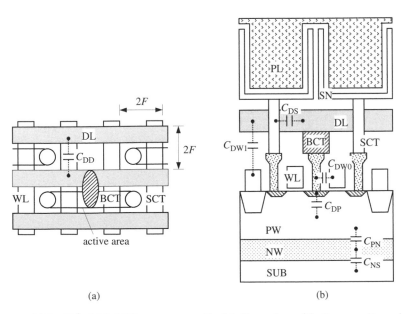

FIGURE 2.25. $8F^2$ folded-DL memory cell. (a) Top view. (b) Cross-section view. Reproduced from [12] with permission; © 2006 IEEE.

TABLE 2.3. Data-line related capacitances of the folded-DL array in 0.13-μm technology. Reproduced from [12] with permission; © 2006 IEEE.

			(aF/cell)	simulated
C_{DW}	C_{DW0}		DL to WL (w. MC)	104
	C_{DW1}		DL to WL (w/o.MC)	4
C_{DS}			DL to SN	119
C_{DP}			DL to PW	34
C_{DD}			DL to DL	18
C_{PN}			PW (-0.7 V) to NW (3.1 V)	53
C_{NS}			NW to SUB (0 V)	15
Cell size				0.135 μm^2
C_S				28 fF
V_t of cell transistor				1.1 V @ $V_{BS} = -0.7$ V

the low-impedance 512-n array was expected to reduce the array noise to 22 mV, which was comparable to 23 mV for the twisted-folded-data line cell.

2.5.3. Data-Line Shielding Circuits

An arising problem with device scaling is the ever-increasing C_{DD}-coupling noise that cannot be cancelled with the open-data-line cell, although the above-described example fortunately has a small enough C_{DD} due to a wider DL pitch and the shielding effect of the storage-node contacts (SCT). Thus, a data-line shielding circuit is needed.

Figure 2.26 shows a good example applied to a 4-Gb DRAM [32] using a stacked-capacitor $8F^2$ cell that is probably similar to that in Fig. 2.25. Two data-lines (D_i, $\overline{D_i}$), each of which is in the two adjacent arrays, are twisted at the sense amplifier to create a pair of data lines. Alternate selection of data lines in the same array allows each selected data line to be shielded with quiescent adjacent data lines which are biased at the half-V_{DD} precharge voltage. Here, the equalizing MOST in the conventional scheme is removed since the DL equalization signals can be merged with the isolation signals (ISO$_i$ and ISO$_j$) using the boosted voltage (V_{DH} or V_{PP}) without degradation of the DL precharge time. The circuit has been reported to achieve a coupling noise of 3 mV, which is quite small compared with 17 mV of the non-twisted folded-data-line cell.

2.6. Design of the 2-T Cell

The 2-T cell (twin cell) is ideal as the highest-speed cell circuit despite the large cell size, if the low-noise characteristics and the differential sensing and driving capabilities are fully utilized. In fact, a 4.8-ns access 144-Mb DRAM with 16-F^2 cells [20] and a 9.6-ns cycle 288-Mb DRAM with 12-F^2 cell [21] have been reported. It is also ideal as a low-voltage cell circuit, because noise

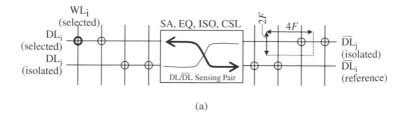

(a)

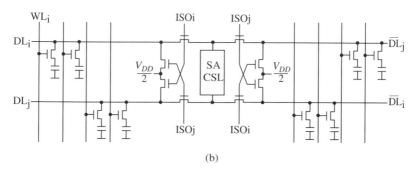

(b)

FIGURE 2.26. Concept of the twisted open data line (a) and actual circuits (b). Reproduced from [32] with permission; © 2006 IEEE.

can be minimized and the reference level is always equal to the signal voltage on another data line.

The data-line voltages for a 'H' read on DL and a 'L' read on $\overline{DL}$ are given as $V(DL) = V_{DD}/2 + v_S(H)$, and $V(\overline{DL}) = V_{DD}/2 - v_S(L)$, assuming that the differential noise $\Delta(C_{DD}) = 0$ with the shielded data line [34], $\Delta(V_{DD}/2) = 0$ after a laser level-trimming, $\Delta(C_P) = \Delta(C_{WD})$ with adjusting the parameters and/or voltage levels, and $\Delta V_t = 0$ with compensation circuits. Thus, the effective signal voltage v_{Seff} and the minimum gate-over-drive (V_{Gmin}) of a turned-on nMOST that is realized by $v_S(H)$ and the highest V_t are given as

$$v_{Seff} = v_S(H) + v_S(L) - \delta V_t, \tag{2.17}$$

$$V_G(L)_{min} = V_{DD}/2 + v_S(H) - (V_{t0} + \delta V_t), \tag{2.18}$$

$$v_S(H) = (Q_{SO} - Q_L - Q_C)/(C_D + C_S), \tag{2.19}$$

$$v_S(L) = Q_{SO}/(C_D + C_S). \tag{2.20}$$

If $v_S(H) \cong v_S(L)$, $V_{min}(D)$ and $V_{min}(C)$ are given as

$$V_{min}(D) = \delta V_t(C_D + C_S)/C_S, \tag{2.21}$$

$$V_{min}(C) = 2(V_{t0} + \delta V_t)/\{1 + C_S/(C_D + C_S)\}. \tag{2.22}$$

Thus, they are eventually determined by three parameters, δV_t, $(C_D + C_S)/C_S$, and V_{t0}. For example, for $\delta V_t = 50\,\text{mV}$, $(C_D + C_S)/C_S = 5$, and $V_{t0} = 200\,\text{mV}$, $V_{\min}(D) = 0.25\,\text{V}$ and $V_{\min}(C) = 0.42\,\text{V}$. Thus, the $V_{\min}$ of this DRAM is 0.42 V.

2.7. Design of Double-Gate Fully-Depleted SOI Cells

The ultra-thin buried-oxide (BOX) double-gate FD-SOI MOST [29] shown in Fig. 2.27(a) is one type of ultra-low voltage DRAMs. The ultra-thin, lightly-doped channel suppresses V_t fluctuations caused by deviations in dopant distribution in the channel region and SOI-layer thickness, as shown in Fig. 2.19. Hence, the offset voltage of sense amplifiers, which is derived from the intrinsic V_t variation (σ_{int}) and is reportedly as large as 30–40 mV even in the 100-nm bulk MOST [27], is negligible. Since the wells are isolated from the diffusion regions by the BOX layer, very little current flows from the diffusion layer to the well, even when forward bias is applied, as shown in Fig. 2.27(b). This design has the additional expected advantage of drastically reducing the soft-error rate, as shown in Fig. 2.27(c) [38]. This is because, in addition to a few collected charges generated by the extremely thin silicon layer, most of the electron-hole pairs generated by alpha-particle or cosmic-ray irradiations are blocked by the BOX and do not affect the extremely thin silicon layers. These characteristics are extremely important in designing ultra-low DRAMs.

The FD-SOI 2-T cell [34] minimizes $V_{\min}(D)$ because $Q_L = 0$, $Q_C = 0$, and because of the already small δV_t and noise. For example, $V_{\min}(D)$ is as low as 25 mV for $V_{t0} = 200\,\text{mV}$, $\delta V_t = 5\,\text{mV}$, and $(C_D + C_S)/C_S = 5$, although $V_{\min}(C)$ is still as high as 0.34 V, as seen in Eqs. (2.21) and (2.22). If the full-V_{DD} date-line precharge (Fig. 2.28) is applied to a data-line-shielded FD-SOI 2-T cell [37], $V_{\min}(C)$ can be minimized by maximizing the gate over-drive of the sense amplifiers. In this precharge, a complementary voltage, 0 and V_{DD}, is stored at two nodes, and, therefore, the signal voltage

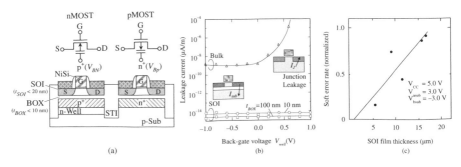

FIGURE 2.27. Structures of double-gate FD-SOI and its characteristics [29, 38].

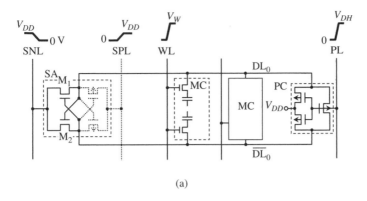

(a)

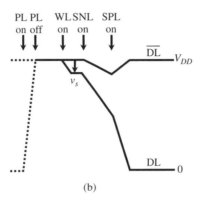

(b)

FIGURE 2.28. Data-line relevant circuits (a) and waveforms (b) of the 2-T cell using the full-V_{DD} data-line precharge.

from one of the twin cells is always negative for the reference level of V_{DD}. Since $v_{Seff} = (Q_{S0} - Q_L - Q_C)/(C_D + C_S) - \delta V_t$, $Q_{S0} = C_S V_{DD}$, and $V_G(L)_{min} = V_{DD} - (V_{t0} + \delta V_t)$, $V_{min}(D)$ and $V_{min}(C)$ are given as

$$V_{min}(D) \cong (1 + C_D/C_S)\delta V_t + (Q_L + Q_C)/C_S, \tag{2.23}$$

$$V_{min}(C) \cong V_{t0} + \delta V_t. \tag{2.24}$$

Figure 2.29 compares $V_{min}(D)$ and $V_{min}(C)$ for the bulk CMOS and the FD-SOI. For the bulk CMOS, δV_ts for a repairable percentage (r) of 0.1 % in Fig. 5.6(a) are used, and the tolerable decay of the stored voltage $(Q_L + Q_C)/C_S$ is assumed to be $0.2V_{DD}$. Thus, $V_{min}(D) \cong (1 + C_D/C_S)\delta V_t/0.8$. For the FD-SOI, $Q_L = 0$, $Q_C = 0$, and $\delta V_t = 5\,\mathrm{mV}$ are assumed. Obviously, for the bulk CMOS, the V_{min} of the chip is determined by $V_{min}(D)$ in the 65-nm generation and beyond, resulting in a V_{min} of 0.43 V in the 32-nm generation. For the FD-SOI, however, V_{min} is taken over by $V_{min}(C)$, and is as low as about 0.2 V. This is due to a drastic reduction in $V_{min}(D)$ with negligible Q_L and Q_C and small δV_t.

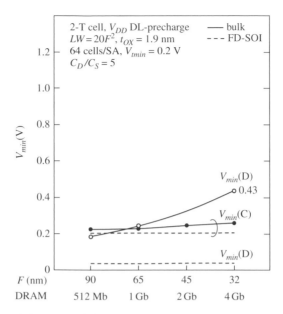

FIGURE 2.29. $V_{min}(\mathrm{D})$ and $V_{min}(\mathrm{C})$ for the bulk CMOS and the FD-SOI [37] using the 2-T cell and the full-V_{DD} DL-precharge.

It should be noted that $V_{min}(\mathrm{D})$ can be increased by increasing C_D/C_S until it is equal to $V_{min}(\mathrm{C})$, while maintaining the V_{min} of DRAMs at the same level. This leads to $(C_D/C_S)\delta V_t = V_t$, and thus $C_D/C_S = V_{t0}/\delta V_t = 40$ for $V_{t0} = 0.2\,\mathrm{V}$, and $\delta V_t = 5\,\mathrm{mV}$. Thus, the FD-SOI makes sensing possible even with an extremely small C_S (e.g., one-eighth that of conventional designs) for a fixed C_D, making it possible to use a simple logic-compatible planar capacitor. Figure 2.30 shows the structure of such a cell structure [34, 37], which fully exploits the advantages of the FD-SOI. It uses a MOS planar capacitor, the plate (PLT) of which is usually

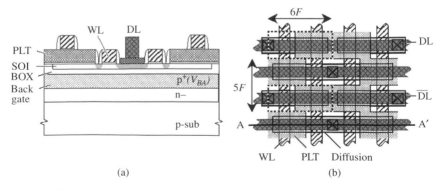

FIGURE 2.30. Cross section and layout of a FD-SOI 2-T cell [34, 37].

at a half-V_{DD}. To maintain extremely low junction-leakage characteristics in the MOST, no implant is added to the stored node. Thus, the work function of the capacitor plate material is different from that of the MOST gate materials, so the MOS capacitor is depleted even when a full V_{DD} is written from the data line. The V_t of the cell MOST can be adusted by implanting ions under the BOX and/or well bias voltages V_{BA}. To avoid coupling noise from adjacent data lines, each data line is shielded with quiescent data lines by alternate activation of paired data lines, allowing Eq. (2.23) without noise components to be justified. The SER is negligible because the collection area of the SOI is very small. The cell size is 30 F^2, making possible a C_S of 0.5 fF for a 1.3-nm t_{OX}(electrical) 32-nm technology. Thus, for a C_D/C_S of 40, at least 128 cells can be connected to one sense amplifier thanks to accepting a large C_D of 20 fF.

References

[1] K. Itoh, *VLSI Memory Chip Design*, Springer-Verlag, NY, 2001.

[2] Y. Nakagome, M. Horiguchi, T. Kawahara, K. Itoh, "Review and prospects of low-voltage RAM circuits," IBM J. R & D, vol. 47, no. 5/6, pp. 525–552, Sep./Nov. 2003.

[3] K. Itoh, K. Osada, and T. Kawahara, "Reviews and prospects of low-voltage embedded RAMs," CICC2004 Dig. Tech. Papers, pp.339–344, Oct. 2004.

[4] K. Itoh, "Low-voltage embedded RAMs in the nanometer era," ICICDT Dig. Tech. Papers, pp.235–242, May 2005

[5] S. Rusu, S. Tam, H. Muljono, D. Ayers, and J. Chang, "A dual-core multi-threaded Xeon processor with 16 MB L3 cache," ISSCC Dig. Tech. Papers, pp. 102–103, Feb. 2006.

[6] M. Khellah, N.S Kim, J. Haward, G. Ruhl, et al., "A 4.2 GHz 0.3mm² 256kb dual-V_{CC} SRAM building block in 65nm CMOS," ISSCC Dig. Tech. Papers, pp. 624–625, February 2006.

[7] J. Davis, D. Plass, P. Bunce, Y. Chan, et al., " A 5.6GHz 64KB dual-read data cache for the POWER6™ processor," ISSCC Dig. Tech. Papers, pp. 622–623, Feb. 2006.

[8] K. Hardee, F. Jones, D. Butler, M. Parris, M. Mound, H. Calendar, G. Jones, L. Aldrich, C. Gruenschlaeger, M. Miyabayashi, K. Taniguchi, and T. Arakawa, "A 0.6V 205MHz 19.5ns tRC 16Mb embedded DRAM," ISSCC Dig. Tech. Papers, pp. 200–2001, February 2004.

[9] M. Iida, N. Kuroda, H. Otsuka, M. Hirose, Y. Yamasaki, K. Ohta, K. Shimakawa, T. Nakabayashi, H. Yamauchi, T. Sano, T. Gyohten, M. Maruta, A. Yamazaki, F. Morishita, K. Dosaka, M. Takeuchi, and K. Arimoto, "A 322 MHz random-cycle embedded DRAM with high-accuracy sensing and tuning," ISSCC Dig. Tech. Papers, pp. 460–461, Feb. 2005.

[10] C. J. Radens et al., "A 0.135μm² 6F² trench-sidewall vertical device cell for 4Gb/16Gb DRAM," Symp. VLSI Technology Dig. Tech. Papers, pp. 80–81, June 2000.

[11] F. Hofmann et al., "Surrounding gate select transistor for 4F² stacked Gbit DRAM," Proc. ESSDERC, pp. 131–134, Sept. 2001.

[12] T. Sekiguchi, K.Itoh, T.Takahashi, M.Sugaya, H.Fujisawa, M.Nakamura, K.Kajigaya and K.Kimura, "A low-impedance open-bitline array for multigigabit DRAM," IEEE J. Solid-State Circuits, vol. 37, pp. 487–498, April 2002.

[13] W. Leung et al., "The ideal SoC memory: 1T-SRAM," 13th Annual IEEE Int. ASIC/SOC Conference, Sept. 2000.

[14] D. Somasekhar et al., "1T-cell DRAM with MOS storage capacitors in a 130nm logic technology for high density microprocessors caches," Proc. ESSCIRC, pp. 127–130, Sept. 2002.

[15] T. Ohsawa, K. Fujita, T. Higashi, Y. Iwata, T. Kajiyama, Y. Asao, and K. Sunouchi, "Memory design using a one-transistor gain cell on SOI," IEEE J. Solid-State Circuits, Vol.37, No.11, pp. 1510–1522, Nov. 2002.

[16] R. Ranica, A. Villaret, P. Mazoyer, D. Lenoble, P. Candelier, F. Jacquet, P. Masson, R. Bouchakour, R. Fournel, J.P. Schoellkopf, and T. Skotnicki, "A one transistor cell on bulk substrate (1T-bulk) for low-cost and high density eDRAM," Symp. VLSI Tech. Dig. Tech. Papers, pp. 128–129, June 2004.

[17] T. Ohsawa, K. Fujita, K. Hatsuda, T. Higashi, M. Morikado, Y. Minami, T. Shino, H. Nakajima, K. Inoh, T. Hamamoto, and S. Watanabe, "An 18.5 ns 128 Mb SOI DRAM with a floating body cell," ISSCC Dig. Tech. Papers, pp. 458–459, Feb. 2005.

[18] F. Morishita et al., "Leakage mechanism due to floating body and countermeasure on dynamic retention mode of SOI-DRAM," Symp. VLSI Tech. Dig. Tech. papers, pp. 141–142, 1995.

[19] F. Morishita, H. Noda, T. Gyohten, M. Okamoto, T. Ipposhi, S. Maegawa, K. Dosaka, and K. Arimoto, "A capacitorless twin-transistor random access memory (TTRAM) on SOI," CICC Dig. Tech. Papers, pp. 435–438, 2005.

[20] H. Noda, S. Miyatake, T. Sekiguchi, R. Takemura, T. Sakata, K. Saino, Y. Kato, E. Kitamura, and K. Kajigaya, "A 4.8-ns randomaccess144-Mb twin-cell-memory fabricated using 0.11-μm cost-effective DRAM technology," Symp. VLSI Circuits Dig. Tech. Papers, pp. 188–189, June 2004.

[21] K-H Kim, U. Kang, H-J Chung, D-H Park et a., "An 8Gb/pin 9.6ns row-cycle288Mb deca-data rate SDRAM with an I/O error-detection scheme," ISSCC Dig. Tech. Papers, pp. 156–157, Feb. 2006.

[22] Kinam Kim and G. Jeong, "Memory technologies in the nano-era: challenges and opportunities," ISSCC Dig. Tech. Papers, pp. 576–577, Feb. 2005.

[23] M. Asakura, T. Ohishi, M. Tsukude, S. Tomishima, et al., "A 34ns 256Mb DRAM with boosted sense-ground scheme," ISSCC Dig. Tech. Papers, pp. 140–141, Feb. 1994.

[24] T. Ooishi, Y. Komiya, K. Hamade, M. Asakura, K. Yasuda, K. Furutani, H. Hidaka, H. Miyamoto, and H. Ozaki, "An automatic temperature compensation of internal sense ground for subquarter micron DRAM's," IEEE J. Solid-State Circuits, Vol. 30, No. 4, pp. 471–479, April 1995.

[25] M. Aoki, Y. Nakagome, M. Horiguchi, H. Tanaka, S. Ikenaga, J. Etoh, Y. Kawamoto, S. Kimura, E. Takeda, H. Sunami, and K. Itoh, "A 60-ns 16M-bit CMOS DRAM with a transposed data-line structure," IEEE J. Solid-State Circuits, vol. 23, no. 5, pp. 1113–1119, Oct. 1988.

[26] J. Okamura, Y. Okada, M. Koyanagi, Y. Takeuchi, M. Yamada, K. Sakurai, S. Imada, and S. Saito, "Decoded-source sense amplifier for high-density DRAM's," IEEE J. Solid-State Circuits, vol. 25, No.1, pp. 18–23, Feb. 1990.

[27] J. Y. Sim, K. W. Kwon, J. H. Choi, S. H. Lee, D. M. Kim, H. R. Hwang, K. C. Chun, Y. H. Seo, H. S. Hwang, D. I. Seo, C. Kim, and S. I. Cho, "A 1.0 V 256Mb SDRAM with offset-compensated direct sensing and charge-recycled precharge schemes," ISSCC Dig. Tech. Papers, pp. 310–311, Feb. 2003.

[28] H. L. Kalter et al., "A 50-ns 16-Mb DRAM with a 10-ns data rate and on-chip ECC," IEEE J. Solid-State Circuits, vol. 25, pp. 1118–1128, Oct. 1990.

[29] R. Tsuchiya et al., "Silicon on Thin BOX: A new paradigm of the CMOSFET for low-power and high-performance application featuring wide-range back-bias control," IEDM Dig. Tech. Papers, pp. 631–634, Dec. 2004.

[30] S. Kuge, F. Morishita, T. Tsurude, S. Tomishima, M. Tsukude, T. Yamagata, and K. Arimoto, "SOI-DRAM circuit technologies for low power high speed multigiga scale memories," IEEE J. Solid-State Circuits, Vol. 31, No. 4, pp. 586–591, April 1996.

[31] K. Shimomura, H. Shimano, N. Sakashita, F. Okuda, T. Oashi, Y. Yamaguchi, T. Eimori, M. Inuishi, K. Arimoto, S. Maegawa, Y. Inoue, S. Komori, and K. Kyuma, "A 1-V 46-ns 16-Mb SOI DRAM with body control technique," IEEE J. Solid-State Circuits, Vol. 32, No. 11, pp. 1712–1720, Nov. 1997.

[32] H. Yoon, J. Y. Sim, H. S. Lee, K. Nam Lim et al., "A 4Gb DDR SDRAM with gain-controlled pre-sensing and reference bitline calibration schemes in the twisted open bitline architecture," ISSCC Dig. Tech. Papers, pp. 378–379, Feb. 2001.

[33] H. Yamauchi, T. Suzuki, A. Sawada, T. Iwata, T. Tsuji, M. Agata, T. Taniguchi, Y. Odake, K. Sawada, T. Ohnishi, M. Fukumoto, T. Fujita, and M. Inoue, "20ns battery-operated 16Mb CMOS DRAM," ISSCC Dig. Tech., Papers, pp.44–45, Feb. 1993.

[34] R. Takemura, K. Itoh, and T. Sekiguchi, "A 0.5-V FD-SOI twin-cell DRAM with offset-free dynamic-V_T sense amplifier," ISLPED Dig. Tech. Papers, pp. 123–126, Oct. 2006.

[35] M. Shirahama, Y. Agata, T. Kawasaki, R. Nishihara, W. Abe, N. Kuroda, H. Sadakata, T. Uchikoba, K. Takahashi, K. Egashira, S. Honda, M. Miura, S. Hashimoto, H. Kikukawa, and H. Yamauchi, "A 400 MHz random-cycle dual-port interleaved DRAM with striped-trench capacitor," ISSCC Dig. Tech. Papers, pp.462–463, Feb. 2005.

[36] K. Itoh, "Analog circuit techniques for RAMs-present and future-," Analog VLSI Workshop, 2005 IEEJ, Dig. Tech. Papers, pp. 1–6, Bordeaux, Oct. 2005.

[37] K. Itoh, M. Yamaoka, and T. Kawahara, "Impact of FD-SOI on deep sub-100-nm CMOS LSIs-a view of memory designers-" Int'l SOI Conf. Dig., pp. 103–104, Oct. 2006.

[38] H. Gotou, Y. Arimoto, M. Ozeki, and K. Imaoka, "Soft error rate of SOI-DRAM," IEDM Dig. Tech. Papers, pp. 870–871, 1987.

3

Ultra-Low Voltage Nano-Scale SRAM Cells

3.1. Introduction

Low-voltage nano-scale SRAMs are becoming increasingly important [1–4] to meet the needs of the rapidly growing mobile market, and to offset the sky-rocketing increase in the power dissipation of high-end MPUs (Microprocessor Units), while ensuring the reliability of miniaturized devices. Thus, sub-1-V SRAMs have been actively researched and developed, exemplified by a large 26.5-MB SRAM cache in a 0.8-V to 1-V MPU [5]. To design such SRAMs, however, the following challenges concerning the memory cell [1–4] must be accomplished. First, leakage currents of SRAM cells must be reduced because, for example, the subthreshold current exponentially increases with reducing threshold voltage (V_t) of MOST, eventually dominating even the active current of an SRAM chip. Second, the voltage margin of SRAM cells must be improved by high signal-to-noise-ratio (S/N) designs because it dramatically decreases with device and voltage scaling, causing degradations of the soft-error charac-teristics and sensing margin. Third, performance variations of memory cells caused by variations in processes, voltage, and temperature (PVT variations) must be reduced, because, for example, the ever-increasing V_t variation with device scaling increases leakage and degrades the voltage margin of SRAM cells. Fourth, to accomplish the above-described challenges, management of internal power-supply voltages with on-chip voltage converters is required. Fifth, the memory cell size, especially for low-voltage embedded memories, must be reduced since the memory block will dominate the chip size of various LSIs. In this sense, a comparison between SRAM cells and DRAM cells in terms of low-voltage potential and cell size is a serious concern.

This chapter describes ultra-low voltage nano-scale SRAM cells, focusing on the six-transistor full-CMOS cell (6-T cell). After explaining trends in SRAM cell-developments, the 6-T cell is discussed in detail in terms of leakage currents, the voltage margin with respects to the lowest necessary V_t and V_t variations for the cell, and the signal charge that is closely related to the soft error rate. After that, improvements of the voltage margin of the 6-T cell are outlined. Finally, a comparison between the 6-T SRAM cell and the one-transistor one-capacitor (1-T) DRAM cell is made, focusing on the low-voltage limitation. Here, the leakage, variability, and power management issues for the SRAM peripheral circuits are described in Chapters 4 to 9 in detail.

3.2. Trends in SRAM-Cell Developments

In the past, many kinds of SRAM cells, such as an enhancement MOST-load cell, a depletion MOST-load cell, a full CMOS cell (i.e., 6-T cell), a high-resistive poly-silicon-load cell, and a thin-film-transistor (TFT) load cell, have been proposed [1]. The 6-T cell, however, has become vital for use in cache memories. This is because it offers a stable operation–that is, a wide voltage margin and noise immunity even for high speed operation–and ease of fabrication, with a process compatible with CMOS logic in MPUs. Although a cache SRAM does not need a density as high as that of a DRAM, the memory cell size is still a prime concern to reduce the fabrication cost, especially for MPUs that have a large on-chip cache. Denser cells have been realized by using the following technologies: multi-level interconnects to form a flip-flop, source-drain silicidation, and trench isolation. Recently, advanced TFT and four-transistor (4-T) SRAM cells have been proposed to reduce the cell size.

Stacked-TFT SRAMs cells [6–8] are promising for tiny cells comparable to DRAM cells. In the conventional TFT (pTFT) load cell, the polycrystalline Si channel film of the pTFTs made it difficult to operate the cells below 3 V due to a high V_t of over 2.5 V that comes from a large S-factor and a poor on-off current ratio. Using an almost single crystal TFT is one solution, although details of the fabrication process remain unknown. However, the drain current is still two-thirds that of the bulk, and the S-factor of pTFT is as large as 140 mV/decade [6]. Thus, it seems that the TFT is not suitable to make a low-voltage high-speed embedded (e-) SRAM. Nevertheless, it has the great advantage of having high density for stand-alone SRAMs. A conventional 84-F^2 (F: feature size) 6-T SRAM cell using a high-density SRAM process was reportedly reduced to 45 F^2 by using load pTFTs stacked over the bulk driver and transfer nMOSTs [7], eliminating the n-well. It was further reduced to 25 F^2 by using load pTFTs and transfer nTFTs double-stacked over bulk driver n-MOSTs [8] in different levels of layers, as shown in Fig. 3.1. Resultant cell sizes were comparable to the sizes of 1-T embedded (e-)DRAM cells. The cells made it possible to realize record-setting high-density stand-alone SRAMs, with 80-nm SRAM processes, such as a 1.3-V 28.5-mm² 49.2-ns access 64-Mb chip with 0.288-μm² (45 F^2) cells [7], and a 1.8-V 61.1-mm² 144-MHz 256-Mb chip with 0.16-μm² (25 F^2) cells [8]. For both SRAM cells, the voltages for word-line and/or cell power supply were selectively boosted to compensate for the limited current driving capability of the TFT described above.

The 4-T cell [9, 10] consists of cross-coupled nMOSTs and transfer pMOSTs working as loads during standby, as shown in Fig. 3.2. When the cell is accessed, the selected word line goes down from V_{DD} to 0, so a differential read signal is outputted on the data lines which were precharged to V_{DD}. The cell size is smaller by about 30% than the 6-T cell. However, the voltage margin and stability may not be sufficient for ultra-low voltage operations: To ensure the data retention characteristics, the leakage current (subthreshold current, I_{offn}) of the off-nMOST must be compensated for by a large enough load current

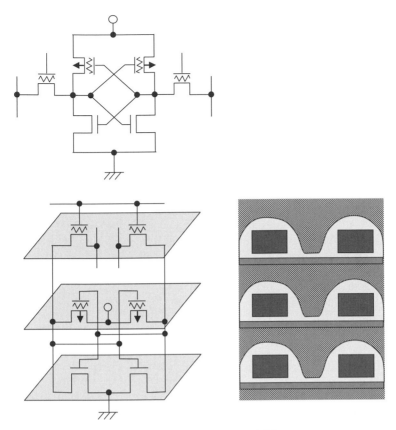

FIGURE 3.1. SRAM cell using thin-film-transistor (TFT) [8].

of the off- pMOST (I_{offp}). This requirement must be met even for the worst cell with the largest V_t of pMOS and the smallest V_t of nMOST within their V_t distributions in a chip. If their V_ts independently vary by $\pm 100\,$mV, and

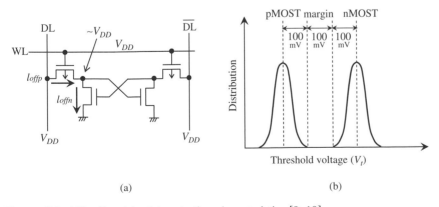

FIGURE 3.2. 4-T cell and its data-retention characteristics [9, 10].

a V_t-difference of 100 mV is needed for the worst cell to hold data, the average V_t of pMOST must be smaller by 0.3 V than that of nMOST, as shown in the figure. This ensures a 10-fold difference in the leakage even for the worst cell with an assumption of $S = 100$ mV/dec. Note that to ensure a retention current of 0.1 μA for a low-power 1-Mb array, the smallest necessary average V_t of the nMOS is around 0.7 V and thus 0.4 V for pMOST, as explained later (see Figure 3.4). Such high V_ts eventually limit the low-voltage operation. In addition, the cell may result in a slow cycle because the precharge time must be long enough so that non-selected cells are 'refreshed' within every cycle. Otherwise their pMOSTs cannot supply large enough load currents while the data line stays low, eventually causing data destructions of the cells during successive activations of other cells along the same data line. Moreover, any voltage coupling to non-selected word lines may degrade the retention characteristics with increased or decreased subthreshold currents. This may occur via the word-line to data-line capacitance when many data lines simultaneously swing at a large voltage.

Note that fully-depleted SOI SRAM cells [14, 15] enable lower-V_{DD} operations with a much lower subthreshold current thanks to reduced V_t variation and the back-gate controls of MOSTs. The 7-T cell [11] has also been proposed to enable a 0.5-V V_{DD} operation with cutting the inverter loop dynamically despite increased area.

3.3. Leakage Currents in the 6-T SRAM Cell

The data retention current of the 6-T SRAM cell consists of many components, as shown in Fig. 3.3, which is an example of a 0.13-μm 1.5-V SRAM cell [12]. Two gate-tunneling currents, five GIDL (gate-induced drain leakage) currents, and three subthreshold currents are developed in each cell. The gate-tunneling and GIDL currents are reduced by using new high-k gate insulator materials and lowering V_{DD}, respectively. The subthreshold current is reduced by increasing V_t. Because the subthreshold current usually dominates the data-retention current, increasing V_t is the key to achieving a small data retention current. The increase, however, requires raising V_{DD} to maintain the S/N of the cell constant. Thus, to determine the low-voltage limitation, the lowest necessary V_t (V_{tmin}) specification on the data-retention current must be investigated.

Subthreshold currents from the cross-coupled MOSTs are unavoidable because the MOSTs always form feed-back loops. These loops dramatically increase the data retention current with decreasing V_t and increasing temperature, as exemplified by the 1-Mb array current in Fig. 3.4 [3]. This identifies the V_{tmin} that satisfies the specification [1–4]. If a low-power 1-Mb SRAM needs a leakage of 0.1 μA at junction temperature $T_{jmax} = 75$ °C, the V_{tmin} at 25 °C might be as high as 0.71 V, while if a high-speed 1-Mb SRAM tolerates a leakage of 10 μA at $T_{jmax} = 50$ °C, it can be as low as 0.49 V. Such high-V_t MOSTs, however, degrade the read signal voltage and static noise margin [34–36], eventually preventing downscaling of the V_{DD}. Note that the V_t used in the Fig. is close to the average V_t (i.e., V_{t0}) because it is the average that usually determines the retention current of the chip. For transfer MOSTs, a low-actual V_t may cause read

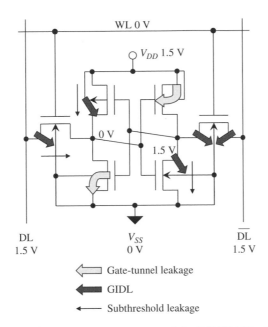

FIGURE 3.3. Standby-leakage current components of the 6-T SRAM cell [12].

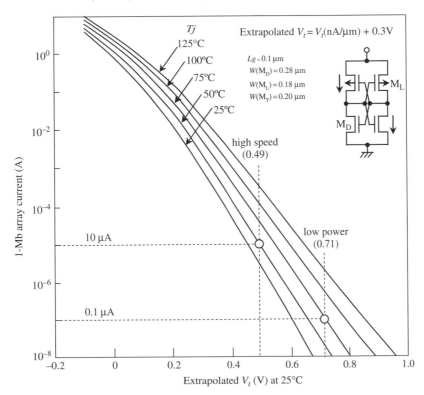

FIGURE 3.4. V_t of cross-coupled MOSTs versus subthreshold current of 1-Mb SRAM [1–4]. Reproduced from [3] with permission; © 2006 IEEE.

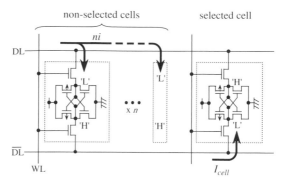

FIGURE 3.5. Read failure of SRAM cell [13].

failures [13], as shown in Fig. 3.5, if the subthreshold current accumulated(ni) from the transfer MOSTs of many non-selected cells along the same DL becomes larger than the read current(I_{cell}) of the selected cell. The subthreshold current, however, can easily be reduced because the MOSTs work as switches, as in DRAM cells. The use of a high enough or effectively high V_t, as in DRAM cells (Fig. 2.11), resolves the problem by isolating the selected cell from other non-selected cells.

3.4. The Voltage Margin of the 6-T SRAM Cell

The voltage margin of the SRAM cell for a given V_{tmin} must be clarified in terms of the S/N (i.e., read/write and soft-error characteristics) to determine the lowest necessary V_{DD} (V_{min}) for a successful operation.

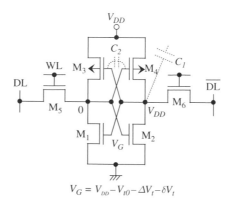

$$V_G = V_{DD} - V_{t0} - \Delta V_t - \delta V_t$$

FIGURE 3.6. SRAM cell circuit. V_{t0}; average V_t in a chip, ΔV_t; V_t variation in a chip, δV_t; V_t-mismatch between paired MOSTs.

3.4.1. Read and Write Voltage Margin

There are two kinds of voltage margins of the cell (Fig. 3.6) that we have to pay attention to; the read margin and write margin [34–36]. The rapidly-reduced read current of the cell with device and voltage scaling decreases the read voltage margin. For example, the read current of the turned-on nMOST(e.g., M_1) in the cell is proportional to the gate-over drive voltage (V_G) that is approximately expressed as $V_G = V_{DD} - V_{t0} - \delta V_t - \delta V_t$, if we can assume that the voltage margin is determined only by cross-coupled driver nMOSTs. Here, V_{t0}, δV_t, and δV_t are the average V_t and V_t variation in a chip, and the V_t-mismatch between paired nMOSTs in the cell, respectively. Thus, even for a fixed $(V_{DD} - V_{t0} - \delta V_t)$, each cell can have a different read current that depends on δV_T. Here, this difference becomes most prominent when V_{DD} reaches $V_{t0} + \delta V_t$, which occurs at quite a high V_{DD} since V_{t0} must be quite high (e.g., 0.49–0.71 V) and δV_t is also high in the nanometer era. For example, for $V_t = 0.71$ V and $\delta V_t = 0.1$ V, the lowest necessary V_{DD} (V_{min}) for a successful operation must always be higher than 0.81 V. In other words, the signal voltage is rapidly reduced as V_{DD} is lowered to below 1 V, entailing a large cell-to-cell variation of the read current. Since both δV_t and δV_t increase as the device gets smaller, as shown in Fig. 3.7(a), the V_{min} that is defined here as the V_{DD} corresponding to a static noise margin (SNM at read) of 0 becomes higher with device scaling, as shown in Fig. 3.7(b) [14, 15]. Note that a 1-V operation of a 90-nm bulk SRAM is expected to be impossible even for a low V_t of 0.4 V, if 6σ of V_t variation is taken into consideration. The write margin is determined by various ratio operations of MOSTs in a cell, as discussed later (see 3.5 and 3.6).

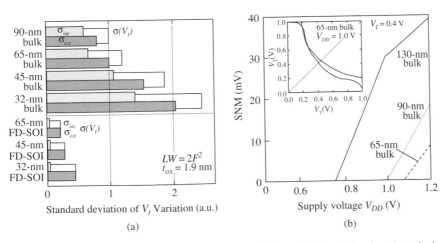

(a)

(b)

FIGURE 3.7. Standard deviations of V_t variation, $\sigma(V_t)$, and intrinsic (σ_{int}) and extrinsic (σ_{ext}) V_t variations (a) [3, 14, 15], and SNM of the 6-T bulk-CMOS SRAM cell taking 6σ of V_t variation into consideration (b) [14]. Reproduced from [14] and [15] with permission; © 2006 IEEE.

3.4.2. Signal Charge

The signal charge (Q_S) of the SRAM cell must be maintained to ensure good soft-error characteristics [1, 2]. Unfortunately, Q_S is usually reduced with device and voltage scaling, causing a rapid increase in the soft error rate (SER). This is because Q_S is given by $(C_1 + 2C_2) V_{DD}$ by using the cell-node capacitance (C_1) and node-to-node capacitance (C_2) [16] (Fig. 3.6), and it is almost equal to the soft-error critical charge in a low-V_{DD} region where the feedback and pull-up speeds in a SRAM cell are much slower than the soft-error phenomena. Here, C_1 and C_2 are usually parasitic and thus small, and they decrease at a rate of $1/k$ with a scaling factor k (Table 1.2). For alpha-particle strikes, chip coating and purification of materials are effective in reducing the number of alpha-particles from materials in a chip and the package. Another remedy is to add a large capacitance to $(C_1 + 2C_2)$, even though this requires more complicated processes and results in a slow write speed of the cell [17]. Fewer charge collection structures, such as p+barriers and triple well structures for shielding the cell array, as a soft-error barrier, are also effective [1]. Neutrons are the main contributors to the soft-error phenomena in recent SRAMs as well as DRAMs. This is because neutrons can generate about ten times as many electron-hole pairs as alpha-particles generate (see Fig. 1.54). Thus, it is almost impossible to reduce the resultant soft errors by means of increasing the signal charge. The ECC circuit is most effective not only for alpha-particle-induced soft errors, but also for cosmic-ray neutron-induced soft errors. The ECC is capable of correcting one error at each address, as discussed in Chapter 1.

A single cosmic-ray incidence often induces soft errors in multiple cells. The resultant multicell (multibit) error, if all of the affected cells belong to the same address, is not corrected by the ECC circuit. However, the circuit does correct a multibit error if all the affected cells belong to different addresses. To increase the rate of multibit-error correction by the ECC circuit, we need to arrange the cell

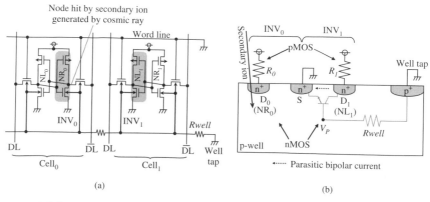

FIGURE 3.8. Models used in circuit simulation (a) and device simulation (b). Reproduced from [18] with permission; © 2006 IEEE.

addresses so that simultaneous cosmic-ray-induced multiple errors are most likely to occur at different physical addresses. Thus, the mechanism responsible for multibit errors, which means that multiple cells fail due to a cosmic ray, must be clarified. Figure 3.8 shows the models [18] used in circuit and device simulations for two adjacent memory cells (Cell$_0$, Cell$_1$). Nodes NR$_0$ and NL$_1$ are initially at a high voltage. The pMOST in one inverter (INV$_0$) is replaced by a resistance (R_0) and the nMOST is replaced by n$^+$ diffusion regions (D$_0$ and S). Analogous replacements are made in the other circuit. Substrate nodes of nMOSTs are tied together and connected to a parasitic p-well resistance (R_{well}), which is connected to a well tap. The parasitic bipolar device between D$_1$ and S is formed. Figure 3.9 depicts simulated waveforms [18] at nodes in Cell$_0$ and Cell$_1$, and p-well, when

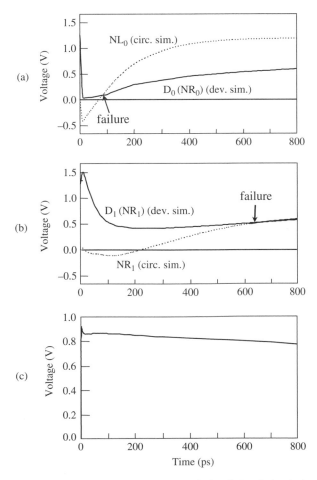

FIGURE 3.9. Nodes and well voltages in device and circuit level simulation. Cell hit by the cosmic ray, Cell$_0$ (a), the adjoining cell (b), Cell$_1$, and p-well voltage, i.e., voltage at V_p between D$_1$ and S (c). Reproduced from [18] with permission; © 2006 IEEE.

a secondary ion generated by a cosmic ray hits the node NR_0. The voltage relation between NL_0 and NR_0 is inverted within 100 ps of the cosmic-ray strike, implying a soft-error occurrence. The voltage relation between NL_1 and NR_1 is also inverted within 700 ps, implying a further soft error occurrence. The mechanisms can be explained as follows [18]. After the secondary ion hits NR_0, the funneling effect [40] leads to the very rapid collection of electrons in NR_0. This lasts for 10 ps and lowers the NR_0-voltage to zero. On the other hand, the holes generated by the ray remain in the p-well. The p-well voltage (V_p) thus floats up to 0.9 V, as shown in the figure. This V_p switches the parasitic bipolar element on, which leads to a current flow from NL_1 to node S. Thus, NL_1 voltage slowly falls. The maximum number of multibit errors has been reported to depend on the number (N_C) of cells between well taps in the word-line direction, as shown in Fig. 3.10 [18]. This assumes $R_{well} = 1 \, k\Omega$/cell and $C_p = 1 fF$/cell. Here, the cell is a conventional one (Fig. 3.12(a)), in which the p-well and n-well run along the word line. For $N_C = 16$, three errors occur. For example, when a cosmic ray hits the center cell, $\langle 7 \rangle$, funneling [40] causes an error in cell $\langle 7 \rangle$, and the parasitic bipolar effect induces errors in the adjoining cells (Fig. 3.10(b)). The key to correcting multicell errors through an ECC circuit is to avoid simultaneous reading out from multiple cells in which a multicell error might occur. For $N_C = 16$, for example, the maximum number of errors

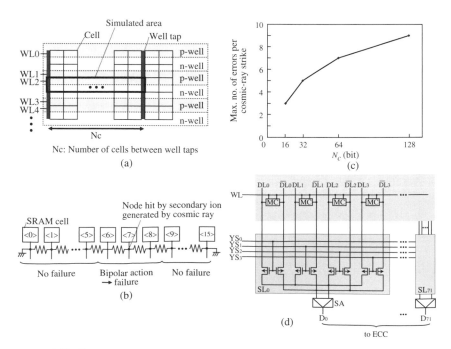

FIGURE 3.10. Error generation along the word line (a, b), the maximum number of errors as a function of N_C (c), and column selectors applied to the read path (d) [18].

in the word-line direction is three, so an ECC circuit can correct almost all of the errors if the data are simultaneously read out from the cells in every third column. This is done by 3-to-1 column selectors in order to select one of three pairs of data lines. Figure 3.10(d) shows actually used 4-to-1 column selectors for the read path, which have been reported to reduce the SER to about 100% [18].

In the LS cell shown in Fig. 3.12(b), in which a p-well is shared with the adjacent cell, a two-cell error might occur, so 2-to-1 column selectors [41] can be used.

The SER reduction by using a single-error correcting code is as follows. The reduction is expressed by the equation $E = (N^2T/2N_0^2W) E_0^2$, [19], where E is SER with ECC, E_0 is SER without ECC, N is the number of bits of an ECC word including check bits, N_0 is the number of data bits of an ECC word, T is the correction period, and W is the number of ECC words in a RAM (i. e., $W = M/N_0$, where M is the memory capacity). Figure 3.11 shows the SER reduction using a code of $N = 136$ and $N_0 = 128$ [43]. Even if SER without ECC is as high as 10^6 FIT (i.e., one upset per 1,000 hours) and the errors are not corrected at all during a ten-year period ($T = 10$ years), the SER is improved by four to five orders of magnitude through ECC. If periodic error correction (one ECC word every 7.8 µs) is performed like a DRAM refresh operation ($T = 7.8$ µs $\times W$), the resulting SER becomes as low as 10^{-6} FIT.

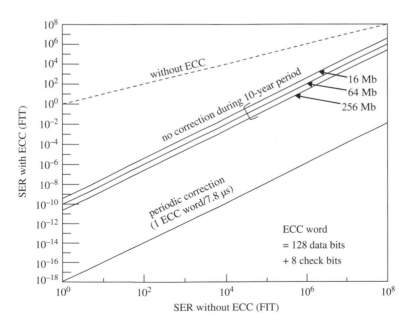

FIGURE 3.11. SER reduction through on-chip ECC [19, 43]. Reproduced from [43] with permission; © 2006 IEEE.

3.5. Improvements of the Voltage Margin of the 6-T SRAM Cell

For non-selected cells, the soft error problem is resolved with the ECC although it becomes ineffective when soft error events excessively increase. For the selected cell, since the write margin is improved by the power supply controls during write operations, as discussed later, the read margin eventually determines the minimum V_{DD} (V_{min}) for a successful operation, which is approximately given by $V_{min} = V_{t0} + \delta V_t + \delta V_t$. For a given leakage, that is, for a given V_{t0}, there are several approaches to improve the voltage margin of the selected cell. First, ECC circuits and redundancy [20, 42] can avoid the fatal cells, enabling a lower V_{min} for the whole chip, as discussed in Chapter 5 (Fig. 5.6). Second, increasing the memory-cell size lowers V_{min} despite losing bit density. Note that the smallest cell using minimum feature sizes available at the process generation makes V_{min} highest with the largest $\delta V_t + \delta V_t$. This is clearly seen in the recently presented dual-V_{DD} processor [21] using a 65-nm CMOS technology, in which V_{min} is 1.0–1.1 V for a 136Mb/cm^2 array density, while it is 0.6–0.7 V for a 95Mb/cm^2 density, as discussed later (Fig. 3.26). Third, an on-chip substrate bias (V_{BB}) control reduces the inter-die δV_t, as explained in Chapter 5. Fourth, a lithographically symmetric cell layout reduces the extrinsic variations, δV_t and δV_t. Fifth, controlling the cell power supply widens the voltage margin and/or reduces the subthreshold current, although there is a limitation to lowering V_{min}. Sixth, fully-depleted SOI greatly reduces inter-and intra-die V_t variations. The following describes the lithographically symmetric cell layout, power-supply controlled cells, and fully-depleted SOI cells.

3.5.1. Lithographically Symmetric Cell Layout

In addition to the intrinsic δV_t, there is the extrinsic δV_t. Pattern deformation after processing, mask-misalignment, and size fluctuation are major sources of extrinsic δV_t in the conventional cell layout shown in Fig. 3.12 (a). Unfortunately, they become larger with device scaling, causing a larger δV_t. The simple patterns of the lithographically symmetric layout cell (LS cell) [22] (Fig. 3.12(b)), the so-called 'thin' cell, almost solves the problems. As shown in Fig. 3.13, it enables an advanced lithography, called optical proximity correction (OPC), to be applied to every layer to reduce pattern deformation. The cell is immune to mask misalignments and the resulting excellent electrical balance and scalability make low voltage operations possible. Moreover, it reduces the data (bit)-line capacitance and the capacitive coupling noise from adjacent data lines. This is due to the short and wide data-line pitch, and shielding effects by V_{DD} and V_{SS} (ground) power lines running along the data line with the same metal layer. Thus, the LS cell has become a de-facto standard cell layout.

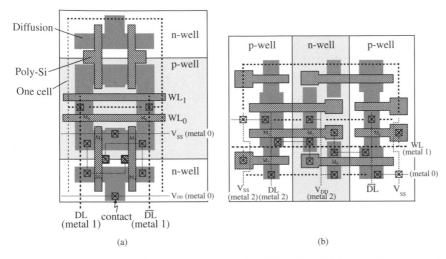

FIGURE 3.12. Comparisons between the conventional SRAM cell (a) and a lithographical symmetrical (LS) SRAM cell [22].

3.5.2. Power-Supply Controlled Cells

(1) Raised Power Supply Cell

Figure 3.14(a) shows a cell with raised power supply ($V_{DH} = V_{DD} + \delta V_D$) and dual-$V_t$ cell [3, 23] that maintains a low data-line voltage (i.e., V_{DD}). It has been reported that the originally proposed V_{DD} [23] was reduced to as low as 0.3 V for a full-V_{DD} data (or bit)-line precharging, while the V_{DH} was dynamically raised up to 1.2 V with a dynamic load. The cell features static high-V_t cross-coupled MOSTs that reduce the subthreshold currents. The V_{DH} is necessary to offset the

	Conventional cells	LS cells
Cell aspect ratio (height/width)	>1	<1
Well layer	Parallel to WL	Parallel to DL
Diffusion layer	Bended	Straight and parallel to DL
Poly–silicon layer	Two directions	Straight and parallel to WL
Application of advanced lithography (phase shift or OPC*)	Difficult	Easy
Pattern fluctuation	Large	Small
Immunity to mask misalignment	Poor	Good
Electrical balance	Poor	Good
Scalability	Poor	Good
Low voltage operation	Difficult	Possible

*OPC: Optical Proximity Correction

FIGURE 3.13. Comparisons between conventional SRAM cell and LS cell [22].

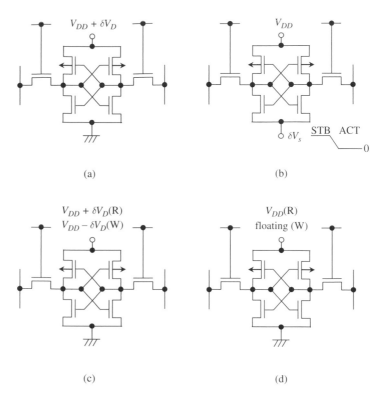

(a)

(b)

(c)

(d)

FIGURE 3.14. Power controls of SRAM cells [4]. Reproduced from [43] with permission; © 2006 IEEE.

high-V_t and the V_t-mismatch (δV_t) of cross-coupled MOSTs, so the signal charge Q_S and drivability of cross-coupled MOSTs are maintained, despite the need for high-stress-voltage-immune MOSTs. Here, the V_{DH} is generated by a charge pump or a voltage down-converter using the I/O power supply. It also features the negative word-line scheme (NWL) (Fig. 3.15) coupled with low-V_t transfer-MOSTs to increase the cell read current, as in DRAMs, to cut leakage during non-selected periods. The word boosting applied to high-V_t transfer MOSTs [37] also produces the same effect. Note that the δV_t not only makes the cell read current (I_{cell}) small, but also makes the static noise margin (SNM) imbalanced and thus narrower on the so-called butterfly curve for the worst combination of V_ts, as exemplified by a δV_t of 0.1 V in Fig. 3.15. Even with a small boost of $\delta V_D = 0.1$ V for $\delta V_t = 0.1$ V both I_{cell} and SNM are greatly increased for both a high-speed design of $V_t = 0.49$ V (Fig. 3.16(a)) and a low-power design of $V_t = 0.71$ V (Fig. 3.16(b)) [3]. Thus, the V_{DH} scheme is likely to make 1-V V_{DD} operation possible even under a design condition of a 100-mV SNM and 20-μA cell current, while the conventional V_{DD} scheme (i.e., no raised power supply and fixed high-V_t transfer MOSTs) makes 1-V V_{DD} operation at $V_t = 0.71$ V impossible. This V_{DH} scheme can maintain the data-line (DL) voltage low (V_{DD}).

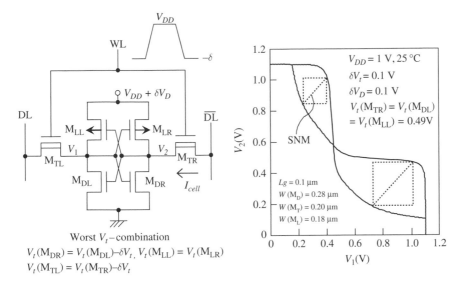

FIGURE 3.15. Raised power-supply 6-T SRAM cell and the butterfly curve [3, 23]. Reproduced from [3] with permission; © 2006 IEEE.

Thus, even if a high voltage V_{DH} is supplied to cells, increase in power dissipation of the chip is negligible. This is because the data-line voltage determines the power of column relevant circuits which are major sources of SRAM power.

(2) Offset Source Driving Cell

Figure 3.14(b) shows a typical offset source driving cell, in which the ground line is raised to δV_s. A raised cell-source line by δV_s during standby periods increases the V_t of off-nMOST due to the body effect. This is an application

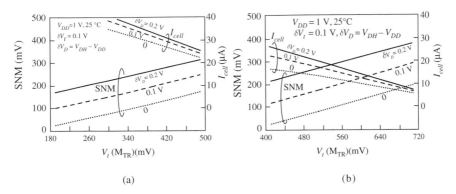

FIGURE 3.16. Static noise margin (SNM) and cell read current (I_{cell}) of raised power-supply cell at 1-V V_{DD} for low V_t (a) and high V_t (b). Reproduced from [3] with permission; © 2006 IEEE.

of (B2) in Table 4.1 to the SRAM to achieve a dynamic high-V_t cell, since the same circuit configuration is realized with the gate-source connection through a turned-on MOST. It was applied to a 1.5-V, 27-ns access, $6.42 \times 8.76 \, mm^2$, 16-Mb SRAM [12]. At ambient temperature, the measured total current per cell of the conventional scheme was 95 fA, as shown in Fig. 3.17. The largest component was the sum of the subthreshold current and the gate-induced drain leakage (GIDL) from nMOSTs and pMOSTs, although the V_ts were as large as 0.7 V and -1 V. The gate-tunneling current of the nMOST was comparable to this, despite an electrical t_{OX} as thick as 3.7 nm. The scheme greatly reduced the total current to 17 fA. Offset source driving by 0.5 V applied to the driver and transfer nMOSTs, and relaxing the electric field of all MOSTs by 0.5 V were responsible for the reduction. The reduction was more remarkable at a higher temperature. At 90 °C, the total current of conventional scheme was drastically increased to 1,244 fA because the subthreshold current is quite sensitive to temperature, unlike GIDL and gate-tunneling current. The scheme reduced the total current to 102 fA. A drawback of the cell, however, is that the signal charge is reduced by δV_s in the standby mode. To cope with the resultant increased SER, an ECC was incorporated with a speed penalty of 3.2 ns and an area penalty of 9.7%.

Various offset source driving schemes to control the common source-line of the cells have been proposed. Of the proposals, switched-source impedance schemes (SSI) [1, 2, 21, 24], as shown in Fig. 3.18, are noteworthy. The SSI shown in Fig. 3.18(a) is composed of a switch (MS), a diode (MD), and a MOS resistor (MR). Here, MR has a long gate and is normally on, so a constant resistance is available regardless of process variations. When the cell V_t is low and the memory-cell leakage is thus large, the cell-source-line voltage (δV_S) is raised, so the subthreshold current is reduced. However, an excessive δV_S is

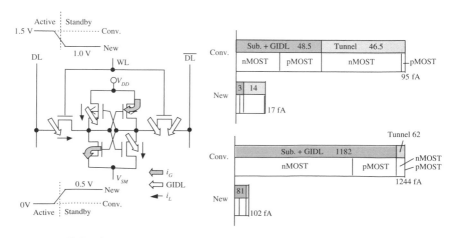

FIGURE 3.17. Leakage current components (a) and measured retention current per cell (b) [12]. i_G and i_L: gate tunneling current and subthreshold current.

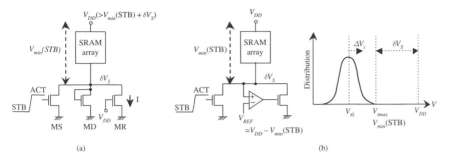

FIGURE 3.18. Switched-source impedance schemes applied to the common source-line of SRAM cells, using a resistor (a) [24] and a comparator (b) [21].

hazardous since there is the minimum V_{DD} in the standby mode, $V_{min}(STB)$, to retain the stored data of all cells in the SRAM array. Although δV_S is determined by the total leakage of the array and thus by the average V_t (i.e., V_{t0}), the voltage margin of each cell is determined by its own $(V_{t0} + \delta V_t + \delta V_t)$. Thus, an excessive δV_S may destroy the stored data of the cells with a high value for $V_{t0} + \delta V_t + \delta V_t$. It is MD that clamps the voltage to a low enough level. The SSI applied to a 0.13-μm 1.2-V 1-Mb SRAM module achieved 25μA/Mb in the standby mode and 50μA/Mb in the low-leakage-active mode at 100 MHz. Another SSI, shown in Fig. 3.18(b), was used for a 1.2-V 65-nm 512-Kb SRAM core [21]. The circuit configuration is quite similar to the boosted sense ground (BSG) scheme for DRAMs, as discussed in Chapter 2 (Fig. 2.11(c)). It uses a comparator to ensure $V_{min}(STB)$. If the reference voltage V_{REF} is set to be $V_{DD} - V_{min}(STB)$, δV_S is always well regulated to $V_{DD} - V_{min}(STB)$. If the V_t distribution of cells in the array is represented by the average $V_t (= V_{t0})$ and the maximum $V_t (= V_{tmax})$, the $V_{min}(STB)$ is set to be nearly equal to V_{tmax} with $V_{min}(STB) > V_{tmax}$. The difference between V_{DD} and V_{tmax} is δV_S. Thus, a lower V_{t0} results in a larger δV_S and deeper body bias, enabling to reduce the leakage with increased V_t.

(3) Dynamic Control of Cell-Power Line

Figure 3.14(c) shows a power-supply switching cell [25] for a wider write margin. A low level of the power supply during write widens the write margin with reduced conductance of load MOSTs, while maintaining the wide voltage margin by using a high level during read. The voltage difference between read and write was reportedly 200 mV at $V_{DD} = 1.1$ V. Figure 3.14(d) shows another one [26] that leaves the supply line to a floating level during write.

3.5.3. Fully-Depleted SOI Cells

A fully-depleted (FD) SOI structure [14, 15, 30] widens the voltage margin or lowers V_{min} of SRAM cells with reduced V_t variation. It also reduces the soft error rate and pn-junction currents. FD-SOI MOSTs using the traditional poly-silicon gate, however, become depleted (i.e., normally on) and thus loses advantages of flexible circuit designs provided by enhancement MOSTs, although it is applicable to DRAM word lines with the negative word-line scheme (NWL) [31]. Thus, different gate materials [32] that can adjust the V_t to be normally off (i.e., enhancement) have been proposed. The following FD-SOI is a good example to achieve an enhancement MOST with multi-static V_t.

Double-gate thin-BOX FD-SOI MOSTs [14, 15] are shown in Fig. 3.19(a). The MOSTs consist of an upper thin-film SOI MOST and a lower thin BOX MOST, the well of which is the gate. The features are as follows. First, the ultra-thin and lightly doped silicon channel achieves a higher mobility and suppresses the short channel effects and V_t variations that come from deviations in dopant distribution in the channel region, and from variations in SOI-layer thickness. Second, a combination of a fully silicided NiSi gate and ion implantation into the substrate beneath the thin BOX achieves the desired V_t values. Third, since the BOX layer is very thin, V_t can be controlled by changing the voltage of the underlying well (back gate, p^+ well for nMOST and n^+ well for pMOST), if the gate and well are disconnected. Since the wells are separated from each other by STI (shallow trench isolation), the back-gate voltage of each MOST is individually controllable. Since the wells are also isolated from the diffusion regions by the BOX layer, very little current flows from the diffusion layer to the well even when forward bias is applied. This differs from a partially depleted (PD-) SOI [27]. The PD-SOI DT MOST always involves a rapidly increasing forward pn-junction current, especially at a higher temperature, as suggested in Fig. 1.18. This strictly limits V_{DD} to an extremely low value of around 0.1 V at 120 °C.

One of the exceptional features of this design is that it creates many innovative low-voltage SRAM cells. If the back-gate bias voltage is applied so as to turn

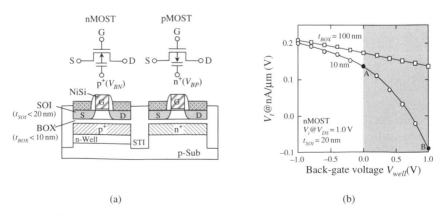

(a)   (b)

FIGURE 3.19. An ultra-thin BOX double-gate FD-SOI MOST (a) and its V_t-characteristics [14, 15]. Reproduced from [14] with permission; © 2006 IEEE.

on the lower BOX MOST, the sum of currents of the upper SOI and lower BOX MOSTs is available. This implies that the V_t is effectively lowered, as shown in Fig. 3.19(b). If it is applied so as to turn off the MOST, only the upper SOI MOST's current is available, implying effectively raised V_t. The bias control is especially useful for flip-flop circuits such as SRAM cells. Figure 3.20 shows how this is applied to 6-T SRAM cells [28]. In Fig. 3.20(a), turned-on driver MOSTs in a cell increase their currents (that is, effectively increase their V_ts or channel widths) due to the gate-well connections, while the currents of the transfer MOSTs remain the same because they have no back-gate bias. Thus, the read voltage margin of the cell is widened by the increased conductance ratio of the transfer and driver MOSTs. Boosting the power and word-line [29] (Fig. 3.20(b)) further enhances performance. The SNM and cell read current are increased, enabling the data line (DL) to be more quickly discharged. Figure 3.21 shows another SRAM cell [29] to widen the read and write voltage margin. During read, the well biases of the accessed

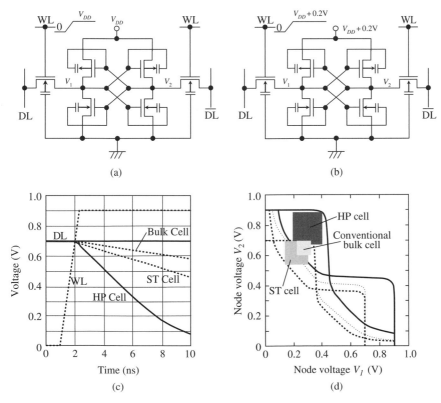

FIGURE 3.20. Double-gate FD-SOI SRAM cells [28]; ST cell (a) and HP cell (b), and DL-discharging waveform (c) and SNM (d). 90-nm process, 128 cells/DL, $V_t = 0.4$V and 3σ of V_t variation are assumed.

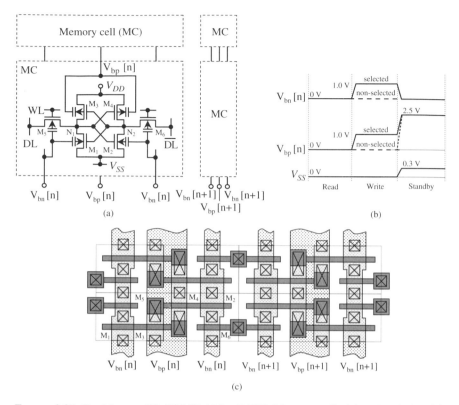

FIGURE 3.21. Double-gate FD-SOI SRAM cell [29]. Memory cells (a), pulse timing (b), and cell layout (c).

column (Vbn[n], Vbp[n]) are controlled to 0 V, so the currents of pMOSTs (M_3, M_4) are increased. For example, when a cell with an N_1 voltage of 0 V and N_2 voltage of V_{DD} is read, the N_1 voltage that is increased as a result of a ratio operation of M_5 and M_1 tends to decrease the N_2 voltage. However, M_4 prevents the decease by increasing current, enabling SNM to improve. During write, the well biases are controlled to a high voltage, so the currents of nMOSTs (M_1, M_2, M_5, M_6) are increased, while the currents of M_3 and M_4 are decreased. Thus, the above voltage combination for N_1 and N_2 is more easily changed to the opposite one when 0 V and V_{DD} are applied to $\overline{DL}$ and DL, respectively, because N_2 is more easily discharged with increased M_6 current and decreased M_4 current. During standby, the nMOST source line (V_{SS}) and the Vbp[n] are raised to 0.3 V and 2.5 V (the I/O supply), respectively, so subthreshold currents of nMOSTs and pMOSTs are reduced with increased V_t. (see Fig. 3.19(b)). The examples described above are for a BOX thickness of 10 nm. The performance is further improved with an even thinner BOX.

3.6. The 6-T SRAM Cell Compared with the 1-T DRAM Cell

This section summarizes essential differences between the 6-T SRAM cell and the 1-T DRAM cell, focusing on ultra-low voltage circuit designs. Low-voltage limitations of peripheral logic circuits that are common to DRAMs and SRAMs are also discussed.

3.6.1. Cell Operations

Figure 3.22 compares cell operations of the 6-T SRAM cell and the 1-T DRAM cell. The 6-T cell has unique features despite a large cell size; gain and feed-back functions, differential-operation capability, and non-destructive read-out (NDRO) capability, which completely differ from DRAMs. The gain function

	DRAM Cell	SRAM Cell
Circuit	(1-T/1-C) 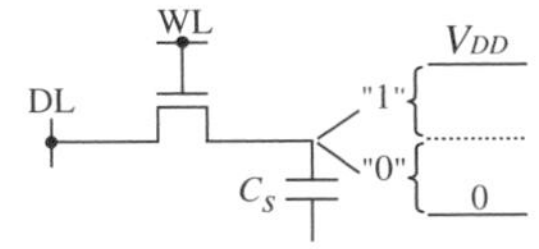	(6-T flip flop) 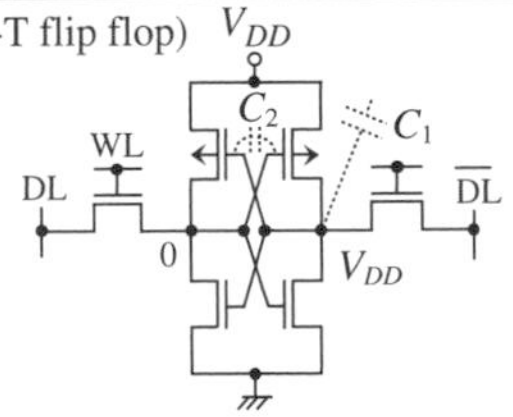
Cell size	Small	Large
Gain function	No → Unreliable sensing small and floating signal subject to noise	Yes → Reliable sensing (static or large signal)
Feedback function	No → Refresh required with a CMOS sense amp on each pair of DLs	Yes → No refresh required self refresh with the load current
Differential operations	No Possible with helps of paired-DL configuration, word boosting, mid-point sensing, and CMOS sense amp	Yes Utilizing the built-in paired-DL configuration → Fast access and cycle
DRO/NDRO	DRO → Restoring required (slow cycle)	NDRO → No restoring required (fast cycle)

FIGURE 3.22. Comparisons of cell operations between DRAM and SRAM cells.

of driver MOSTs in the cell enables a reliable sensing with a static or large read signal voltage developed on the data line. The feedback function of the cross-coupled MOSTs requires no refresh operation because the data is retained by a leakage-compensation current from the load in the cell. In other words, the cell is continuously "self-refreshed". The differential operation capability utilizing the built-in paired data-line configuration enables fast read/write operations even if the word voltage is not boosted. Despite a V_t drop at the transfer MOST, a new write data from the data line is quickly amplified in the cell to the power supply (V_{DD}) level, with the help of the feedback function of cross-coupled MOSTs. Moreover, NDRO characteristics achieve a fast cycle time by eliminating a restore or rewrite operation, and allow a CMOS sense amplifier on each data line to be eliminated. Some of these features eventually restrict low-voltage operations of the cell, as discussed later.

The 1-T DRAM cell has no gain, no feedback function, and no differential-operation capability in the cell itself despite a small cell size. In addition, the cell does not have non-destructive read-out (NDRO) characteristics. No gain function tends to cause unreliable sensing with a small and floating read signal voltage on the data line that is subject to various noises, as discussed in Chapter 2. No feedback function in the cell requires refresh operations to retain the data using one sense/restoring circuit connected to each pair of data lines. The sense/restoring circuit is composed of a cross-coupled CMOS sense amplifier similar to the SRAM cell in circuit configuration. Note that a huge number of data lines and sense amplifiers must simultaneously be activated to reduce the refresh busy rate. Although differential operation is impossible as the cell, it is possible with the help of the paired data-line configuration, word-bootstrapping (i.e. raised word voltage), mid-point sensing (i.e., half-V_{DD} data-line precharge), and CMOS sense amplifiers. DRO necessitates the sense/restoring with the CMOS sense amplifier on each data line, causing a slow cycle time. A write operation is quickly performed with the help of the feedback function of the CMOS sense amplifier.

3.6.2. Flip-Flop Circuits

The flip-flop circuit eventually determines the voltage margin and thus the minimum V_{DD} (V_{min}) for a successful operation of RAMs. RAMs use flip-flop circuits that SRAMs use for cells, while DRAMs use for CMOS sense amplifiers, each of which cascades an nMOS amplifier and a pMOS amplifier, as shown in Fig. 3.23. Their flip-flop circuits are similar in circuit configuration. However, they are quite different in operation: The flip flop in SRAMs is always turned on, independent of active or standby periods, operating in static and ratio modes. Thus, all MOSTs in a cell affect the voltage margin of the cell. In contrast, the flip flop in DRAMs is turned on only during active periods, operating in dynamic mode by an activation of the nMOS amplifier, followed by an activation of the pMOS amplifier. Thus, only paired nMOSTs affect the voltage margin of the

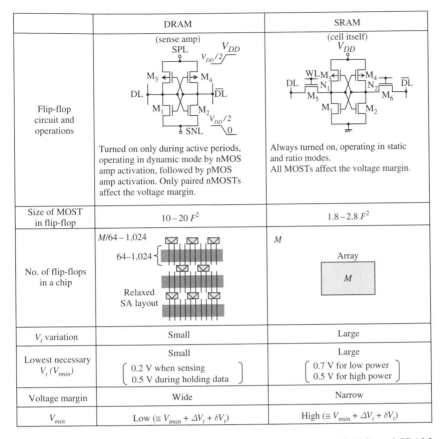

	DRAM	SRAM
Flip-flop circuit and operations	(sense amp)	(cell itself)
	Turned on only during active periods, operating in dynamic mode by nMOS amp activation, followed by pMOS amp activation. Only paired nMOSTs affect the voltage margin.	Always turned on, operating in static and ratio modes. All MOSTs affect the voltage margin.
Size of MOST in flip-flop	$10 - 20\,F^2$	$1.8 - 2.8\,F^2$
No. of flip-flops in a chip	$M/64 - 1{,}024$ $64-1{,}024$ Relaxed SA layout	M Array M
V_t variation	Small	Large
Lowest necessary $V_t\,(V_{tmin})$	Small ⎰ 0.2 V when sensing 0.5 V during holding data ⎱	Large ⎰ 0.7 V for low power 0.5 V for high power ⎱
Voltage margin	Wide	Narrow
V_{min}	Low ($\cong V_{tmin} + \Delta V_t + \delta V_t$)	High ($\cong V_{tmin} + \Delta V_t + \delta V_t$)

FIGURE 3.23. Comparisons of flip-flop circuits in a chip between DRAMs and SRAMs. M; memory capacity.

cell. For SRAMs, the flip flop must be small for a high density because the number of flip-flops equal the number of cells, while for DRAMs the size can be large because the required number of flip-flops is less. The differences in the size and number make the V_t variation of the SRAM flip-flop larger than that of the DRAM flip-flop. Furthermore, the SRAM flip flop needs a large V_t to reduce the data retention current, unlike the DRAM flip flop. Eventually, the static/ratio-mode operations, the large V_t variation, and the high V_t of SRAM cells prevent SRAMs from voltage down-scaling. The details are in what follows.

(1) Voltage Margin

SRAMs: The voltage margin of the 6-T SRAM cell (Fig. 3.23) is quite sensitive to design parameters, such as V_{DD}, V_t, W/L, temperature, and their variations. For example, during non-selected periods with M_5 and M_6 off, and N_1 low and N_2 high, it is worsened when one inverter has a high-$V_t M_1$ and low (small)-$V_t M_3$

and another inverter has a low-V_tM$_2$ and high (large)-V_tM$_4$, since the combination tends to raise the N$_1$ voltage. During read/write, the margin is further degraded by additional ratioed operations of the transfer MOSTs to driver MOSTs or load MOSTs. During read operation, for example, the margin is degraded by a low-V_tM$_5$due to a raised N$_1$ voltage. During write operation, the situation is more complicated. For example, when N$_1$ is changed from low to high with a high (V_{DD}) to DL and a low (0 V) to $\overline{DL}$, N$_2$ quickly becomes low, so M$_3$ is turned on for charging up N$_1$, because the conductance of M$_4$ is usually much smaller than that of M$_6$. Since M$_5$ is always in the source-follower mode, it cannot quickly charge up N$_1$ because of its small conductance. However, a combination of one inverter with a high-V_t M$_6$ and a low (small)-V_t M$_4$ and the other with a low-V_t M$_1$ and a high-V_t M$_3$ worsens the voltage margin, because it tends to prevent N$_2$ and N$_1$ from changing. In addition, the voltage margin strongly depends on variations in W/L and V_t, and the mismatch between the two of the six MOSTs in a cell, making the worst-case design of SRAM cells more complicated.

DRAMs: The operation of the DRAM cell and its related circuits are fully dynamic without static and ratioed operations. Each DRAM cell during standby or non-selected periods can continue to store the data at the cell capacitor without any current compensation, unlike the SRAM cell, although an extremely slow read/rewrite (restore) cycle of 8–64 ms is necessary for refresh operations of each cell. Except during refresh operations the sense amplifiers continue to be cut off for a long time. Fortunately, each sense amplifier operates almost dynamically: A CMOS sense amplifier never operates in static and ratioed modes at the initial sensing. The well-known mid-point sensing is performed by a successive activation of nMOS and pMOS amplifiers, as discussed in Chapters 1 and 2. After a small cell signal has been amplified to a certain level by the preceding activation of the nMOST amplifier, the signal is further amplified to a high enough voltage (i.e., V_{DD}) by subsequent activation of the pMOST amplifier. Thus, the sensing is not influenced by the V_t variation and offset voltage (V_t-mismatch) of the pMOST amplifier, but only by those of the nMOS amplifier. After amplifying the signal to V_{DD}, the CMOS sense amplifier consumes subthreshold currents, as the 6-T SRAM cell does, despite a short period within one cycle time. For write, the node voltage of the selected cell can be changed more easily, with the help of word-boosting, than the SRAM cell. Note that the SRAM cell suffers from a narrow write voltage margin due to the existence of load MOSTs.

(2) Variations in V_t-Relevant Parameters

The size of each flip flop and the necessary number in a chip are quite different for SRAMs and DRAMs, as mentioned previously. Due to the differences, variations in V_t-relevant parameters of SRAM cells are larger, physically and statistically, than those of DRAM sense amplifiers. This is because SRAMs must use the smallest flip-flop (i.e., cell) possible with small MOSTs for a high density, exemplified by channel areas (LWs) of about 2.0 F^2 for transfer MOSTs (M$_5$, M$_6$), 2.8 F^2 for driver MOSTs (M$_1$, M$_2$), and 1.8 F^2 for load

MOSTs (M_3, M_4) in a cell by usually using the minimum channel length $L = F$ (F: the feature size) to all MOSTs. Thus, a wide V_t variation is involved in a chip. Moreover, the range of V_t variations is statistically large because a large number of flip-flops are used in a chip. In contrast, DRAMs sense amplifiers can use the largest MOST possible, as long as the data line (DL) capacitance does not increase intolerably, exemplified by channel areas of about $10–20\,F^2$ with a larger channel width of $1.5\,F$ for all MOSTs. This is because only one sense amplifier suffices for a pair of DLs and the layout can be relaxed with the alternate placement [1]. Thus, a narrow V_t variation is realized. Moreover, the range of V_t variations is statistically small because a smaller number of flip-flops are used due to one sense amp shared by 64 to 1,024 cells. Therefore, the V_t variation (δV_t) and the V_t-mismatch (δV_t) for SRAM cells are larger than those for DRAM sense amplifiers, thus more preventing SRAM cells from voltage down-scaling. Unfortunately, δV_t and δV_t increase as the physical size of MOST decreases. More details are explained in Chapter 5.

3.6.3. The Minimum V_{DD} of RAMs

There are differences in the minimum $V_{DD}(= V_{min})$, which is defined as the lowest necessary V_{DD} for a successful operation, between DRAMs, SRAMs, and peripheral logic circuits.

(1) Cell and Sense Amplifier

Investigating V_{min} is vital, although the actual operating voltage V_{DD} must be set to be higher than the V_{min}, since the speed must be taken into account. The V_{min}s for DRAMs and SRAMs are different. The V_{min} for DRAMs is usually determined by the larger one of the two V_{min} values; $V_{min}(D)$ derived from the S/N of the cell and $V_{min}(C)$ derived from the gate-over-drive of turned-on MOST in the sense amplifier, as discussed in Chapter 2. $V_{min}(D)$ and $V_{min}(C)$ are different for memory cells, data-line precharging schemes, sensing schemes, and S/N. They are drastically reduced by data-line-shielded twin cells [38, 39] if these cells are coupled with a full-V_{DD} data-line precharge [1], due to the increased S/N and gate-over drive. They are minimized using the FD-SOI design, as discussed in Chapter 2 and shown in Fig. 3.24(a), for a repair percentage of 0.1%.

For the 6-T SRAM (Fig. 3.6), the V_{min} is expressed as $V_{min} = V_{tmin} + \delta V_{tmax}$, assuming that the voltage margin is determined only by cross-coupled nMOSTs. Here, V_{tmin} is 0.49 V for a high-speed (HS) design, and 0.71 V for a low-power (LP) design, as discussed earlier. the δV_{tmax} is also given in Fig. 5.6. Thus, the V_{min} is as high as around 0.7 V for HS design and 0.9 V for LP design in the 32-nm generation (Figure 3.24(b)), because of $\delta V_{tmax} = 180\,mV$ for $r = 0.1$ %. In practice, however, the actual V_{min} becomes much higher because the margin of the cell is determined by combinations of V_ts of all MOSTs in the cell. In fact, the V_{min} has been reported to be as high as 1.1 V even for a 65-nm high-speed

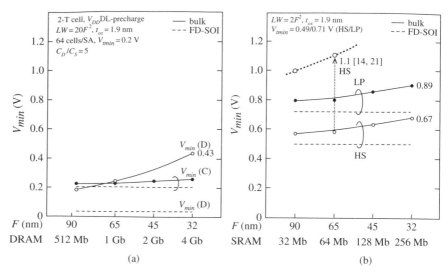

FIGURE 3.24. The minimum V_{DD} ($= V_{min}$) of DRAMs (a) and SRAMs (b) [38]. $V_{min}(D)$; V_{min} derived from the SN of cell, $V_{min}(C)$; V_{min} derived from the gate-over-drive of sense amplifier. HS; high-speed design, LP; low-power design. A repairable percentage (r) of 0.1 % is assumed in Fig. 5.6.

design, as also discussed later. For the FD-SOI, a smaller V_t variation reduces the V_{min}, as shown in the figure. The FD-SOI also widens the voltage margin of SRAM cells with well-voltage controls, as mentioned previously. In any event, the V_{min} of SRAMs is much higher than that of DRAMs.

To be more exact, the V_{min} of DRAMs is not equal to $V_{min}(D)$ or $V_{min}(C)$, but equal to the lowest necessary word voltage, V_{Wmin}, because V_{Wmin} is usually higher than $V_{min}(D)$ or $V_{min}(C)$. Thus, the V_{min} is expressed as

$$V_{min} = V_{Wmin} = (V_{min}(D) \text{ or } V_{min}(C)) + V_{tFW}, \qquad (3.1)$$

where V_{tFW} is the lowest necessary V_t of the transfer MOST (not for sense amplifiers) for a full V_{DD} write, as discussed in Section 2.4.1. V_{tFW} needs to increase in each successive generation of memory capacity to preserve the refresh busy rate constant. Thus, the V_{min} of DRAMs given by Eq. (3.1) increases more rapidly than that in Figure 3.24(a). Even if such a high V_{Wmin} is needed, DRAMs have solved the high voltage problem by using a small transfer MOST with a thick gate oxide coupled with high-voltage tolerant word drivers, as discussed in Chapter 9. Eventually, DRAM designs focusing on $V_{min}(D)$ or $V_{min}(C)$ in Figure 3.24(a) are essential.

In general, the V_{tmin} of SRAM cells is lower than the V_{tFW} because, unlike DRAM cells, SRAM cells accept a larger subthreshold current, and have no source-follower mode in the cells. The V_{tmin}, however, must also be gradually increased with memory capacity to maintain the retention current of the chip to the same, regardless of memory capacity, for low-power designs. This stems from the data retention current specification. Thus, the V_{min} of SRAMs also increases more rapidly than that in Figure 3.24(b).

(2) Peripheral Logic Circuits

Figure 3.25 compares the V_{min}, of bulk-CMOS SRAMs and bulk-CMOS logic gates in peripheral circuits, using Figs. 3.24(b) and 5.8. This assumes that the average V_t ($= V_{t0}$) of logic gates is kept at a constant value of 0.3 V to ensure a low enough subthreshold current and that the acceptable intra-die speed variation $\Delta \tau$ is between 1.2 and 1.3. Obviously, the V_{min} of SRAMs is higher and increases more slowly as device size decreases than the V_{min} of the logic gate does. This is mainly because, in contrast with logic gates, the lowest necessary V_t (i.e., V_{tmin}) of SRAM cells is much higher than the V_{t0} described above, and increase in δV_{tmax} is suppressed by repair techniques. As a result, the V_{min} of logic gates may surpass the V_{min} of SRAMs as device size decreases.

Figure 3.26 shows a dual-V_{DD} approach applied to a 65-nm MPU [21]. An MPU core consisting of logic gates and three level caches (L0-L2) that use smaller capacity memories with lower density cells operates at a low voltage (V_{CORE}) that can change from 0.6 (standby)-0.7 V (V_{min}) to 1.2 V (V_{max}), based on the dynamic V_{DD} and frequency (DVF) approach. In contrast, the last level

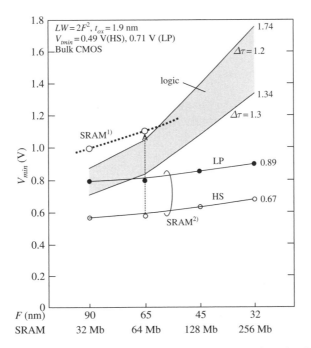

FIGURE 3.25. Expected V_{min} of SRAMs and logic gates using the bulk CMOS. SRAM[1];V_{min} in the worst case combination of all MOSTs in a cell. SRAM[2];V_{min} determined only by cross-coupled nMOSTs.

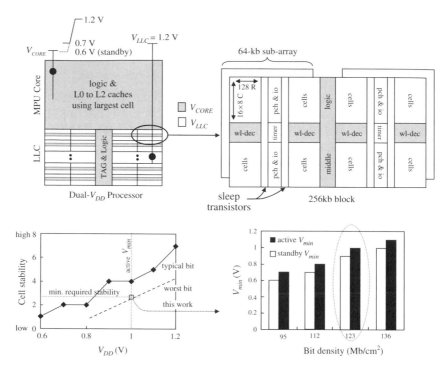

FIGURE 3.26. Dual V_{DD} processor. Reproduced from [21] with permission; © 2006 IEEE.

cache (LLC) using the largest capacity memory with the highest density cell ($136\,\text{Mb/cm}^2$) operates at a high constant V_{DD} (V_{LLC}) of 1.2 V. The difference in the operating voltage is as follows. The active V_{min} of the SRAM array, which is susceptible to speed variations, depends on the bit density (i.e., cell size), as exemplified by 0.7 V and 1.1 V for $95\,\text{Mb/cm}^2$ and $136\,\text{Mb/cm}^2$, respectively. This is because a smaller SRAM cell for a high density needs a higher V_{DD} to offset the higher V_{min} caused by the wider variation in V_t and V_t mismatch. Here, a 1.2-V V_{DD} is the reliability-limited maximum V_{DD} (V_{max}) for 65-nm devices. As device size continues to decrease, the V_{min} for the logic gate might become higher than that for L0-L2 and LLC, as described above.

3.6.4. Soft-Error Immunity

There are two kinds of soft-error modes; the data-line mode (or bit-line mode) and the cell mode. The data-line mode occurs only on the selected data lines for only a short period holding a floating signal, thus depending on the cycle time. It never occurs during standby and on many non-selected data lines, since the data lines are fixed to a given voltage. In contrast, the cell mode can always occur for all the non-selected cells. Thus, the data-line mode is not so serious as the cell mode. The soft-error rate (SER) of the data-line mode is reduced with

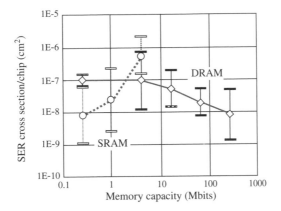

FIGURE 3.27. Soft error rate of RAMs [33].

reducing diffused area of the data lines. The cell mode strongly depends on the signal charge Q_S. The Q_S of SRAM cells has rapidly decreased with device and voltage scaling, as mentioned previously, while that of DRAM cells has been gradually decreased despite device scaling to maintain a large enough signal voltage, as mentioned in Chapter 2. Hence, SRAM's SER rapidly increases with device down-scaling despite spatial scaling that reduces the collected charge, as shown in Fig. 3.27 [33], while DRAM's SER gradually decreases. Thus, ECC circuits are particularly important for SRAMs, as mentioned earlier.

3.6.5. Memory Cell Size

The memory cell size of the above-described twin (2-T) DRAM cell is $12–37F^2$, as expected from the previously reported DRAM cells [39, 42]. The FD-SOI might accept a planar capacitor, as discussed in Chapter 2, and is reduced to $6–8F^2$ if a folded or open data-line arrangement is used, although vertical capacitors may be needed in some cells to offset noise despite using the FD-SOI. For SRAMs, the size is about $120\,F^2$ for the standard cell (Figure 3.12(b)) for the bulk and FD-SOI. Because of this, existing SRAMs might be replaced by DRAMs in large-capacity low-voltage applications.

References

[1] K. Itoh, *VLSI Memory Chip Design*, Springer-Verlag, NY, 2001.
[2] Y. Nakagome, M. Horiguchi, T. Kawahara, K. Itoh, "Review and future prospects of low-voltage RAM circuits," IBM J. R&D, vol. 47, no. 5/6, pp. 525–552, Sep./Nov. 2003.
[3] K. Itoh, K. Osada, and T. Kawahara, "Reviews and future prospects of low-voltage embedded RAMs," CICC Dig. Tech. Papers, pp. 339–344, Oct. 2004.

[4] K. Itoh, "Low-voltage embedded RAMs in the nanometer era," ICICDT Dig. Tech. Papers, 235–242, May 2005.

[5] S. Naffziger, B. Stackhouse, and T. Grutkowski, "The implementation of a 2-core multi-threaded Itanium®-family processor," ISSCC Dig. Tech. Papers, pp. 182–183, Feb. 2005.

[6] S-M Jung, J. Jang, W. Cho, J. Moon, K. Kwak, B. Choi, B. Hwang, H. Lim, J. Jeong, J. Kim, and K. Kim, "The revolutionary and truly 3-dimensional 25F^2 SRAM technology with the smallest S^3 (Stacked Single-crystal Si) cell, 0.16μm^2, and SSTFT (Stacked Single-crystal Thin Film Transistor) for ultra high density SRAM," Symp. VLSI Tech. Dig. Tech. Papers, pp. 228–229, June 2004.

[7] H-J An, H-Y Nam, H-S Mo, J-P Son, B-T Lim, S-B Kang, G-H Han, J-M Park, K-H Kim, S-Y Kim, C-K Kwak, and H-G Byun, "64Mb mobile stacked single-crystal Si SRAM(S^{3}RAM) with selective dual pumping scheme (SDPS) and multi cell burn-in scheme (MCBS) for high density and low power SRAM," Symp. VLSI Circuits Dig. Tech. Papers, pp. 282–283, June 2004.

[8] Y. H. Suh, H. Y. Nam, S. B. Kang, B. G. Choi, H. S. Mo, G. H. Han, H. K. Shin, W. R. Jung, H. Lim, C. K. Kwak, and H. G. Byun, "A 256 Mb synchronous-burst DDR SRAM with hierarchical bit-line architecture for mobile applications," ISSCC Dig. Tech. Papers, pp. 476–477, Feb. 2005.

[9] K. Noda, K. Matsui, K. Imai, K. Inoue, K. Takashiki, H. Kawamoto, K. Yoshida, K. Takada, N. Nakamura, T. Kimura, H. Toyoshima, Y. Koishikawa, S. Maruyama, T. Saitoh, and T. Tanigawa, "A 1.9-μm^2 loadless CMOS four-transistor SRAM cell in a 0.18-μm logic technology," IEDM Dig. Tech. Papers, pp. 643–646, Dec. 1998.

[10] K. Takeda, Y. Aimoto, N. Nakamura, H. Toyoshima, T. Iwasaki, K. Noda, K. Matsui, S. Itoh, S. Masuoka, T. Horiuchi, A. Nakagawa, K. Shimogawa, and H. Takahashi, "A 16-Mb 400-MHz loadless CMOS four-transistor SRAM macro," IEEE J. Solid-State Circuits, Vol. 35, no. 11, pp. 1631–1640, Nov. 2000.

[11] K. Takeda, Y. Hagihara, Y. Arimoto, M. Nomura, Y. Nakazawa, T. Ishii, and H. Kobatake, "A read-static-noise-margin-free SRAM cell for low-V$_{DD}$ and high-speed applications," ISSCC Dig. Tech. Papers, pp. 478–479, Feb. 2005.

[12] K. Osada, Y. Saitoh, E. Ibe, and K. Ishibashi, "16.7fA/cell tunnel-leakage-suppressed 16-Mbit SRAM based on electric-field-relaxed scheme and alternate ECC for handling cosmic-ray-induced multi-errors", ISSCC Dig. Tech Papers, pp. 302–303, Feb. 2003.

[13] K. Itoh and H. Mizuno, "Low-voltage embedded-RAM technology: present and future," Proc. 11th IFIP Int'l Conf. VLSI, pp. 393–398, Dec. 2001.

[14] R. Tsuchiya, M. Horiuchi, S. Kimura, M. Yamaoka, T. Kawahara, S. Maegawa, T. Ipposhi, Y. Ohji, and H. Matsuoka, "Silicon on thin BOX: a new paradigm of the CMOSFET for low-power and high-performance application featuring wide-range back-bias control," IEDM Dig. Tech. Papers, pp. 631–634, Dec. 2004.

[15] M. Yamaoka, K. Osada, R. Tsuchiya, M. Horiuchi, S. Kimura, and T. Kawahara, "Low power SRAM menu for SOC application using yin-yang-feedback memory cell technology," Symp. VLSI Circuits Dig. Tech. Papers, pp. 288–291, June 2004.

[16] P. M. Carter and B. R. Wilkins, "Influences on soft error rates in static RAM's," IEEE J. Solid- State Circuits, vol. sc-22, No. 3, pp. 430–436, June 1987.

[17] S-M Jung, H. Lim, W. Cho, H. Cho, H. Hong, J. Jeong, S. Jung, H. Park, B. Son, Y. Jang, and K. Kim, "Soft error immune 0.46μm^2 SRAM cell with MIM node capacitor by 65 nm CMOS technology for ultra high speed SRAM," IEDM Tech. Dig., pp. 289–292, Dec. 2003.

[18] K. Osada, K. Yamaguchi, Y. Saitoh, and T. Kawahara, "SRAM immunity to cosmic-ray-induced multierrors based on analysis of an induced parasitic bipolar effect," IEEE J. Solid-State Circuits, vol. 39, No. 5, pp. 827–833, May 2004.

[19] M. Horiguchi, M. Aoki, Y. Nakagome, S. Ikenaga, and K. Shimohigashi, "An experimental large-capacity semiconductor file memory using 16-level/cell storage," IEEE J. Solid-State Circuits, Vol. 23, no. 1, pp. 27–33, Feb. 1988.

[20] H. L. Kalter, C. H. Stapper, J. E. Barth, Jr., J. DiLorenzo, C. E. Drake, J. A. Fifield, G. A. Kelley, Jr., S. C. Lewis, W. B. V. D. Hoeven, and J. A. Yankosky, "A 50-ns 16-Mb DRAM with a 10-ns data rate and on-chip ECC," IEEE J. Solid-State Circuits, vol. 25, pp. 1118–1128, Oct. 1990.

[21] M. Khellah, N.S. Kim, J. Haward, G. Ruhl, M. Sunna, Y. Ye, J. Tschanz, D. Somasekhar, N. Borkar, F. Hamzaoglu, G. Pandya, A. Farhang, K. Zhang, and V. De, "A 4.2 GHz 0.3 mm² 256kb dual-V_{CC} SRAM building block in 65nm CMOS," ISSCC Dig. Tech. Papers, pp. 624–625, Feb. 2006.

[22] K. Osada, J-U Shin, M. Khan, Y-de Liou, K. Wang, K. Shoji, K. Kuroda, S. Ikeda, and K. Ishibashi, "Universal-V_{DD} 0.65–2.0V 32kB cache using voltage-adapted timing-generation scheme and a lithographical-symmetric Cell," ISSCC, Dig. Tech. Papers, pp. 168–169, Feb. 2001.

[23] K. Itoh, A. R. Fridi, A. Bellaouar, and M. I. Elmasry, "A deep sub-V, single power-supply SRAM cell with multi-Vt, boosted storage node and dynamic load," Symp. VLSI Circuits Dig. Tech. Papers, pp. 132–133, June 1996.

[24] M. Yamaoka, Y. Shinozaki, N. Maeda, Y. Shimazaki, K. Kato, S. Shimada, K. Yanagisawa, and K. Osada, "A 300MHz 25µA/Mb leakage on-chip SRAM module featuring process-variation immunity and low-leakage-active mode for mobile-phone application processor," ISSCC Dig. Tech. Papers, pp. 494–495, Feb. 2004.

[25] K. Zhang, U. Bhattacharya, Z. Chen, F. Hamzaoglu, D. Murray, N. Vallepalli, Y. Wang, and M. Bohr, "A 3-GHz 70Mb SRAM in 65nm CMOS technology with integrated column-based dynamic power supply," ISSCC Dig. Tech. Papers, pp. 474–475, Feb. 2005.

[26] M. Yamaoka, N. Maeda, Y. Shinozaki, Y. Shimazaki, K. Nii, S. Shimada, K. Yanagisawa, and T. Kawahara, "Low-power embedded SRAM modules with expanded margins for writing," ISSCC Dig. Tech. Papers, pp. 480–481, Feb. 2005.

[27] F. Assaderaghi, S. Parke, P. K. Ko, and C. Hu, "A novel silicon-on-insulator (SOI) MOSFET for ultralow voltage operation," Symp. Low Power Electronics Dig. Tech. Papers, pp. 58–59, 1994.

[28] M. Yamaoka, K. Osada, K. Itoh, R. Tsuchiya, and T. Kawahara, "Dynamic-Vt, dual-power-supply SRAM cell using D2G-SOI for low-power SoC application," Int'l SOI Conf. Dig., pp. 109–111, Oct. 2004.

[29] M. Yamaoka, R. Tsuchiya, and T. Kawahara, "SRAM circuit with expanded operating margin and reduced stand-by leakage current using thin-BOX FD-SOI transistors," A-SSCC Dig. Tech. Papers, pp. 109–112, Nov. 2005.

[30] T. C. Chen, "Where CMOS is going: trendy hype vs. real technology," ISSCC Dig. Tech. Papers, pp. 22–28, Feb. 2006.

[31] C-Ho Lee, J-M Yoon, C. Lee, K. Kim, S. B. Park, Y. J. Ahn, H. S. Kang, and D. Park, "The application of BT-FinFET technology for sub 60nm DRAM integration," ICICDT Dig. Tech. Papers, pp. 37–41, May 2005.

[32] J. Kedzierski, E. Nowak, T. Kanarsky, Y. Zhang, D. Boyd, R. Carruthers, C. Cabral, R. Amos, C. Lavoie, R. Roy, J. Newbury, E. Sullivan, J. Benedict, P. Saunders,

K. Wong, D. Canaperi, M. Krishnan, K.-L. Lee, B. A. Rainey, D. Fried, P. Cottrell, H.-S. P. Wong, M. Ieong, and W. Haensch, "Metal-gate FinFET and fully-depleted SOI devices using total gate silicidation," IEDM Dig. Tech. Papers, pp. 247–250, Dec. 2002.

[33] E. Ibe, "Current and future trend on cosmic-ray-neutron induced single event upset at the ground down to 0.1-micron-devices," *The Svedberg Laboratory Workshop on Applied Physics*, Uppsala, May 3, 2001.

[34] E. Seevinck, F. J. List, and J. Lohstroh, "Static-noise margin analysis of MOS SRAM cells," IEEE J. Solid-State Circuits, Vol. SC-22, No.5, pp.748–754, Oct.1987.

[35] T. Douseki and S. Mutoh, "Static-noise margin analysis for a scaled-down CMOS memory cell," IEICE Trans. on Electronics, Vol. J75-C-II, No. 7, pp. 350–361, July 1992 (in Japanese).

[36] T. Sakurai, A. Matsuzawa, and T. Douseki, *Fully-Depleted SOI CMOS Circuits and Technology for Ultralow-Power Applications*, Springer, 2006.

[37] R. V. Joshi, Y. Chan, D. Plass, T. Charest, R. Freese, R. Sautter, W. Huott, U. Srinivasan, D. Rodko, P. Patel, P. Shephard, and T. Werner, "A low power and high performance SOI SRAM circuit design with improved cell stability," Int'l SOI Conf. Dig., pp. 4–7, Oct. 2006.

[38] K. Itoh, M. Yamaoka, and T. Kawahara, "Impact of FD-SOI on deep sub-100-nm CMOS LSIs-a view of memory designers-" Int'l SOI Conf. Dig., pp. 103–104, Oct. 2006.

[39] R. Takemura, K. Itoh, and T. Sekiguchi, "A 0.5-V FD-SOI twin-cell DRAM with offset-free dynamic-V_T sense amplifiers," ISLPED Dig. Tech. Papers, pp. 123–126, Oct. 2006.

[40] C. M. Hsieh, P. C. Murley, and R. R. O'brien, "A Field-funneling effect on the collection of alpha-particle-generated carriers in silicon devices," IEEE Electron Device Letters, Vol. EDL-2, No.4, pp.103–105, April 1981.

[41] K. Osada, Y. Saitoh, E. Ibe, and K. Ishibashi, "16.7-fA/cell tunnel-leakage-suppressed 16-Mbit SRAM for handling cosmic-ray-induced multi-errors," IEEE J. Solid-State Circuits, Vol. 38, no. 11, pp. 1952–1957, Nov. 2003.

[42] K.-H. Kim, U. Kang, H.-J. Chung, D.-H. Park, W.-S. Kim, Y.-C. Jang, M. Park, H. Lee, J.-Y. Kim, J. Sunwoo, H.-W. Park, H.-K. Kim, S.-J. Chung, J.-K. Kim, H.-S. Kim, K.-W. Kwon, Y.-T. Lee, J. S. Choi, and C. Kim, "An 8Gb/s/pin 9.6ns row-cycle 288Mb deca-data rate SDRAM with an I/O error-detection scheme," ISSCC Dig. Tech. Papers, pp. 156–157, Feb. 2006.

[43] K. Itoh, M. Horiguchi, and T. Kawahara, "Ultra-low voltage nano-scale embedded RAMs," ISCAS Proc., pp.25–28, May 2006.

4

Leakage Reduction for Logic Circuits in RAMs

4.1. Introduction

Low-voltage RAMs, especially embedded RAMs (e-RAMs) using nano-meter technology, are becoming increasingly important because they play critical roles in reducing power dissipation of LSIs for power-aware systems. Thus, research and development aimed at sub-1-V RAMs has actively been done, as exemplified by the development of a 0.6-V 16-Mb e-DRAM [1]. However, we face many challenges to achieving such e-RAMs [2–4]. Of the challenges, reduction of the subthreshold leakage current of MOSTs is a key. Thus, circuit developments for leakage reduction even for the active mode have been investigated since the late 1980's. To the best of our knowledge, almost all basic reduction-circuit concepts had been proposed by around 1993 mainly through developments in exploratory DRAMs [5–14]. The proposals are summarized as static and/or dynamic realizations of high threshold-voltage(V_t) MOSTs or effectively high-V_t MOSTs through various reverse-biasing schemes such as substrate (V_{BB}) reverse biasing [8], gate-source (V_{GS}) reverse biasing [9], or V_{GS} self-reverse biasing [10, 11]. The resulting high-V_t was proposed for use not only in RAMs but also in logic LSIs in the form of applications to all MOSTs in a chip, a limited number of MOSTs on non-critical paths in a chip [12], a power-switch (i.e., power gating) MOST [8], a buffer [9], various logic gates and an inverter chain [11], and iterative circuit blocks [10, 11]. All these applications were initially to reduce leakage in the standby mode. As early as 1993, however, they were quickly followed by attempts to reduce leakage in the active mode of a hypothetical 16-Gb DRAM [13, 14, 25]. Numerous attempts have subsequently been made to reduce leakage in logic LSIs [15], independently of RAMs, although RAM and logic designers must cooperate to develop reduction schemes for memory-rich LSIs.

To realize low-voltage RAMs, subthreshold leakage currents of both RAM cells and peripheral circuits must be reduced. For RAM cells, the leakage is reduced by raising the threshold voltage (V_t) of MOSTs, as discussed in Chapters 2 and 3. For peripheral circuits, it cannot be reduced unless the reduction circuit meets some requirements, which are similar to those for LSI logic designs. First, the reduction circuit must be fast enough to be able to control even the leakage in the active mode [2–4]. Once such a high-speed circuit is available, it can also be applied to the standby mode, making its design much easier as it can be applied to both. Even if the scheme can be applied to the standby mode, it may not be able to be applied to the active mode if it is too slow. Therefore, the leakage in each block must be reduced within one active cycle, ready to reduce the leakage of any block in the next active cycle. This necessitates a better reduction efficiency that allows a smaller voltage swing for a given leakage reduction, smaller load capacitance, and simpler control. Second, the reduction circuit must confine activated circuitry to a minimum fraction of total peripheral circuits at a given time, so a faster control is possible. It is desirable if the circuit simultaneously reduces the leakage in the remaining inactive (i.e., non-selected) circuitry, which dominates the total leakage in peripheral circuits. Fortunately, peripheral circuits in RAMs accept such schemes so as to reduce the leakage easily and effectively, whereas random logic circuits in MPU/MCU/SoC do not. This stems from the features of RAMs we describe later. Third, the circuit must minimize the area penalty. Fourth, it must compensate for leakage variations as well as speed variations caused by V_t-variations, as discussed in Chapter 5.

This chapter describes subthreshold-current reduction circuits for peripheral logic circuits in RAMs. First, basics of leakage reduction in MOSTs are described. Second, basic circuit concepts that have been proposed to date for RAMs and logic LSIs are presented and compared. Third, applications to peripheral circuits in standby and active modes are discussed in detail.

4.2. Basic Concepts for Leakage Reduction of MOSTs

The leakage current (i.e., subthreshold current) of a MOST is expressed by

$$i \propto \exp\left[\pm \frac{V_{GS} - V_t - K\left(\sqrt{\mp V_{BS} + 2\psi} - \sqrt{2\psi}\right) + \lambda V_{DS}}{S/\ln 10} \right]$$
$$\times \left\{ 1 - \exp\left[\mp \frac{qV_{DS}}{kT} \right] \right\}, \tag{4.1}$$

where the upper signs refer to nMOSTs and the lower signs to pMOSTs. V_t is the actual threshold voltage, while S is the subthreshold swing, K is the body-effect coefficient, and λ is the drain-induced barrier lowering (DIBL) factor [18]. Here, q is the electronic charge, k is the Boltzmann constant, and T is the absolute temperature. Leakage is usually reduced to 1/10 with a V_t increment of only 0.1 V for a bulk MOST with $S \cong 100\,\text{mV/decade}$ at 100 °C. To reduce the leakage current, using a high-V_t MOST through increasing V_t is the best way. Two ways of obtaining the high-V_t MOST from a low-actual V_t MOST is enhancing the "actual" V_t by increasing the doping level of the MOST substrate and raising the "effective" V_t by applying reverse biases as discussed in the following.

Although there have been many attempts to develop reverse-biasing schemes, the basic concepts can still be categorized into the three in Table 4.1. These are

TABLE 4.1. Concepts to create effective high V_t. Reproduced from [3] with permission of IBM.

Modified voltage(s)		nMOST	pMOST
(A) V_{GS} reverse biasing	(A1) V_S: self-reverse biasing	$D\|V_{DD}$; 0—G B—0; $S\|+\delta$	$S\|V_{DD}-\delta$; V_{DD}—G B—V_{DD}; $D\|0$
	(A2) V_G: offset gate driving	$D\|V_{DD}$; $-\delta$—G B—0; $S\|0$	$S\|V_{DD}$; $V_{DD}+\delta$—G B—V_{DD}; $D\|0$
(B) V_{BS} reverse biasing	(B1) V_B: substrate driving	$D\|V_{DD}$; 0—G B—$-\delta$; $S\|0$	$S\|V_{DD}$; V_{DD}—G B—$V_{DD}+\delta$; $D\|0$
	(B2) $V_S=V_G$: offset source driving	$D\|V_{DD}$; G B—0; $S\|+\delta$	$S\|V_{DD}-\delta$; G B—V_{DD}; $D\|0$
(C) V_{DS} reduction		$D\|V_{DD}-\delta$; 0—G B—0; $S\|0$	$S\|V_{DD}$; V_{DD}—G B—V_{DD}; $D\|+\delta$

(A) gate-source (V_{GS}) reverse biasing, (B) substrate-source (V_{BS}) reverse biasing, and (C) drain-source voltage (V_{DS}) reduction. Here, the V_{GS} reverse biasing scheme can further be categorized as V_S-control with a fixed V_G (A1) [10, 11] and V_G-control with a fixed V_S (A2) [9]. The V_{BS} reverse biasing schemes can be categorized as V_B-control with a fixed V_S(B1) [8, 26] and V_S-control with a fixed V_B (B2) [27, 28]. The efficiencies for reducing leakage for offset voltage δ applied to a low-actual V_t MOST are plotted in Fig. 4.1 using 0.1-µm MOST parameters. The reduction efficiency of (A2) is the leakage current-ratio without and with V_{GS} reverse bias:

$$r_1 = \frac{i(V_{GS}=0)}{i(V_{GS}=-\delta)} = \exp(\frac{\delta}{S/\ln 10}) \tag{4.2}$$

This is quite large because δ has been directly added to the low-actual V_t, realizing an effective high V_t ($V_t+\delta$). The reduction efficiency of (B1) is calculated in the same manner:

$$r_2 = \exp[\frac{K(\sqrt{\delta+2\Psi}-\sqrt{2\Psi})}{S/\ln 10}] \tag{4.3}$$

This is smaller than r_1 because of the square-root dependence on δ and the small K. (C) has quite a small reduction efficiency of

$$r_3 = \exp(\frac{\lambda\delta}{S/\ln 10}) \tag{4.4}$$

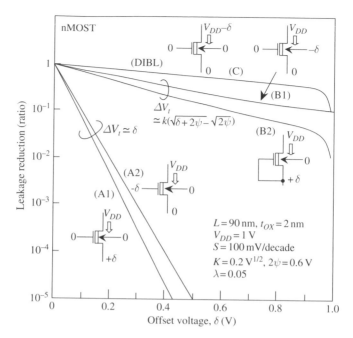

FIGURE 4.1. Leakage reduction efficiency. Reproduced from [3] with permission of IBM.

because of the small λ, unless V_{DS} approaches the thermal voltage (kT/q), where i is drastically reduced due to the second factor of Eq. (4.1). Scheme (A1) has the largest reduction efficiency of $r_1 r_2 r_3$ because all three effects are combined. (B2) has a reduction efficiency of $r_2 r_3$, which is larger than that of (B1) because of the additional effect of reducing V_{DS}.

Note the inherently small offset voltage required to reduce the given leakage provided by scheme (A). This effectively reduces not only subthreshold current in low-power mode, but also achieves a faster recovery time in high-speed mode, as explained later.

4.3. Basics of Leakage Reduction Circuits

4.3.1. Basic Concepts

Reverse biasing involves two types of biasing, static and dynamic. The selective use of the resulting high-V_t MOSTs in low-actual-V_t circuits decreases the leakage current of circuits. The static high V_t scheme, or the so-called dual-V_t scheme, is used to statistically combine low-V_t MOSTs and the resulting high-V_t MOSTs in circuits. A CMOS dual-V_t scheme [19, 20] in which a low V_t is applied only to the critical path occupying a small portion of the core is quite effective in simultaneously achieving high speed and low leakage current, although the basic scheme was proposed for an nMOST 5-V 64-Kb DRAM [21].

A difference in V_t of 0.1 V reduces the standby subthreshold current to one-fifth its value for a single low V_t although an excessive V_t difference might cause a race condition problem between low- and high-V_t circuits. The dual-V_t scheme is also applied to SRAMs [20, 22]. A combination of dual V_t and dual V_{DD} has been reported to achieve a high-speed low-power 1-V e-SRAM [22]. Another application of the dual-V_t scheme is in a high-V_t power switch [8, 10–12, 14, 15] that can cut the subthreshold current of an internal low-V_t core in the standby mode, as discussed in the section 4.4. High-V_tMOSTs can easily be produced in DRAMs [17] by using the internal supply voltages (i.e., V_{BB} and V_{DH}) that DRAMs require, as will be explained in the section 4.5. The high V_t, however, eventually restricts the lower limit of V_{DD} as the transconductance of a MOST degrades at a lower V_{DD}. The dynamic high-V_t scheme changes the V_t so that it is sufficiently low in high-speed modes, such as the active mode with no reverse bias, while in low-power modes, such as the standby mode, it is increased by changing bias conditions (see Fig. 1.64).

This section reviews the dynamic high-V_t scheme, assuming that all MOSTs in the circuit have low actual V_ts.

Gate-Source Self-Reverse Biasing (A1). Figure 4.2(a) shows self-reverse biasing. It features a low-V_t switch M_{SP} inserted between the source of the MOST M_P and V_{DD}. The MOST M_{SP} stacked to M_P is a kind of power switch that works as a source impedance turning on and off during active and standby modes, respectively. A subthreshold current flowing from M_{SP}, when M_{SP} and M_P are off in the standby mode, generates an offset voltage (δ) on V_{DL}, automatically providing a reverse bias δ to M_P so that the current is eventually reduced. This biasing is a combination of V_{GS} reverse biasing, V_{BS} reverse biasing and V_{DS} reduction. The primary effect is by the V_{GS} reverse biasing and the secondary effect is by the V_{BS} reverse biasing and the V_{DS} reduction, as described above. The gate voltage must be V_{DD}, not V_{DL}, to take advantage of the V_{GS} reverse bias. Note that no matter how large the original leakage current at M_P is, it is eventually

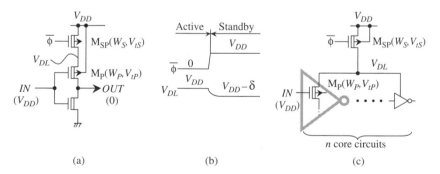

FIGURE 4.2. Circuits for self-reverse biasing (A1) [10, 11]: (a) Principle; (b) operating waveforms; (c) application to iterative circuits. W_S and W_P denote the respective channel widths of M_{SP} and M_P. V_{tS} and V_{tP} denote the respective threshold voltages of M_{SP} and M_P. Reproduced from [3] with permission of IBM.

confined to the constant current of M_{SP} through the automatic adjustment of the offset voltage δ. Here, δ is expressed as $V_{tS} - V_{tP} + S \log (W_P/W_S)$, and the current reduction ratio is expressed as $10^{-\delta/S}$, if secondary effects are neglected [2], as explained later. Thus, the reduction is adjustable with δ, that is, V_{tS} and W_S. If V_{tS} is high enough, the current is completely cut off with a larger δ, creating a perfect switch. A large δ, however, results in slow recovery time, large charging/discharging current and spike noise at mode transients. If V_{tS} is lower, δ becomes smaller (allowing leakage flow), causing an imperfect (leaky) switch, but the above problems are reduced. Moreover, a low-V_t switch is favorable to reduce the necessary channel width of M_S, since the increased transconductance can supply the accumulated current of the logic core with a smaller channel width, especially at a lower V_{DD}. Sharing a low-V_t switch through iterative circuits in RAMs (Fig. 4.2(c)) [10, 11, 13, 14] is quite effective. The input voltage of each circuit must be V_{DD}, not V_{DL}, to take advantage of the V_{GS} reverse bias. Because a feature of RAM circuits is that only one of the iterative circuits is active, W_S can be comparable to W_P with little speed penalty in the active mode, while $\delta = S/ \log (nW_P/W_S)$ in the standby mode for $V_{tS} = V_{tP}$. Therefore, both the leakage and area penalty as a result of adding M_{SP} are negligible with increasing n (i.e. δ). To be more precise, secondary effects must be taken into consideration: The substrate connection of M_P to V_{DD} creates substrate reverse bias. The effect of reduced V_{DS} is also added if δ is large (i.e. a small V_{DS}).

Gate-Source Offset Driving (A2). Figure 4.3(a) shows offset gate driving, where the input voltage is "overdriven" by δ. This is difficult to apply to random logic circuits because the logic swing of the output must be smaller than that of the input. However, it is useful to reduce the leakage (I_{leak}) in bus drivers [9], in power switches that have a low actual V_ts (Fig. 4.3(b) [3]), and RAM cells (negative word-line scheme called NWL, Fig. 4.3(c) [3]), as explained in Chapter 2. Offset gate driving applied to an imperfect switch reduces I_{leak} in standby, realizing an effectively perfect switch. However, the problems of a perfect switch described above arise.

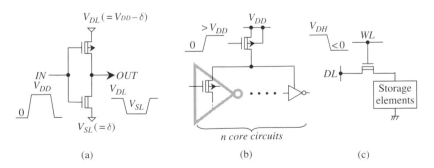

FIGURE 4.3. Circuits for offset gate driving (A2) [2, 9]: (a) Principle; (b) application to power switch; (c) application to RAM cells (negative word line). Reproduced from [3] with permission of IBM.

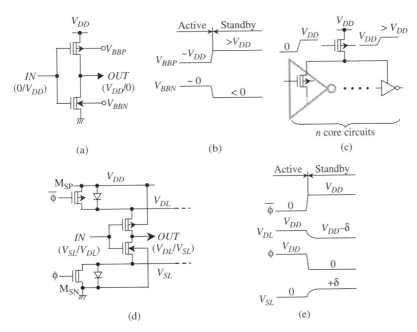

FIGURE 4.4. Circuits for substrate-source voltage (V_{BS}) reverse biasing [3]: (a) Substrate (well) driving (B1); (b) its operating waveforms; (c) application to power switch; (d) offset source driving (B2); (e) its operating waveforms. Reproduced from [3] with permission of IBM.

Substrate (Well) Driving (B1). Figure 4.4(a) shows substrate (well) driving [3], where the substrate voltages of MOSTs in core circuits change between active and standby modes. Figure 4.4(b) shows the operating waveforms. This scheme can also be applied to reduce I_{leak} in power switches (Fig. 4.4(c)).

Offset Source Driving (B2). Figure 4.4(d) has the circuit for offset source driving, with switches M_{SP} and M_{SN} inserted between the MOST sources and power supplies. Note that this is quite different from (A1), though both utilize source switches. The input (gate) voltage of (A1), which is the output of the previous stage, is "full swing" (V_{DD}), while that of (B2) is not (i.e. V_{DL} or V_{SL}). This difference results in the large discrepancy in I_{leak} reduction efficiency as shown in Fig. 4.1. From this viewpoint, power switches [15] applied to logic circuits can be categorized as (B2). Another application of this scheme is to reduce I_{leak} in SRAM cells [3], as discussed in Chapter 3.

4.3.2. Comparisons between Reduction Circuits

There is a big difference between the two schemes (A) and (B) in mode-transient time, especially recovery (standby-to-active) time [2, 3]. In V_{GS} reverse biasing, the small voltage swing (δ) enables quick recovery (several nanoseconds). In V_{BS} reverse biasing, however, it takes more than 100 ns for recovery when it is applied to a power line with a heavy capacitance, because V_{BS} reverse biasing requires a large V_B swing (ΔV_B) or V_S swing (ΔV_S), which is usually more than 1.5 V for a given change in V_t (ΔV_t). The necessary voltage swing imposes different requirements on substrate driving (B1) and offset source driving (B2). In (B1), the necessary voltage is significantly larger than V_{DD}, which is the sum of V_{DD} and ΔV_B. For example, existing MOSTs with a 0.2-$V^{1/2}$-body-effect coefficient (K) require a ΔV_B as large as 2.5 V to reduce the current by two decades with a 0.2-V ΔV_t. A larger-K MOST is thus needed to reduce the swing. However, this slows down the speed in stacked circuits, such as NAND gates. In contrast, the K value decreases with MOST scaling, implying that the necessary ΔV_B for a given ΔV_t will continue to increase further in the future (i.e., unscalable ΔV_B). Unfortunately, a need for the ever-larger ΔV_t reflecting the low-V_t era accelerates the increase. Eventually, this will enhance short-channel effects and increase other leakage current, such as GIDL current [23]. A shallow reverse V_B setting, or even a forward V_B setting is thus required to realize a larger ΔV_t with a smaller ΔV_B, because V_t changes more sharply with V_B [2]. However, the requirements to suppress V_B noise will instead become more stringent. In fact, a connection between the substrate and source every 200 μm [24] to reduce noise has been proposed, despite the area penalty. In addition, problems inherent in LSIs with an on-chip substrate bias (V_{BB}) generator, which DRAM designers have experienced since the late 1970s, may occur even with a low V_{DD}. These problems include spike current and CMOS latch-up during power-on and mode transitions, and V_{BB} degradation caused by increased substrate current in high-speed modes and screening tests at high stress V_{DD}. They also include the ever-increasing pn-junction current (see Fig. 1.18.) and slow recovery time as a result of poor current drivability of the on-chip charge pump [2].

In offset source driving (B2), the necessary voltages and voltage swing at any node are smaller than V_{DD}. This control becomes ineffective as V_{DD} is lowered owing to a smaller substrate bias. However, the problems accompanied by an on-chip V_{BB} generator are not expected. The energy overhead for offset source driving (B2) through mode transitions is usually larger than that for substrate driving (B1). This is because the parasitic capacitances of source lines are larger than those of substrate lines, though the necessary δ is smaller. The parasitic capacitances of substrate lines consist mainly of junction capacitances between the substrate (well) and source/drain of MOSTs, while those of source lines include the gate capacitances of on-state MOSTs as well as junction capacitances. The energy overhead for self-reverse biasing (A1) is quite small because of the small and self-adjusted δ .

4.4. Gate-Source Reverse Biasing Schemes

This section describes gate-source reverse (or back) biasing (i.e., (A) in Table 4.1) in detail because it is the most promising for LSI designs. It can be categorized as gate-source self-reverse biasing (A1), and gate-source offset driving (A2). In particular, self-reverse biasing is most effective because it features fast and simple control so as to be applicable even to the active mode, a small area penalty, and confinement to minimum active circuitry if applied to an iterative circuit block in RAMs, as explained later.

4.4.1. Gate-Source Self-Reverse Biasing

Switched-Source Impedance (SSI). A typical example of gate-source self-reverse biasing is the switched-source impedance (SSI) scheme applied to input predictable logic [10, 11]. In Fig. 4.5, SSI is stacked at the source of the MOST in the subthreshold region to provide gate-source self-reverse biasing. For example, SSI is at the nMOST source with the switch off for an inverter with low-level input during inactive periods, while for high-level input it is at the pMOST source. The switch is on during active periods so that the inverter operates normally. SSI features can be explained using the circuit in Fig. 4.6, assuming that an SSI is equivalent to one low-V_t MOST (M_2). During inactive periods with M_2 off, no matter how large the original subthreshold current (i'_1) is, it is eventually confined to the M_2 constant current (i_2) with self-adjustingδ. The δ and leakage reduction ratio $\gamma (= i_1/i'_1 = i_2/i'_1)$ are simply expressed by making i_1 equal to i_2 as

$$i_1 = aW_1 10^{-(\delta+V_{t1})/S}, \quad a:\text{ current density,} \tag{4.5}$$

$$i_2 = aW_2 10^{-V_{t2}/S}, \tag{4.6}$$

$$\delta = (V_{t2} - V_{t1}) + (S/\ln 10)\ln(W_1/W_2), \tag{4.7}$$

$$\gamma = 10^{-\delta/S}. \tag{4.8}$$

Primary Stacking Effect: The additional MOST M_2 makes δ and thus γ adjustable with its channel width (W_2) and/or threshold voltage (V_{t2}), yielding the primary

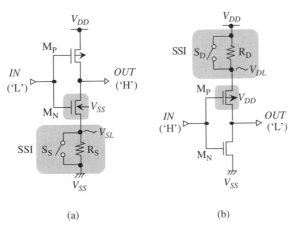

(a)   (b)

FIGURE 4.5. SSI applied to an inverter with a low input (a) and a high input (b). Reproduced from [11] with permission; © 2006 IEEE.

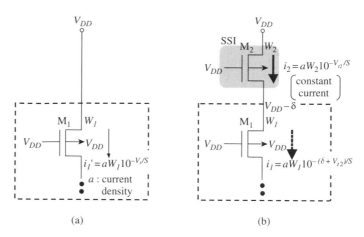

(a)   (b)

FIGURE 4.6. Circuit without SSI (a) and with SSI (b) [10, 11]. Reproduced from [34] with permission of Springer Science and Business Media.

stacking effect. A larger δ and thus a smaller γ are attained with a smaller W_2 and/or a higher V_{t2}. In an extreme case where V_{t2} is high so that i_2 becomes almost zero, γ is almost zero with a large δ. This is just the power switch with a sufficiently high V_t. Therefore, there is no substantial difference between the power switch and a 'leaky' power switch with a lower V_t, although a higher V_t reduces leakage more, but needs a longer recovery time instead. For the same V_t, γ is simply expressed as the channel-width ratio of W_2/W_1. For example, a small voltage swing (δ) of only 0.1 V reduces leakage by one order of magnitude when $W_2 = W_1/10$ and $S = 100\,\mathrm{mV/decade}$. Hence, fast control is achieved with such small δ, small load capacitance of the node, and simple self-reduction control.

Secondary Stacking Effects: If $W_2 = W_1$ and $V_{t2} = V_{t1}$, no reduction in leakage is expected because of $\delta = 0$ and $\gamma = 1$. This is true only because the above equations neglect the secondary effects. Actually, the secondary effects deriving from offset source driving ((B2) in Table 4.1) to M_1, and the DIBL effect ((C) in Table 4.1) of M_2 actually reduce leakage, although these reductions are quite small. As a first approximation, leakage (I_{leak}) is at least halved because M_1 and M_2 are equivalent to one MOST with halved channel width. To be more exact, about one-third reduction is expected [3], as seen in Fig. 4.7. This is because the node-voltage lowering V_M at the connection and the I_{leak}-reduction efficiency are determined by the equilibrium of the two currents and expressed by the crossing point of the two curves. Since the reduction efficiency becomes larger with the number of series MOSTs, the I_{leak} of NAND gates or NAND decoders with more series-connected nMOSTs is efficiently reduced.

Many applications of SSI to logic gates [11], such as inverters, NAND gates, NOR gates, clocked inverters, and R-S latches have been proposed.

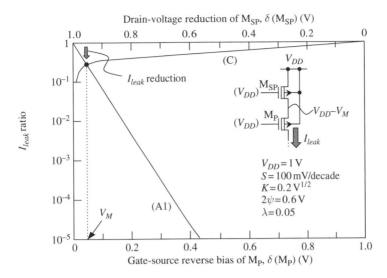

FIGURE 4.7. Leakage reduction due to stacking MOSTs using concepts (A1) and (C) in Table 4.1. The same parameters as in Figure 4.1 are used. Reproduced from [3] with permission of IBM.

SSI Sharing. One of the most effective SSI applications is SSI sharing [2–4, 10, 11, 13, 14, 25]. Figure 4.8 shows examples of SSI sharing applied to an inverter chain and an iterative circuit block to minimize the area penalty. For a 0-V input CMOS inverter chain (Fig. 4.8(a)), in which all outputs of inverters have settled, leakage flows from the V_{DD} supply to a pMOST in each V_{DD} input

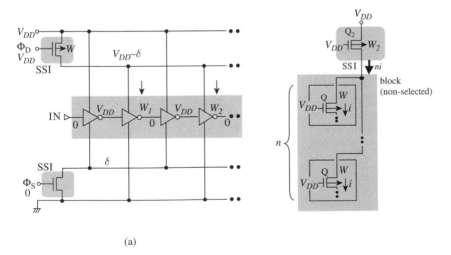

(a)

FIGURE 4.8. SSI sharing for inverter chain (a) and iterative (parallel) circuit (b) [11]. Reproduced from [34] with permission of Springer Science and Business Media.

inverter. Thus, there is accumulated leakage flowing into the pMOST SSI so that it is reduced with resulting δ. The accumulated leakage is proportional to the total channel width of relevant pMOSTs in the inverter chain. Thus, the leakage reduction ratio is expressed by the ratio of the channel width (W) of pMOST SSI to the total channel width of the pMOSTs, i.e., $\gamma = W/\Sigma W_k$ $(k = 1$ to $n)$ for a $2n$-stage inverter chain. Here, the W can be comparable to each pMOST channel width without a speed penalty, because each inverter switches with different timing during active periods. Thus, γ is eventually expressed as $1/n$. Therefore, with a large n, leakage and area penalties become smaller, because γ is almost 0, and W is much narrower than the total channel width. As a result, the leakage in the inverter chain can be reduced to that of pMOST SSI. Such is the case for nMOST SSI.

The n iterative (parallel) inverter block (Fig. 4.8(b)), with each inverter represented by one pMOST, coupled with SSI connected to the common source line, also minimizes the area penalty. During non-selected periods at the V_{DD} gate voltage of each MOST, the block is equivalent to one pMOST with W_1 equal to nW, so that the leakage reduction ratio (γ) is simply expressed as W_2/nW. Here, W_2 can be comparable to M-channel width (W) without a speed penalty, because only one MOST is activated during selected periods with SSI on. Thus, γ is eventually expressed as $1/n$. Therefore, the leakage and area penalties are greatly reduced with a large n.

SSI Applications to Multi-Divided Block. Minimizing the amout of active circuitry further reduces leakage in the active mode, if the large leakage from the remaining inactive circuitry is sufficiently reduced by SSI. This can be done by partial activation of a multi-divided block using SSI. Table 4.2 compares various SSI applications to an iterative circuit block [13, 14, 25]. Configuration (b) corresponds to Fig. 4.8(b). Configuration (c) is a multi-divided block with m sub-blocks with n/m circuits each, coupled with SSI connected to the common source to select each sub-block. Configuration (d) is hierarchical SSI with an additional SSI (i.e., SSI_2). Here, the same V_t, the same S, and the same leakage for pMOSTs and nMOSTs have been assumed. The channel-width ratios of the SSI MOST and the additional SSI MOST to the circuit MOST that flows leakage (i) with a channel width of W have also been assumed to be a and b, respectively. The leakage currents for configurations (b), (c), and (d) in the standby mode are confined to their SSI constant currents, ai, mai, and bi, respectively, no matter how large the total circuit leakage and the total SSI leakage are. In the active mode, leakage, ni, flows for (b) with the SSI on. Leakage, $n/mi + (m-1)ai$, flows for multi-divided block (c) because one selected SSI and one selected circuit in the selected subblock are turned on, while other SSIs and other circuits remain off. For another multi-divided block (d), i.e., hierarchical SSI, the same leakage flows with SSI_2 on. The standby leakage is minimized by architecture (d) because the mai current of (c) is eventually confined to bi, and active leakage is reduced by multi-divided blocks (c) and (d). Active leakage is minimized by the condition of $m = \sqrt{n/a}$ for $m \gg 1$, although this reduction must be compromised

TABLE 4.2. Various SSI applications to iterative circuit block [13, 14, 34].

	(a) No SSI	(b) SSI to one block	(c) SSI to multi-divided block	(d) Hierarchical SSI to multi-divided block
Configuration[*1]				
Penetrating[*2] current — Standby	ni	ai	mai	bi
Penetrating[*2] current — Active	ni		$(n/m)i + (m-1)ai$	
Drivability[*3]	1		$\dfrac{1}{1 + 1/a}$	$\dfrac{1}{1 + 1/a + 1/b}$
Charging to get into[*4] active mode	0		$C_1 \Delta V_1$	$C_1 \Delta V_1 + C_2 \Delta V_2$

[*1] Gate voltages shown are for the active mode. pMOSTs in iterative circuits have a gate width W.

[*2] All the transistors have the same threshold voltage. Channel widths are tailored as $a << n$ and $b << ma$.

[*3] Relative value to the conventional ignoring wiring resistance.

[*4] C_1 and C_2 are wiring capacitances and ΔV_1 and ΔV_2 are voltage drops from V_{DD} in the standby mode.

with speed. The three dimensional selection of subblocks [2, 13, 14, 25] is the most effective for leakage reduction.

Compensations for V_t-variations. Whatever SSI scheme is applied, the constant leakage current expressed by Eq. (4.6) still remains. It widely varies with V_t and temperature. Thus, compensation for these variations is key, and they are described in Chapter 5 (see Fig. 5.7).

4.4.2. Gate-Source Offset Driving

A negative supply voltage ($-\delta$ or V_{BB}), a low supply voltage raised from ground ($+\delta$), or a supply voltage (V_{DH}) higher than V_{DD} is essential to substantially cut the subthreshold current from a low-actual V_t nMOST or pMOST during inactive periods, while providing a large ON-current during active periods. This is because a high-V_t is effectively achieved as the sum of δ and a low-actual V_t, or as the sum of ($V_{DH} - V_{DD}$) and a low-actual V_t. The basic scheme that corresponds to (A2) in Fig. 4.1 was first proposed for a bus architecture [9], followed by power switches and word driving schemes (NWL and BSG, Fig. 2.11).

Bus Architecture. Figure 4.9 shows the bus architecture [9]. It has two supply voltage systems (external V_{DD} and V_{SS}, and internal V_{DL} and V_{SL}), and the bus signal swing (V_{DL} to V_{SL}) is smaller than the swing in the logic stage ($V_{DD} - V_{SS}$).

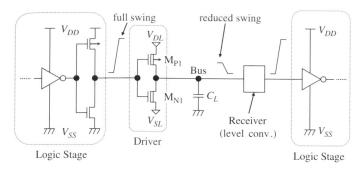

FIGURE 4.9. Bus architecture. Reproduced from [9] with permission; © 2006 IEEE.

The bus driver consists of a conventional CMOS inverter using lower-V_t MOSTs (M_{P1} and M_{N1}) fed from V_{DL} and V_{SL}. Figure 4.10 shows a bus receiver which converts the reduced-swing signal to a full-swing signal for the logic stage. It has a symmetric configuration with two level converters ($V_{DL} \rightarrow V_{DD}$ and $V_{SL} \rightarrow V_{SS}$), each consisting of a transmission gate and a cross-coupled MOST pair. This efficiently converts the signal swing from $V_{DL} - V_{SL}$ to $V_{DD} - V_{SS}$ without increasing the standby current. The reduced voltage swing on the bus line with a heavy capacitance and the gate-source offset driving are responsible for low-power high-speed operations and low standby current.

DRAM Word Drivers. The negative word driving (NWL) and the boosted sense ground (BSG) schemes use gate-source offset driving. NWL has been more popular for DRAMs (see Fig. 2.11). The driving is stable despite using a charge-pump generator, because the necessary pumping current is small due to only one word-line activation and the slow row cycle of DRAMs. To our knowledge, BSG has not been used in products because of reduced signal voltage resulting from reduced data-line voltage by δ.

Power Switch. The power switch shuts off the power supply of internal low-V_t core. Two options for implementing the switch are a low actual-V_t power switch and a high actual-V_t power switch. Although the low-V_t pMOST switch (see

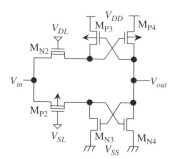

FIGURE 4.10. Bus receiver. Reproduced from [9] with permission; © 2006 IEEE.

Fig. 1.65) can increase the drive current in the active mode, it needs a raised power supply (V_{PP} or V_{DH}) to cut leakage in the standby mode with G-S offset driving. The high-V_t switch also needs the V_{DH} to increase the drive current, although it cuts standby-leakage. In the same manner, low-V_t and high V_t -nMOS power switches need a negative supply (V_{BB}) and a raised power supply. The power switches, however, always involve the following problems, as exemplified by a charge-pump V_{DH} generator as shown in Fig. 4.11.

Problem (1) Unregulated floating V_{PP} (i.e., V_{DH}) and increased pumping power. The gate voltage (V_{PP}) of the power-switch MOST must be well regulated against random operations of the internal core. For example, unless the high V_t is set to be high enough with a high V_{PP}, the leakage of the turned-off switch MOST is unexpectedly large due to unregulated gate voltages, because it is quite sensitive to the gate voltage. For example, the setting accuracy of the gate voltage must be less than 30 mV if a leakage variation must be less than 50 % for $S = 100$ mV/decade. To ensure such a regulation, the generator must compensate for the large randomly-changing load current that is maximized at the highest operation frequency of the power switch (i.e., internal core). Hence, the dissipated charge, $C_2 V_{PP} f$, at the load (C_2) must be much smaller than the pumping charge, $C_P V_{DD} f_P$, at C_P. In addition, C_1 must be much larger than C_P and C_2 for small ripples, and a level monitor and a feedback circuit may be necessary. To reduce the pumping power, the pumping charge must be reduced, thus calling for reduction in dissipated charge. Therefore, a slower frequency operation of the switch and a smaller load capacitance C_2 are essential.

Problem (2) An area penalty caused by a large power-switch MOST. Thus, the internal core must be small.

Problem (3) A long recovery time or a large spike current of the power line of the internal core if the internal capacitance and voltage swing are large. Thus, a small internal core is preferable.

Problem (4) Malfunctions due to floating nodes. Fixing the internal node voltages in the core after turning off the power switch is extremely important, since any floating node is the source of malfunctions. This is done by various high-V_t level

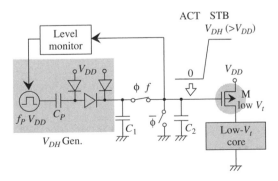

FIGURE 4.11. A low-V_t pMOS power switch using a charge-pump V_{DH} generator.

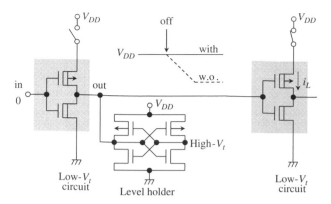

FIGURE 4.12. A high-V_t level holder connected to the floating output node [3, 14].

holders, exemplified by a high-V_t flip-flop type as shown in Figure 4.12 [3, 14]. In this scheme, the input of a low-V_t circuit is evaluated, and then the evaluated output is maintained by a small level holding latch. After that, the power switches are turned off to stop leakage in the low-V_t circuit, while preventing output from unnecessary discharge. Thus, the switches can be turned on quickly at the necessary timing to prepare for the next evaluation, ensuring fast random logic operations. Note that unnecessary discharging of the node prevents the output from quick recovery, especially for a heavy output capacitance, and thus prevents fast active operation even if the power switch itself can quickly be switched. In addition, the degraded output voltage may generate leakage currents at pMOSTs in the succeeding blocks if their power switches are still turned on.

Special attention must be paid to unusual operating conditions such as power on/off or burn-in test, since the previously mentioned detrimental effects are more enhanced. Possible problems [2, 3] include spikes or a large leakage current during power-on or during mode transitions. Such abnormal currents may occur not only for circuits involved in on-chip V_{BB} generators [2], but also for the power-switch relevant circuits while the gate voltage of the power switch has not yet settled. In addition, voltage degradation of V_{BB} or V_{PP} caused by increased load current at high stress V_{DD} in screening tests may be hazardous, especially for low-V_t circuits.

4.5. Applications to RAMs

Some of the above-described schemes can drastically reduce the leakage current of RAMs thanks to their matrix architecture. In this section, first, the leakage sources in RAMs and the features of RAMs in terms of leakage reduction are briefly explained. Then, leakage-conscious designs for peripheral logic circuits are discussed. Here, we assume that the chip (Fig. 4.13) is comprised of a RAM cell array and peripheral circuits such as row/column decoders and drivers, control logic circuits, amplifiers and I/O circuits, and on-chip voltage converters [2] that bridge the supply-voltage gap between the RAM cell array and peripheral circuits.

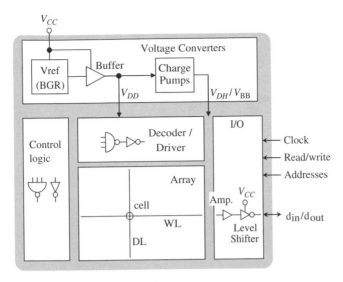

FIGURE 4.13. Architecture of a RAM chip.

4.5.1. Leakage Sources in RAMs

RAMs consist of a memory cell array and peripheral circuits [2], both of which are composed of many iterative (parallel) circuit blocks, due to the matrix architecture. These blocks include the cell array itself and row/column decoder blocks and their relevant driver blocks. Of the iterative circuit blocks, the memory cell array is the largest leakage source because of the largest total channel width despite the small channel width of each cell. Leakage in RAM cells causes different detrimental effects for DRAM and SRAM cells. For DRAM cells, leakage shortens the data retention time (i.e., the refresh time), while for SRAM cells it increases the data retention current (Fig. 1.63(a)). The use of a high-V_t or effectively high-V_t MOST eventually reduces the leakage of DRAM/SRAM cells and thus of the memory cell array, as explained in Chapters 2 and 3.

For peripheral circuits, the above-described iterative circuit blocks are major sources of leakage because each of the blocks usually has a large total channel width. Thus, for example, as we can see from Fig. 1.63(b) [2, 14, 16], the DC subthreshold leakage (I_{DC}) from such iterative circuit blocks in DRAM chips rapidly increases and surpasses AC capacitive current (I_{AC}), eventually dominating the total chip current (I_{ACT}). Note that in some cases leakage reduction of pMOSTs is more important than that of nMOSTs because of the larger channel width and sometimes larger S-factor of pMOSTs [17] (see Fig. 4.18).

Array Driver Blocks. Of the iterative circuit blocks in the peripheral circuits, the row (i.e., word) driver block contributes most to leakage because it has the largest total channel width. Note that each pMOST word driver has a large channel width to quickly drive a heavily capacitive word line that is composed of the gate capacitances of many cells. Here, a pMOST is necessary for the output driver to simplify the design with reduced stress voltage to the MOST [2]. On the contrary, the total channel width of the column driver block is usually narrow because each driver theoretically controls only one column switch when connecting a selected DL to a common DL. Even with multi-divided data-line architectures [2] the number of switches is limited.

Row and Column Decoder Blocks. The total channel width is effectively narrow because each decoder block consists of robust circuits (i.e., leakage-immune NAND gates) [2], as discussed later.

Sense Amplifier Block. The total channel width of DRAM sense amplifiers (SAs) is much larger than that of SRAMs. This is because DRAMs need a larger number of tiny SAs in a chip, because one SA must be placed at each data line due to refresh requirements. In addition, in standard mid-point (half-V_{DD}) sensing of DRAMs, the SA must operate at the lowest voltage (i.e., half-V_{DD}) in the chip, calling for a lower V_t for fast sensing. The many low-V_t SAs resulting from this thus generate large leakage despite small MOSTs being used. In contrast, SRAMs have a small number of SAs usually operating at V_{DD} on the common data input/output line.

Control Logic Block and Others. This block may operate at its highest frequency, making leakage reduction extremely difficult. Fortunately, however, the total channel width of the block is small (usually 10% at most of total channel width of chip). Moreover, the I/O block of the chip, which operates at a high I/O power supply voltage, accepts a high enough V_t to suppress leakage.

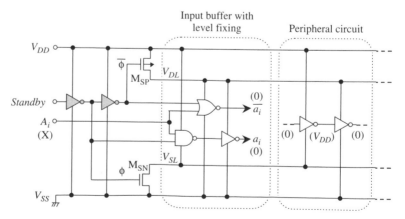

FIGURE 4.14. Method to make the internal nodes of RAMs predictable. Each node voltage during standby mode (standby signal is at high level) is in parentheses. M_{SP}, M_{SN}, and solid inverters consist of high-V_t MOSTs. Other logic gates consist of low-V_t MOSTs. Reproduced from [11] with permission; © 2006 IEEE.

4.5.2. Features of Peripheral Circuits of RAMs

Reducing leakage in the active mode is extremely difficult because of the limited time to control it, while reducing it in the standby mode is rather easy because there is sufficient time available. Fortunately, however, RAM peripheral circuits favor the reduction of subthreshold-current (I_{leak}) in the active mode, compared with random logic LSIs, because of the following inherent features of RAMs.

Use of Multiple Iterative Circuit Blocks. RAMs consist of multiple iterative circuit blocks, such as row/column decoders and drivers, each of which has quite a large total-channel width, as discussed previously. In addition, all circuits except the selected one in each block are inactive even during the active period. This feature is extremely important for reducing the leakage in each block simply and effectively, even in the active mode with a smaller area penalty than for logic LSIs, as will be explained later.

Use of Input-Predictable Circuits. RAMs are composed of many input-predictable circuits, allowing circuit designers to predict all node voltages in the chip and to prepare the most effective subthreshold-current reduction scheme (e.g. V_{GS} self-reverse biasing) in advance. As for input nodes, which are not predictable, the level-fixing input buffer (Fig. 4.14) [11] can force the internal node voltages to be predictable. In standby mode (signal *Standby* is at high level), internal nodes including a_i, $\bar{a}_i$ and the following-stage outputs are forced to be at predetermined levels irrespective of input node A_i. Similar techniques are applied to logic LSIs, though their node voltages are usually unpredictable because they contain registers or latches to retain internal states. Latches (Fig. 4.15) [31] that fix the output level while retaining the latched data are effective in reducing I_{leak} in standby/sleep modes. Level-fixing flip-flops [32] combined with the self-reverse biasing [11], power switches [33], and level holders [14] enable quick recovery from standby/sleep modes. These techniques, in turn, can be applied to RAM peripheral circuits with registers or latches.

Slow RAM Cycle. RAMs feature a slow cycle compared with random logic-gate LSIs. A physically large memory-cell array that occupies over 50% of the chip, a relatively large capacitance and high resistance in the word line and data (or bit) line, and a small memory-cell signal that necessitate succeeding amplification

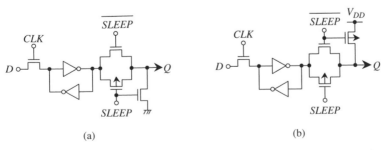

(a)

(b)

FIGURE 4.15. Latches with output fixing in sleep mode: (a) Output fixed low; (b) output fixed high in sleep mode [3]. Reproduced from [31] with permission; © 2006 IEEE.

are all responsible for a slower cycle time of RAMs [2]. This slow cycle makes leakage reduction easier: The circuits are active only for a short period within the "long" memory cycle, allowing additional time to control the subthreshold current. This is true for DRAM row circuits, which are sufficiently slow to accept leakage control. However, the column circuits in modern DRAMs [2, 3] feature a fast burst cycle and unpredictable circuit operation (every column may be selected during the memory cycle). Therefore, it is difficult to reduce leakage in column circuits in the active mode. This is also the case for extremely high-speed SRAMs and logic LSIs. Fortunately, however, even the total channel width of such circuits in RAMs is small, as previously explained.

Use of Robust Circuits. Modern CMOS RAMs do not use leakage-sensitive circuits such as dynamic NOR gates that require a level keeper (see Fig. 1.20) to prevent malfunctions caused by leakage. For example, the decoder block consists of dynamic (for the row) and static (for the column) NAND gates to reduce the power. NAND decoders discharge only one output node in a selected decoder, while the NOR decoders used in the nMOS era discharged all output nodes in decoders, except for the selected one [2]. In addition to low power, they reduce leakage through the stacking effect, as explained earlier. In the standby mode, RAMs have a feature in that all address inputs to each nMOS NAND decoder are low [2], enabling leakage to be dramatically reduced. Even in the active mode, leakage is reduced by the stacking effect because a considerable number of decoders have at least two low-level inputs.

Utilization of Internal Power-Supply Voltages. Multi-static V_t [2, 3] is available for DRAMs because the internal power-supply voltages necessary for DRAM operations can be utilized to achieve a high V_t.

4.5.3. Applications to DRAM Peripheral Circuits

SSI was applied to iterative circuit blocks to reduce leakage current, first in the standby mode and then in the active mode with advanced SSIs.

Standby-Current Reductions in 256-Mb DRAMs. SSI dramatically reduced the data retention current with refresh operations that was dominated by the word-driver block. Figure 4.16 shows a low-V_t pMOS SSI (M_S) shared with the n word-driver block in a 256-Mb DRAM [29]. This is an example of (A1) in Table 4.1 or Fig. 4.1, although a raised supply (V_{DH}) necessary for word-bootstrapping is used. In the standby mode, the off-SSI enables the voltage (V_{DL}) of the common source line to drop by δ from V_{DH} as a result of the total subthreshold current flow of ni. As each pMOS driver (M_P) is self-reverse-biased, the total current eventually decreases. Hence, the V_{DH} is well regulated despite the use of the on-chip charge pump. In the active mode, the selected word line is driven after V_{DL} becomes V_{DH} by turning on M_S. Here, the channel width of M_S can be reduced to several times that of M_P without a speed penalty, as previously discussed. A resulting δ as small as 254 mV reduced the standby subthreshold current of the word-driver and decoder blocks to 1.5×10^{-3} for $n = 256$ and $W(M_S)/W(M_P) = 5$ in a 0.2μm technology, enabling to reduce

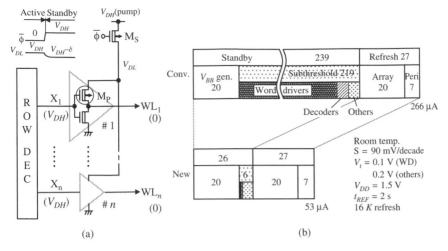

(a)   (b)

FIGURE 4.16. Gate-source (V_{GS}) self-reverse biasing applied to a 256-Mb DRAM (a) and retention current with refreshes [29]. W_S and W_P denote the respective channel widths of M_S and M_P. Reproduced from [34] with permission of Springer Science and Business Media.

the subthreshold current of the chip to 3% (from 219 to 6 μA). The total data retention current was thus reduced to 53 μA. The small δ enabled a fast recovery time to the active mode of 1 ns.

Figure 4.17 shows another 256-Mb DRAM [17] with a hierarchical word-line architecture [2]. The SSI and multi-static high V_t utilizing well biases have been combined. Here, $\overline{\text{MWL}}$ and SWL are the main word line and sub-word line, and CSL is the column select line for the multi-divided data-line architecture [2]. The circled MOSTs in the figure are in the subthreshold region during the standby mode. Here, SSI_1 was only applied to the pMOSTs (open circles) in the inverter chain with an output of RX in the array control circuit, which corresponds to Fig. 4.8(a). This was because pMOSTs have a larger total channel width and a larger S-factor (see Fig. 4.18) due to their buried-channel MOST structure. Since these inverters operate in series in the time domain, this connection does not cause a speed penalty, as previously explained. SSI_1 was also applied to the main-word driver block. Furthermore, the nMOSTs (shaded circles) and pMOSTs (shaded circles) in the column-decoder block had a higher V_t due to the respective well bias V_{BB} and V_{DH}. SSI_2 further reduced the leakage in the column decoder block. By combining both SSI and multi-static high-V_t schemes, the total subthreshold current in the power-down/self-refresh mode was reduced to one-sixth, as Fig. 4.18 shows. The current can further be reduced by applying a multi-static V_t scheme to the peripheral circuits.

Active-Current Reduction for a 16-Gb DRAM. Hierarchical SSI schemes (i.e., Table 4.2(d)) and a power switch with a level holder (Fig. 4.12) have been reported to be effective in reducing the active current of a hypothetical 1-V 16-Gb DRAM [13, 14]. Figure 4.19(a) shows an application to the word-driver

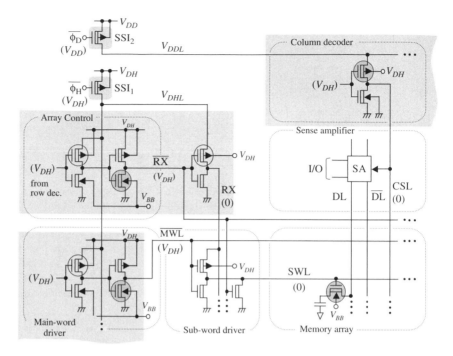

FIGURE 4.17. Various leakage-reduction schemes applied to array associated circuitry in a 256-Mb SDRAM [17]. ( ): voltages in the standby mode. Reproduced from [3] with permission of IBM.

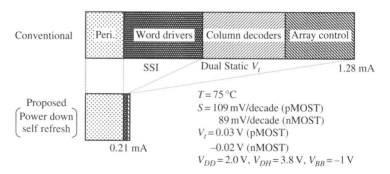

FIGURE 4.18. Leakage in the standby mode of a 256-Mb SDRAM [17]. The peripheral circuits component is from peripheral MOSTs without substrate bias. Reproduced from [3] with permission of IBM.

block, which is divided into m sub-blocks with n/m word drivers each. An SSI$_1$ is connected to the common source line of pMOST word drivers to select the sub-block. In the active mode, while turning on the selected SSI$_1$ and SSI$_2$, the leakage in each non-selected sub-block is confined to the SSI$_1$ small constant

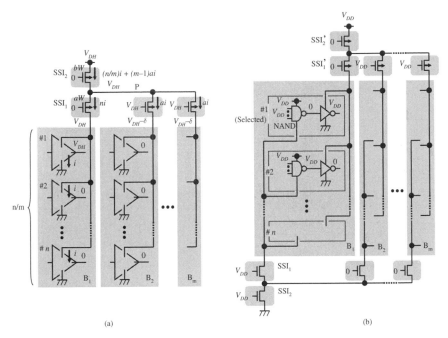

FIGURE 4.19. Word driver block (a) and decoder block (b) with the hierarchical SSI scheme [13, 14]. Reproduced from [34] with permission of Springer Science and Business Media.

current since the SSI_1 channel width is much narrower than the total channel width of driver MOSTs in the corresponding sub-block, as previously discussed. In standby mode, the total leakage confined by SSI_1s (i.e., *mai* in Table 4.2(c)) is further confined to the SSI_2 small constant current since the SSI_2 channel width is much narrower than the total channel width of SSI_1s. Figure 4.19(b) shows another application to the decoder block, which is divided into m sub-blocks $(B_1 - B_m)$ with n drivers each. One decoder unit consists of an nMOST NAND gate with address inputs and a CMOS inverter. The hierarchical SSI scheme is applied to the nMOST common source of the NAND gate block, and to the pMOST common source of the CMOS inverter block. Leakage current in active and standby modes is reduced in the same manner as in the word driver block.

Figure 4.20(a) shows an application to the sense-amplifier (SA) driver block, which is divided into m sub-blocks $(B_1 - B_m)$ with n drivers each. In the active mode, the selected driver in the selected sub-block, e.g., driver #1 in sub-block B_1, drives the SAs in the selected sub-array with M_{D1} and M_{D1}'. This is done by turning on SSI_2 and SSI_2', and only SSI_1 and SSI_1' of selected sub-block B_1, while all SSI_1s and SSI_1's of non-selected sub-blocks remain off. After the signals developed on the half-V_{DD} data lines have been amplified to V_{DD} or 0 by the selected SAs, accumulated leakage (i'), which is the sum of leakage (i_S) flowing into each SA (Fig. 4.20(b)), flows into M_{D1} and M_{D1}'. The accumulated

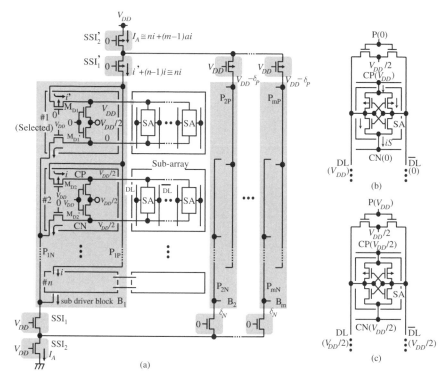

FIGURE 4.20. Sense amp (SA) driver block (a) with a symmetric hierarchical SSI scheme in the active mode, and selected SA-relevant circuits (b) and non-selected SA-relevant circuits (c) [13, 14]. P; precharge. Reproduced from [34] with permission of Springer Science and Business Media.

leakage is proportional to the total channel width of leakage-relevant MOSTs in SAs, which is comparable to the channel width of M_{D1} and M_{D1}' to quickly drive the SAs. For other drivers in the selected block, all SAs remain off because all data lines and all drive lines (CPs and CNs) are at $V_{DD}/2$ (Fig. 4.20(c)). However, leakage (i), which is proportional to the channel width of M_D or M_D', develops at each non-selected (OFF) driver pMOST and nMOST (e.g., M_{D2} and M_{D2}'). Hence, the total leakage in non-selected drivers is much larger than the leakage in the selected driver. However, for non-selected blocks, SSI_1s and SSI_1's cause voltage drops or raised voltages (δ_P or δ_N) on corresponding power lines (e.g., P_{2N} and P_{2P}) to reduce leakage. The total leakage is $(m-1)\ ai$, assuming that the channel width ratio of an SSI_1 MOST to M_D is a. Note that M_{D1} and M_{D1}' are common to plural SAs in this example.

In the conventional design, the simulated active current in a hypothetical 16-Gb DRAM was as large as 1.18 A (see Fig. 4.21). The DC subthreshold leakage was as large as 1.105 A, while the AC capacitive current was as small as 75 mA at a cycle time of 180 ns. Major leakage current came from iterative circuit blocks,

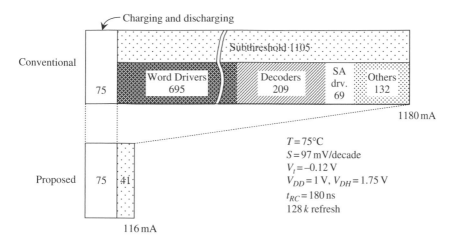

FIGURE 4.21. Active current reduction in the hypothetical 16-Gb DRAM [14]. Reproduced from [13] with permission; © 2006 IEEE.

such as the word driver block, decoder blocks (row and column), and sense-amp driver blocks. Note that depletion MOSTs (e.g., nMOST's $V_t = -0.12\,\text{V}$) were responsible for such large leakage. The above-described SSI schemes and a high-V_t power switch with a high-V_t level holder (Fig. 4.12) [14] reduced the active current to $116\,\text{mA}$.

Sleep-Mode Current Reduction in 90-nm 16-Mb e-DRAM. A variety of SSI schemes and compensation circuits for V_t-variations has enabled a 16-Mb e-DRAM to operate at a record-setting low voltage of 0.6 V using a 0.195-μm^2 trench capacitor ($C_S = 40\,\text{fF}$) 1-T cell [1]. The total operating power at a 0.6-V 20-ns row cycle was only 39 mW, and the standby and sleep-mode currents at 0.9 V and 105 °C were as low as $328\,\mu\text{A}$ and $34\,\mu\text{A}$. The details are in what follows.

Figure 4.22 shows the cell-relevant circuits. An excessively raised word voltage of 3 V is probably needed for high-speed charging of a large C_S rather than for a full-V_{DD} write. A low V_t of 0.2 V is used for SA MOSTs ($M_1 - M_4$) to enable high-speed half-V_{DD} sensing, while a normal V_t of 0.3 V is used for SA-driver MOSTs (M_5, M_6) to reduce subthreshold current in standby mode. The substrate biases of SA and driver MOSTs are independently controlled because their different V_t implants cause different temperature dependencies for V_t. Gate-source offset driving in the sleep mode of nMOSTs and pMOSTs (above V_{DD} and below V_{SS} by 0.3 V) reduces subthreshold current. The RAM data is retained by periodically exiting the sleep mode and performing burst refresh cycles at 20 ns, as we can see from the figure. The refresh scheme minimizes the pump currents of the converters for $-0.3\,\text{V}$ and $V_{DD} + 0.3\,\text{V}$, enabling a simple design to be used for the converters. Figure 4.23 depicts a sense-amplifier driver consisting of an inverter chain that shares SSIs, as shown in Fig. 4.8(a). Figure 4.24 shows row circuits composed of multiple iterative circuit blocks such as a NAND decoder block, an inverter block, a level shifter block, and a word-driver block. Each block has its own SSI MOST. For example, in the sleep

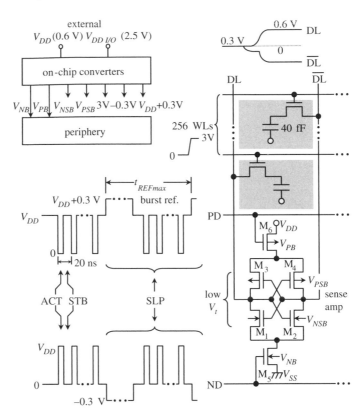

FIGURE 4.22. Cell relevant circuits of 0.6-V 16-Mb e-DRAM [1]. STB: standby mode, ACT: active mode, and SLP: sleep mode. Reproduced from [4] with permission; © 2006 IEEE.

mode, the leakage from each block is reduced by the respective SSI because each circuit in the block and SSI are off. In the active mode, all SSIs are turned on so that the selected word line is activated by the corresponding row circuits. Only one circuit in each block is activated in the active mode, and thus leakage in the sleep mode is reduced without speed and area penalties, as previously described.

The substrate biases of SAs and other periphery circuits must statically be controlled according to variations in process, temperature, and V_{DD} to suppress variations in subthreshold-current as well as speed. The details are discussed in Chapter 5.

4.5.4. Applications to SRAM Peripheral Circuits

Figure 4.25 illustrates a 0.13-μm 1.2-V 300-MHz 1-Mb e-SRAM module [30]. The partial activation of the multi-divided block discussed previously is employed. In the active mode, a four-bank architecture (#1 in the figure) with

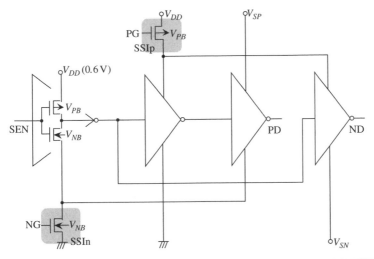

SEN: V_{DD} (ACT), 0(STB/SLP)
NG: V_{DD}+0.3 V(ACT/STB), −0.3 V(SLP)
PG: −0.3 V(ACT/STB), V_{DD}+0.3 V(SLP)

V_{SP}: V_{DD} (ACT/STB), V_{DD}+0.3 V(SLP)
V_{SN}: 0(ACT/STB), −0.3 V (SLP)

FIGURE 4.23. Sense-amp driver of 0.6-V 16-Mb e-DRAM [1]. Reproduced from [4] with permission; © 2006 IEEE.

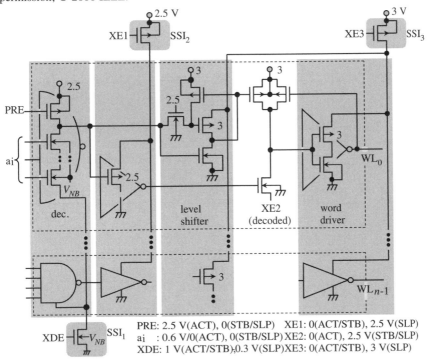

PRE: 2.5 V(ACT), 0(STB/SLP)   XE1: 0(ACT/STB), 2.5 V(SLP)
ai : 0.6 V/0(ACT), 0(STB/SLP) XE2: 0(ACT), 2.5 V(STB/SLP)
XDE: 1 V(ACT/STB), 0.3 V(SLP) XE3: 0(ACT/STB), 3 V(SLP)

FIGURE 4.24. Row circuits of 0.6-V 16-Mb e-DRAM [1]. Reproduced from [4] with permission; © 2006 IEEE.

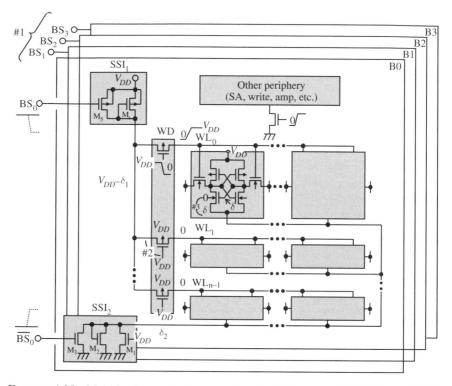

FIGURE 4.25. Multi-bank architecture applied to 0.13-μm 1.2-V 1-Mb e-SRAM module [30]. Reproduced from [34] with permission of Springer Science and Business Media.

only one-bank activation by turning on M_3 and M_5 confines the active circuitry to one-fourth, and thus reduces the AC power of control signals to one-fourth. This also reduces leakage in inactive banks, if SSI is applied to the word-driver block and cell array. SSI_1 reduces the leakage in each off-pMOST in the word driver block [4] by gate-source self-reverse biasing (#2). SSI_1 allows a small voltage drop, δ_1, in the small capacitive power line of the word-driver block, enabling a fast recovery time of 0.3 ns. SSI_2 causes δ_2 on the common cell-source line as a result of accumulated leakage from many cells. Thus, the body effect by δ_2 (#3) increases the V_t of the cross-coupled off-nMOST in each cell so that leakage is reduced. Here, the diode (M_2) clamps the source voltage, so the signal charge of memory cells is not reduced by excessive δ_2.

However, it has drawbacks: A large necessary δ_2 of about 0.4 V due to poor reduction efficiency of source-driving ((B2) in Fig. 4.1), and large source-line capacitance result in a slow recovery time of about 3 ns. In addition, the signal charge of the non-selected cell is reduced by δ_2. In the sleep mode, peripheral circuits such as SAs and write amplifiers are turned off with the power switch off, resulting in a slow recovery time of 3 ns. In the standby mode, all banks are off, causing further reduced leakage. The results indicated that leakage current was reduced by 25% in the high-speed (300 MHz) active mode with only SSI_1 turned on, and by 67% in the slow-speed (100 MHz) active mode with both SSI_1 and SSI_2 turned on. In the sleep mode, leakage in peripheral circuits was reduced by 95%. This source driving of the cell is effective only if V_{DD} is still high (such as 1.2 V as this example). However, the resultant reduced power supply, $V_{DD} - \delta_2$, and thus reduced signal charge of memory cells in the standby mode would pose a stability problem at a lower V_{DD}, strictly limiting low-V_{DD} operations, as discussed in Chapter 3.

References

[1] K. Hardee, F. Jones, D. Butler, M. Parris, M. Mound, H. Calendar, G. Jones, L. Aldrich, C. Gruenschlaeger, M. Miyabayashi, K. Taniguchi, and T. Arakawa, "A 0.6V 205MHz 19.5ns tRC 16Mb embedded DRAM," ISSCC Dig. Tech. Papers, pp. 200–201, Feb. 2004.

[2] K. Itoh, *VLSI Memory Chip Design*, Springer-Verlag, NY, 2001.

[3] Y. Nakagome, M. Horiguchi, T. Kawahara, K. Itoh, "Review and future prospects of low-voltage RAM circuits," IBM J. R & D, vol. 47, no. 5/6, pp. 525–552, Sep./Nov. 2003.

[4] K. Itoh, K. Osada, and T. Kawahara, "Reviews and future prospects of low-voltage embedded RAMs," CICC Dig. Tech. Papers, pp. 339–344, Oct. 2004.

[5] M. Aoki, J. Etoh, K. Itoh, S. Kimura and Y. Kawamoto, "A 1.5V DRAM for battery-based applications," ISSCC Dig. Tech. Papers, pp. 238–239, Feb 1989.

[6] Y. Nakagome, Y. Kawamoto, H. Tanaka, K. Takeuchi, E. Kume, Y. Watanabe, T. Kaga, F. Murai, R. Izawa, D. Hisamoto, T. Kisu, T. Nishida, E. Takeda and K. Itoh, "A 1.5-V circuit technology for 64Mb DRAMs," Symp. VLSI Circuits Dig. Tech. Papers, pp. 17–18, June 1990.

[7] K. Itoh, "Reviews and prospects of deep sub-micron DRAM technology," SSDM Ext. Abst., pp. 468–471, Aug. 1991.

[8] J. Etoh, K. Itoh, Y. Kawajiri, Y. Nakagome, E. Kume and H. Tanaka, "Large scale integrated circuit for low voltage operation," US Patent No. 5297097 (Mar. 1994), RE37593 (Mar. 2002).

[9] Y. Nakagome, K. Itoh, M. Isoda, K. Takeuchi and M. Aoki, "Sub-1-V swing bus architecture for future low-power ULSIs," Symp. VLSI Circuits Dig. Tech. Papers, pp. 82–83, June 1992.

[10] T. Kawahara, M. Horiguchi, Y. Kawajiri, G. Kitsukawa, T. Kure and M. Aoki, "Subthreshold current reduction for decoded-driver by self-reverse biasing," IEEE J. Solid-State Circuits, vol. 28, pp. 1136–1144, Nov. 1993.

[11] M. Horiguchi, T. Sakata and K. Itoh, "Switched-source-impedance CMOS circuit for low standby subthreshold current giga-scale LSI's," IEEE J. Solid-State Circuits, vol. 28, pp. 1131–1135, Nov. 1993.

[12] D. Takashima, S. Watanabe, H. Nakano, Y. Oowaki, K. Ohuchi and H. Tango, "Standby/active mode logic for sub-1-V operating ULSI memory," IEEE J. Solid-State Circuits, vol. 29, pp. 441–447, Apr. 1994.

[13] T. Sakata, M. Horiguchi, and K. Itoh, "Subthreshold-current Reduction Circuits for Multi-gigabit DRAMs," Symp. VLSI Circuits Dig. Tech. Papers, pp. 45–46, May 1993.

[14] T. Sakata, K. Itoh, M. Horiguchi and M. Aoki, "Subthreshold-current reduction circuits for multi-gigabit DRAM's," IEEE J. Solid-State Circuits, vol. 29, No.7, pp. 761–769, July 1994.

[15] S. Mutoh, T. Douseki, Y. Matsuya, T. Aoki, S. Shigematsu and J. Yamada, "1-V power supply high-speed digital circuit technology with multithreshold-voltage CMOS," IEEE J. Solid-State Circuits, vol. 30, pp. 847–854, Aug. 1995.

[16] K. Itoh, "Reviews and prospects of low-power memory circuits" (invited), Low-Power CMOS Design, A. Chandrakasan and R. Brodersen, Eds., Wiley–IEEE Press, NJ, pp. 313–317, 1998.

[17] M. Hasegawa, M. Nakamura, S. Narui, S. Ohkuma, Y. Kawase, H. Endoh, S. Miyatake, T. Akiba, K. Kawakita, M. Yoshida, S. Yamada, T. Sekiguchi, I. Asano, Y. Tadaki, R. Nagai, S. Miyaoka, K. Kajigaya, M. Horiguchi and Y. Nakagome, "A 256Mb SDRAM with subthreshold leakage current suppression," ISSCC Dig. Tech. Papers, pp. 80–81, Feb. 1998.

[18] S. Narendra, S. Borkar, V. De, D. Antoniadis and A. Chandrakasan, "Scaling of stack effect and its application for leakage reduction," Proc. ISLPED, pp. 195–199, Aug. 2001.

[19] C. Akrout, J. Bialas, M. Canada, D. Cawthron, J. Corr, B. Davari, R. Floyd, S. Geissler, R. Goldblatt, R. Houle, P. Kartschoke, D. Kramer, P. McCormick, N. Rohrer, G. Salem, R. Schulz, L. Su and L. Whitney, "A 480-MHz RISC microprocessor in a 0.12-um Leff CMOS technology with copper interconnects," IEEE J. Solid-State Circuits, vol. 33, pp. 1609–1616, Nov. 1998.

[20] H. Morimura and N. Shibata, "A 1-V 1-Mb SRAM for portable equipment," Proc. ISLPED, pp. 61–66, Aug. 1996.

[21] K. Itoh, R. Hori, H. Masuda, Y. Kawajiri, H. Kawamoto and H. Katto, "A single 5V 64K dynamic RAM," ISSCC Dig. Tech. Papers, pp. 228–229, Feb. 1980.

[22] I. Fukushi, R. Sasagawa, M. Hamaminato, T. Izawa and S. Kawashima, "A low-power SRAM using improved charge transfer sense amplifiers and a dual-Vth CMOS circuit scheme," Symp. VLSI Circuits Dig. Tech. Papers, pp. 142–145, June 1998.

[23] A. Keshavarzi, S. Ma, S. Narendra, B. Bloechel, K. Mistry, T. Ghani, S. Borkar and V. De, "Effectiveness of reverse body bias for leakage control in scaled dual Vt CMOS ICs," Proc. ISLPED, pp. 207–212, Aug. 2001.

[24] H. Mizuno, K. Ishibashi, T. Shimura, T. Hottori, S. Narita, K. Shiozawa, S. Ikeda and K. Uchiyama, "A 18-µA standby current 1.8-V 200-MHz microprocessor with self-substrate-biased data-retention mode," IEEE J. Solid-State Circuits, vol. 34, pp. 1492–1500, Nov. 1999.

[25] T. Sakata, M. Horiguchi, M. Aoki and K. Itoh, "Two-dimensional power-line selection scheme for low subthreshold-current multi-gigabit DRAMs," Proc. ESSCIRC, pp. 131–134, Sep. 1993.

[26] K. Seta, H. Hara, T. Kuroda, M. Kakumu and T. Sakurai, "50% active-power saving without speed degradation using standby power reduction (SPR) circuit," ISSCC Dig. Tech. Papers, pp. 318–319, Feb. 1995.

[27] K. Kumagai, H. Iwaki, H. Yoshida, H. Suzuki, T. Yamada and S. Kurosawa, "A novel powering-down scheme for low Vt CMOS circuits," Symp. VLSI Circuits Dig. Tech. Papers, pp. 44–45, June 1998.

[28] M. Mizuno, K. Furuta, S. Narita, H. Abiko, I. Sakai and M. Yamashina, "Elastic-Vt CMOS circuits for multiple on-chip power control," ISSCC Dig. Tech. Papers, pp. 300–301, Feb. 1996.

[29] G. Kitsukawa, M. Horiguchi, Y. Kawajiri, T. Kawahara, T. Akiba, Y. Kawase, T. Tachibana, T. Sakai, M. Aoki, S. Shukuri, K. Sagara, R. Nagai, Y. Ohji, N. Hasegawa, N. Yokoyama, T. Kisu, H. Yamashita, T. Kure and T. Nishida, "256-Mb DRAM circuit technologies for file applications," IEEE J. Solid-State Circuits, vol. 28, pp. 1105–1113, Nov. 1993.

[30] M. Yamaoka, Y. Shinozaki, N. Maeda, Y. Shimazaki, K. Kato, S. Shimada, K. Yanagisawa, and K. Osada, "A 300MHz 25μA/Mb leakage on-chip SRAM module featuring process-variation immunity and low-leakage-active mode for mobile-phone application processor," ISSCC Dig. Tech. Papers, pp. 494–495, Feb. 2004.

[31] J. P. Halter and F. N. Najm, "A gate-level leakage power reduction method for ultra-low-power CMOS circuits," Proc. CICC, pp. 475–478, May 1997.

[32] K-S Min, H. Kawaguchi, and T. Sakurai, "Zigzag super cut-off CMOS (ZSCCMOS) block activation with self-adaptive voltage level controller: an alternative to clock-gating scheme in leakage dominant era," ISSCC Dig. Tech. Papers, pp. 400–401, Feb. 2003.

[33] H. Kawaguchi, K. Nose and T. Sakurai, "A super cut-off CMOS (SCCMOS) scheme for 0.5-V supply voltage with picoampere stand-by current," IEEE J. Solid-State Circuits, vol. 35, pp. 1498–1501, Oct. 2000.

[34] T. Kawahara and K. Itoh, "Memory Leakage Reduction," Chapter 7 in *Leakage in Nanometer CMOS Technologies*, Edited by S. G. Narendra and A. Chandrakasan, Springer, 2006.

5
Variability Issue in the Nanometer Era

5.1. Introduction

To create low-voltage nano-scale LSIs, many challenges remain with the memory-cell array and peripheral logic circuits in a memory chip. Of the challenges, the variability issue is a key because variations in process, voltage, and temperature (PVT variations) decrease chip performance intolerably. For example, as CMOS approaches the 25-nm node, random threshold-voltage (V_t) variation caused by varying dopant implant positions in ultra-small inversion regions has been reported to give rise to more than 100 mV of V_t variation [1]. Additionally, a temperature variation of 100°C causes subthreshold currents to increase by four orders of magnitude, as explained later. Speed variation also becomes intolerable. The voltage margin of V_t-sensitive circuits, such as flip-flop circuits that DRAMs use for sense amps and SRAMs use for cells, also widely varies. These performance variations eventually restrict low-voltage operations.

This chapter describes the variability issue. First, the V_t variation issue in the nanometer era is briefly discussed. And then, the resulting variations in subthreshold currents, logic speed, and voltage margins of the flip-flop circuits are investigated in detail, and several solutions for the problem are outlined. Based on the discussion, two approaches in the nanometer era [5] are envisaged: One is high-V_{DD} bulk-CMOS LSIs for low-cost applications, and the other is low-V_{DD} FD-SOI LSIs for high-speed and low-power applications.

5.2. V_t Variation in the Nanometer Era

The V_t variation (ΔV_t) consists of two components [2, 3]; the extrinsic variation due to implant non-uniformities and channel length/width variations, and the intrinsic variation due to random microscopic fluctuations of dopant atoms in the channel area. Here, the standard deviation of the intrinsic V_t variation [3] is given as

$$\sigma(V_t) = \frac{q}{C_{OX}}\sqrt{\frac{N_A \cdot D}{3L \cdot W}} \tag{5.1}$$

where q is the electronic charge, C_{OX} is the gate-oxide capacitance per unit area, N_A is the impurity concentration, D is the depletion layer width under the gate, L is the channel length, and W is the channel width. It is obvious that the variation

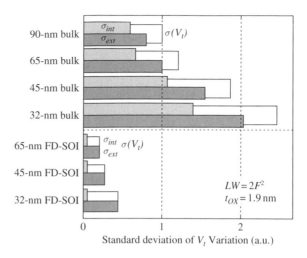

FIGURE 5.1. Standard deviations of V_t variation, $\sigma(V_t)$, and intrinsic (σ_{int}) and extrinsic (σ_{ext}) V_t variations [4, 18]. Reproduced from [4] with permission; © 2006 IEEE.

increases with device scaling because of a smaller LW and the ever-larger N_A for smaller short-channel effects. In practice, for smaller MOSTs, N_A is not uniform in the channel region due to an effect to reduce short channel effects and to suppress the punch-through of nanometer MOSTs. Therefore, empirical data are more useful to evaluate the variation. According to empirical data and device simulations for $2F^2$-size 1.9-nm-t_{OX} MOSTs (F: feature size, t_{OX}: gate oxide thickness) [4], $\sigma(V_t)$ increases with device scaling because the two components, the extrinsic σ_{ext} and intrinsic variation σ_{int}, increase with $\sigma_{ext} > \sigma_{int}$ every generation, as shown in Fig. 5.1. Note that the inter-die V_t variation makes the chip-to-chip characteristics different, while the intra-die V_t variation makes the characteristics of each memory cell and each circuit in a chip different. Here, $\sigma(V_t)$ decreases for fully-depleted SOI (FD-SOI), as shown in the figure, due to an extremely-thin and lightly-doped channel (i.e., small D and N_A).

5.3. Leakage Variations

Subthreshold currents are quite sensitive to V_t and temperature. For example, the leakage of a 1-Mb SRAM array with an average (intra-die) V_t of 0.49 V varies as much as four orders of magnitude for a V_t variation of ±0.1 V and a temperature variation of 100°C, as shown in Fig. 5.2 [5]. Thus, to ensure the lowest necessary retention current for the worst chip in the V_t variation, the V_t of the average chip must be increased to account for V_t variation, causing a degraded voltage margin for a given V_{DD}. Otherwise compensation circuits are needed. This is also the case for peripheral circuits of RAMs.

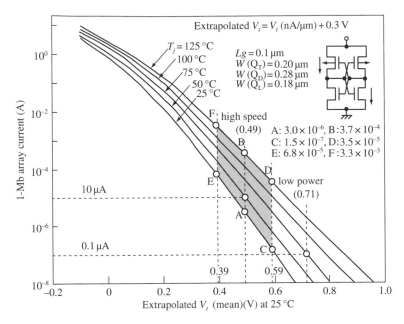

FIGURE 5.2. 1-Mb array current vs. V_t of cross-coupled MOSTs [5].

5.4. Speed Variations of Logic Circuits

The delay τ, which is proportional to $V_{DD}/(V_{DD} - V_t)^{1.25}$ [6], widely varies with voltage and device scaling: For any V_t variation (ΔV_t), the degree of delay variation ($\Delta \tau / \tau$) is expressed as

$$\Delta \tau / \tau = 1.25 \Delta V_t / (V_{DD} - V_{t0}),$$
$$V_t = V_{t0} + \Delta V_t, \tag{5.2}$$

where ΔV_t is the total V_t variation including extrinsic and intrinsic V_t variations, and V_{t0} is the average V_t of the chip. The delay variation is enhanced by two factors, V_{DD} and ΔV_t, in the nanometer era where V_{DD} must be reduced, but ΔV_t becomes larger. In particular, the enhancement becomes prominent as V_{DD} approaches V_{t0}, which does not continue to scale so that subthreshold currents will remain sufficiently small. V_{t0} is usually about 0.3 V for logic circuits. This gives a lower limit for $V_{DD}(V_{min})$ of logic circuits [20]. Note that unless V_{min} is lowered, dynamic voltage scaling [7], in which the clock frequency and V_{DD} vary dynamically in response to the computational load, would be less effective as devices are scaled down. This is due to the ever-smaller difference between V_{DD} and V_{min}.

Figure 5.3 shows delay in a chip versus feature size, F, with assumptions of V_t variations normalized by that for the 90-nm (bulk) generation, and a $6F^2$ for the driver MOST in an inverter. Delay times for $\pm 3\sigma(V_t)$ are normalized by that

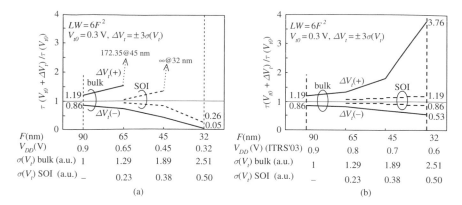

FIGURE 5.3. Speed variations of an inverter for the traditional V_{DD} scaling (a), in which V_{DD} and F are scaled down with the same scaling factor, and the V_{DD} projected by ITRS 2003 [8, 21] (b). The speed is assumed to be proportional to $V_{DD}/(V_{DD} - V_t)^{1.25}$ [6]. Reproduced from [21] with permission; © 2006 IEEE.

for the average V_t (i.e., $V_{t0} = 0.3\,V$) for each generation. Here, $\Delta V_t(+)$ is for $+3\sigma(V_t)$, and $\Delta V_t(-)$ is for $-3\sigma(V_t)$. For the traditional V_{DD}-scaling, in which V_{DD} and F are scaled down with the same scaling factor, as shown in Fig. 5.3(a), the speed spread in a bulk CMOS chip reaches an intolerable level soon, even if such a V_{DD} scaling has been traditional because of advantages of low power and ease of device development. Such a spread is due to un-scalable V_t variations. Even for the more-relaxed V_{DD} scaling projected by ITRS 2003 [8, 21], as shown in Fig. 5.3(b), the speed spread results in an unacceptable increase: For the bulk CMOS, the spread is from 1.19 to 0.86 in the 90-nm generation. However, it rapidly increases with device scaling, reaching as large as 3.76 to 0.53 in the 32-nm generation. A 2.5-time increase in V_t variation and a decrease in V_{DD} from 0.9 V to 0.6 V are responsible for increased speed spread.

5.5. Variations in V_t Mismatch of Flip-Flop Circuits

Voltage margins of flip-flop circuits in RAM chips (i.e., SRAM cells and DRAM sense amplifiers) are strongly affected by V_t variations, as discussed in Chapters 2 and 3. In this section, the variation issue of the V_t-mismatch (δV_t) between a pair of MOSTs is discussed based on the following assumptions.

(1) The inter-die variation is compensated for with V_{BB} control. The intrinsic V_t variation σ_{int} for 45-nm generation is 30 mV [9], and the trend in σ_{int} follows the data shown in Fig. 5.1.
(2) The standard deviation of V_t mismatch is given by $\sigma(\delta V_t) = \sqrt{2}\,\sigma_{int}$.
(3) The thickness of the gate oxide t_{OX} of MOSTs is fixed to 1.9 nm for both SRAM cells and DRAM sense amplifiers, independent of device

scaling. This is because, at present, thin gate materials with a small enough gate-tunneling current are difficult to develop. It is also assumed that DRAMs use dual-t_{OX} technology [10, 11] with thin-t_{OX} MOSTs for sense amplifiers and thicker-t_{OX} MOSTs for memory cells.

(4) The MOST size is $2F^2$ for SRAM cells and $20F^2$ for DRAM sense amplifiers. For DRAMs, the number of cells connected to one sense amplifier is 64–1024, corresponding to a high-speed design with short data lines (HS-DRAM in the following figure) and a low-cost design with long data lines (LC-DRAM), as mentioned in Chapter 2.

(5) The voltage margins of SRAM cells and DRAM sense amplifiers are determined by only the cross-coupled nMOSTs. This assumption is correct for the sense amplifiers because of the preceding activation of an nMOST amplifier, as discussed in Chapter 2. For SRAM cells, however, it underestimates detrimental effects from other MOSTs in a cell due to the static and ratio operations.

Figure 5.4 shows the comparison of $\sigma(\delta V_t)$ of SRAM cells and DRAM sense amplifiers. The mismatch of the former is about three times that of the latter due to the smaller MOST size. The maximum V_t mismatch $|\delta V_t|_{max}$, however, depends not only on the device parameters, but also on the number of MOSTs N used in a chip. The ratio $m = |\delta V_t|_{max}/\sigma(\delta V_t)$ increases with N, and its expectation is expressed by (see Section A5.1 for the detailed calculation)

$$m = \int_0^\infty \left[1 - \left\{ \frac{1}{\sqrt{2\pi}} \int_{-x}^x \exp\left(-\frac{u^2}{2}\right) du \right\}^N \right] dx. \tag{5.3}$$

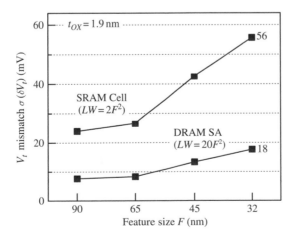

FIGURE 5.4. Trend in standard deviation of V_t-mismatch for the bulk CMOS.

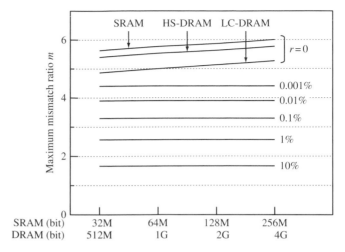

FIGURE 5.5. Maximum V_t-mismatch ratio m. Maximum V_t-mismatch $|\delta V_t|_{max}$ is m times standard deviation.

The calculated ratio m is shown by the lines "$r = 0$" in Fig. 5.5. The ratio m of SRAM cells is slightly larger than that of DRAM sense amplifiers. This is because N corresponds to the number of cells in a SRAM, while it is the number of sense amplifiers in a DRAM, which is much smaller than the number of SRAM cells. The resultant maximum V_t mismatch $|\delta V_t|_{max} = m\sigma(\delta V_t)$ of DRAM sense amplifiers and that of SRAM cells are shown in Fig. 5.6(a) and (b), respectively. The mismatch increases about 2.5 times with feature-size scaling from 90 nm to 32 nm, reaching as much as 100 mV for a 4-Gb DRAM and 330 mV for a 256-Mb SRAM. The δV_t of DRAM sense amplifiers is smaller than that of SRAM cells. However, it seriously affects the S/N and sensing speed

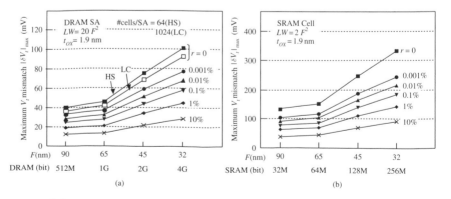

FIGURE 5.6. Trend in maximum V_t-mismatch for the bulk CMOS with a parameter of repairing percentage r: (a) DRAMs and (b) SRAMs. Reproduced from [20] with permission; © 2006 IEEE.

of the amplifiers, especially in the case of the mid-point sensing, as discussed in Chapter 2. In SRAMs, the minimum operating voltage V_{min} must be higher than $V_t + |\delta V_t|_{max}$ for a turned-on MOST to drive a data line. Thus, it is as high as 0.82 V even for a high-speed 256-Mb SRAM necessitating a 0.49-V V_t (see Fig. 3.4), while it is 1.04 V for a low-power design necessitating 0.71-V V_t. One method to solve the V_t-mismatch problem of DRAM sense amplifiers is the mismatch-compensation circuit technique [12, 13], which, however, causes area and access overheads. Therefore, a redundancy technique and/or an on-chip error checking and correcting (ECC) are needed, as described in the next section.

5.6. Solutions for the Reductions

5.6.1. Redundancy and ECC

Redundancy and/or on-chip ECC circuits are essential in avoiding DRAM sense amplifier and SRAM cell failures caused by excessive intra-die V_t mismatches and in allowing for the minimum V_{DD} for successful operations to be lowered (see Fig. 3.25). Repairing a small number of sense amplifiers or memory cells reduces ratio m described in Section 5.5. The relationship between m and repairable percentage r is derived as follows.

Let us consider an N-bit SRAM and assume that the V_t mismatch, δV_t, of each memory cell follows a Gaussian distribution with a mean of zero and a standard deviation of σ. The probability density function is expressed as

$$f(\delta V_t) = \frac{1}{\sqrt{2\pi}\sigma} \exp\left(-\frac{\delta V_t^2}{2\sigma^2}\right). \tag{5.4}$$

Probability p of a memory cell with an excessive (larger than $m\sigma$) δV_t is calculated as

$$p = P(|\delta V_t| > m\sigma) = \int_{-\infty}^{-m\sigma} f(x)\,dx + \int_{m\sigma}^{\infty} f(x)\,dx$$

$$= 1 - \frac{1}{\sqrt{2\pi}} \int_{-m}^{m} \exp\left(-\frac{u^2}{2}\right) du. \tag{5.5}$$

Therefore pN cells are expected to have excessive δV_t's. If all these cells are repaired by redundancy and/or ECC (that is, $r = p$), the maximum V_t mismatch $|\delta V_t|_{max}$ is limited to $m\sigma$, because the remaining cells have smaller δV_t's. The relationship between m and r and the resulting $|\delta V_t|_{max}$ are shown in Figs. 5.5 and 5.6, respectively. Unlike the no-repair case ($r = 0$), m is independent of N, because equation (5.5) does not contain N, while equation (5.3) does. For example, if 0.1% of memory cells are repaired, $|\delta V_t|_{max}$ is limited to 3.3σ. The above discussion is also applied to DRAM-sense-amplifier failures, if N memory cells are interpreted as N sense amplifiers.

Since DRAM sense amplifiers with excessive δV_t's result in column defects [22], they can be efficiently repaired by column redundancy with a small-area penalty. Only 0.1% of redundant columns and 4096 sets of storages for storing defective addresses achieve $r = 0.1\%$ for a 4-Gbit LC-DRAM, extending the memory-capacity limitation by 1–2 generations as shown in Fig. 5.6(a). For HS-DRAMs, however, the required number of storages to achieve as much r is larger (65536 sets for $r = 0.1\%$ of a 4-Gbit HS-DRAM). For SRAMs, the required number of storages is larger still (262144 sets for $r = 0.1\%$ of a 256-Mbit SRAM), because SRAM cells with excessive δV_t's cause random bit defects. It is impractical to repair all such random defects by redundancy because the area for programmable storages such as fuses and antifuses would have to be too large. An on-chip ECC circuit [14, 15] is an efficient method for repairing random defects because it requires no programmable storages. One problem with ECC is the possibility that multiple defects that cannot be corrected by a single-error correction code will occur. However, combining ECC with redundancy drastically increases the number of repairable bits due to the combinations's synergistic effect [14, 19]. Most defects, including hard defects and excessive δV_t's, are repaired by ECC, while a few blocks with multiple defects are repaired by redundancy. Therefore, the combination of ECC and redundancy will be the best solution for the V_t-mismatch problem of future HS-DRAMs and SRAMs.

5.6.2. Symmetric Layouts for Flip-Flop Circuits

Stringent controls of channel length and width, and the use of the largest MOSTs possible are effective to reduce the V_t variation and V_t mismatch. Enlarging MOSTs, however, is fatal for a large-capacity SRAM because of increased SRAM cell area, but it can be tolerated for DRAM sense amplifiers, as discussed in Chapters 2 and 3. Lithographical symmetric layouts for flip-flop circuits (see Figs. 2.19 and 3.13) are indispensable to reduce the V_t mismatch.

5.6.3. Controls of Internal Supply Voltages

Suppression of, and compensation for variations of design parameters (e.g., V_t, V_{DD}, and temperature) through controls of internal voltages with on-chip voltage converters are useful. Controlling the substrate bias voltage (V_{BB}) of sense amplifiers, SRAM cells, and other periphery circuits is thus effective, as exemplified by a MPU [16] and the DRAM discussed next. The controls are useful for inter-die V_t variation but ineffective for intra-die V_t variation.

An nMOS body-bias generator [17] shown in Fig. 5.7(a) was applied to a 0.6-V 16-Mb DRAM. The current (I_{DS}) through the MOST(M_1) is a good indicator of both MOST-OFF current and the switching speed of peripheral circuits. The V_{GS} is $V_{DD}/2$ and is set to approximately V_t, so the current is sensitive to V_{DD} and the body bias (V_{NB}). The drain voltage (V_D) is compared to $V_{DD}/2 + \Delta v$ and $V_{DD}/2 - \Delta v$ using two OP-amps (i.e., comparators) to determine whether the body bias (V_{NB}) should be increased or decreased. When the V_t of peripheral circuits is low due to fast process conditions or high temperatures to the extent of $V_D < V_{DD}/2 - \Delta v$, the lower OP-amp senses the reduced V_D so that PUMP (i.e., on-chip charge pump) starts the built-in ring oscillator to oscillate if M_{P2} is on. Thus, V_{NB} starts to decrease

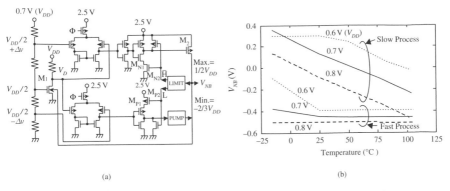

FIGURE 5.7. Body-bias generator for nMOSTs (a) and generated V_{NB} (b) [17].

so that the V_t is increased to compensate. The oscillations continue until the resulting deep V_{NB} increases the V_t to a point where the OP-amp turns M_{P1} off with $V_D > V_{DD}/2 - \Delta v$. This is true as long as the deep V_{NB} does not exceed the acceptable lower limit for V_{NB}. Once V_{NB} exceeds the lower limit for V_{NB} on the way to the deep V_{NB}, LIMIT detects the level of the lower limit for V_{NB} and turns M_{P2} off with the "H" output to inhibit oscillation. The upper OP-amp works in the same manner, comparing V_D to $V_{DD}/2 + \Delta v$. However, when V_t is high due to slow process conditions or low temperatures, the lower OP-amp disables PUMP while the upper OP-amp discharges the M_2 gate for driving the body. Thus, the V_t of nMOSTs is reduced and compensated for. Another feature of the generator is that as V_{DD} increases, body bias V_{NB} becomes negative to raise V_t and reduce standby power with reduced subthreshold current. The positive body bias is limited to a maximum of $V_{DD}/2$ to limit the pn-leakage current into the body-to-source/drain junctions. To be more exact, it is carefully set to about 0 V at 100°C and to below 0.3 V at 0°C not only to reduce a large pn-leakage current, which sharply increases with the forward biasing and temperature, but also to eliminate detrimental effects due to parasitic bipolar transistors that develop more easily at forward bias, as discussed in Chapter 1 (see Fig. 1.18). In addition, the negative body bias is limited to $-2/3V_{DD}$ to avoid over-stressing the MOSTs. Figure 5.7(b) plots V_{NB} versus temperature as a function of V_{DD}. The pMOST body bias generator is a complementary version of the nMOST generator. It has been reported that the negative body bias reduced subthreshold currents by 75% under fast process conditions, and positive body bias improved the speed by 63% under slow process conditions. These significant improvements suggest that operations are very sensitive to substrate noise. Thus, a triple-well structure for isolating the floating DRAM array substrate from floating p/n peripheral substrates, coupled with the mid-point sensing, is necessary for quiet substrates, as discussed in Chapter 2.

Controlling the internal V_{DD} with an on-chip voltage down-converter also compensates for the speed variations without any instability caused by V_{BB} controls. However, it does not compensate for variations in the subthreshold current that is almost independent of V_{DD}. In addition, it entails a power loss at the converter, as discussed in Chapters 1 and 7.

5.6.4. Raised Power Supply Voltage

When intra-die variations in V_t and V_t mismatch become large, each cell, each sense amplifier, and each logic circuit in a chip can have a different speed or voltage margin. Thus, reliable LSIs designs become impossible because the V_{BB} or V_{DD} described above control cannot correct for intra-die V_t variation, and redundancy and ECC are ineffective for logic gates. The best way to cope with these problems is to raise the supply voltage V_{DD}, although the voltage scaling and low-power merits that we take for granted today will be lost. For example, to confine the speed spread in a chip to a tolerable level, V_{DD} must be raised in accordance with the ever-increasing V_t variation. Figure 5.8 shows the expected minimum V_{DD} (i.e., V_{min}) of the logic circuit that is defined as the V_{DD} for a fixed value of the speed ratio, $\Delta\tau$ (i.e., delay for $+3\sigma(V_t)$ normalized by the average delay) [20]. In practice, speed variations in a chip are much larger, since another speed variation coming from $-3\sigma(V_t)$, as shown in Fig. 5.3, must be taken into account. In any event, for the bulk, the V_{min} gradually increases along with each $\Delta\tau$ caused by device scaling, reaching as high as 1.74 V in the 32-nm generation for an acceptable $\Delta\tau$ of 1.2. Instead, however, stress voltage-immune MOSTs must be developed, with special attention paid to the resulting detrimental effects. For example, such high V_{DD}s may increase substrate currents, as shown in Eq. (1.15), to the extent that an on-chip V_{BB} generator cannot

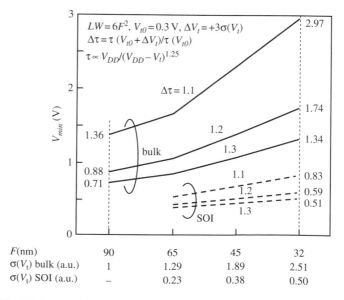

FIGURE 5.8. Minimum V_{DD} (V_{min}) of the logic circuit. Reproduced from [20] with permission; © 2006 IEEE.

manage any more, eventually degrading V_{BB} levels. Another serious problem is that ever-larger V_{DD} causes ever more serious power crises. Thus, new MOSTs that can sufficiently reduce V_t variation are urgently needed.

5.6.5. Fully-Depleted SOI

An ultra-thin BOX double-gate fully-depleted (FD) SOI [4, 18], discussed in Chapters 2 and 3 (see Figs. 2.28 and 3.19), is reportedly a strong candidate for coping with the problems in the nanometer era despite the expense incurred in producing the wafers. One of this design's main advantage is its drastic reductions of the intrinsic V_t variation (Fig. 5.1) dramatically reduce V_t-mismatch, widing the voltage margin of flip-flop circuits such as SRAM cells and DRAM sense amplifiers. Reductions in the total V_t variation also reduce variations in subthreshold currents and speed variations. It should be noted that even in the 32-nm generation the speed spread of the inverter remains in the same range as that for the 90-nm bulk CMOS, as shown in Fig. 5.3(b). This implies that the SOI would extend the low-voltage limitation of bulk CMOSs by at least three generations. This is true for a relatively high V_{DD}. If V_{DD} is further scaled down, as in Fig. 5.3 (a), however, not even the SOI will be able to manage the speed variations, which means new circuits will be required. For a given speed variation, the FD-SOI also reduces the V_{min} of the logic circuit, as shown in Fig. 5.8. For example, for $\Delta\tau = 1.2$ it reduces the V_{min} to one-third of that for most chips in the 32-nm generation (i.e., from 1.74 V to 0.59 V), enabling chips to reduce power dissipation by 90%.

Based on the discussion so far, two approaches in the nanometer era have been envisaged [5]: one is high-V_{DD} bulk-CMOS LSIs for low-cost applications, and the other is low-V_{DD} FD-SOI LSIs for high-speed, low-power applications. A low-cost FD-SOI process would be a unified process technology.

A5.1. Derivation of maximum V_t mismatch

Let us assume the threshold-voltage mismatch δV_t is expressed as a Gaussian distribution with mean value of zero and standard deviation of σ. The probability distribution function is given by

$$f(\delta V_t) = \frac{1}{\sqrt{2\pi}\sigma} \exp\left(-\frac{\delta V_t^2}{2\sigma^2}\right) \tag{A5.1}$$

(see Fig. 5.9). The probability of the δV_t of a memory cell being within $\pm x\sigma$ is the hatched area of Fig. 5.9, and is expressed as

$$Y_1(x) = \int_{-x\sigma}^{x\sigma} f(t)\,dt = \frac{1}{\sqrt{2\pi}\sigma} \int_{-x\sigma}^{x\sigma} \exp\left(-\frac{t^2}{2\sigma^2}\right) dt$$

$$= \frac{1}{\sqrt{2\pi}} \int_{-x}^{x} \exp\left(-\frac{u^2}{2}\right) du. \tag{A5.2}$$

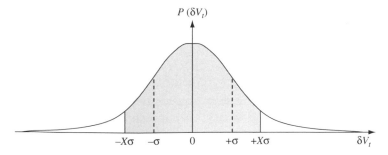

FIGURE 5.9. Distribution of V_t-mismatch δV_t.

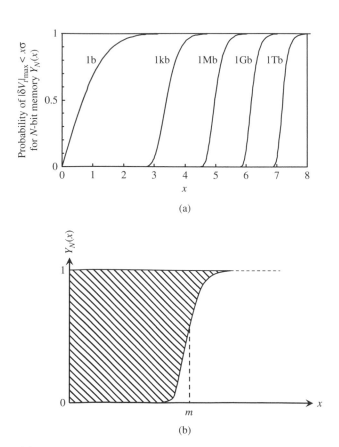

FIGURE 5.10. (a) Probability of the maximum V_t-mismatch of an N-bit memory, $|\delta V_t|_{max}$, being within $x\sigma$, and (b) calculation of the expectation of $|\delta V_t|_{max}$.

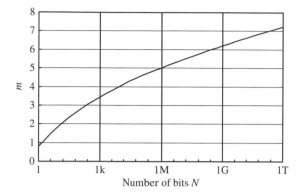

FIGURE 5.11. Relationship between m and the number of bits N. Here, the expectation of $|\delta V_t|_{\max}$ is equal to $m\sigma$.

The probability of all the δV_t's of N memory cells being within $\pm x\sigma$, that is, the probability of $|\delta V_t|_{\max} < x\sigma$ is given by

$$Y_N(x) = \{Y_1(x)\}^N. \tag{A5.3}$$

Plotting equation (A5.3) with N as a parameter results in Fig. 5.10(a). If we denote the expectation of $|\delta V_t|_{\max}$ as $m\sigma$, m is the area of lefthand side of this curve as shown in Fig. 5.10(b), and is expressed as

$$m = \int_0^\infty \{1 - Y_N(x)\}\, dx = \int_0^\infty \left[1 - \left\{ \frac{1}{\sqrt{2\pi}} \int_{-x}^x \exp\left(-\frac{u^2}{2}\right) du \right\}^N \right] dx. \tag{A5.4}$$

Figure 5.11 shows the relationship between m and the number of bits N.

References

[1] T.C. Chen, "Where CMOS is going: trendy hype vs. real technology," ISSCC Dig. Tech. Papers, pp. 22–28, Feb. 2006.
[2] K. Itoh, *VLSI Memory Chip Design*, Springer-Verlag, NY, 2001.
[3] Y. Taur, D.A. Buchanan, W. Chen, D.J. Frank, K.E. Ismail, S.-H. Lo, G.A. Sai-Halasz, R. G. Viswanathan, H.-J. C. Wann, S.J. Wind, and H.-S. Wong, "CMOS scaling into the nanometer regime," Proc. IEEE, vol. 85, pp. 486–504, April 1997.
[4] M. Yamaoka, K. Osada, R. Tsuchiya, M. Horiuchi, S. Kimura, and T. Kawahara, "Low power SRAM menu for SOC application using yin-yang-feedback memory cell technology," Symp. VLSI Circuits Dig. Tech. Papers, pp. 288–291, June 2004.
[5] K. Itoh, K. Osada, and T. Kawahara, "Reviews and future prospects of low-voltage embedded RAMs," CICC Dig. Tech. Papers, pp.339–344, Oct. 2004.

[6]   K. Chen, C. Hu, P. Fang, M. R. Lin, and D. L. Wollesen, "Predicting CMOS speed with gate oxide and voltage scaling and interconnect loading effects," IEEE Trans. Electron Devices, Vol. 44, no. 11, pp. 1951–1957, Nov. 1997.

[7]   T. Burd, T. Pering, A. Stratakos and R. Brodersen, "A dynamic voltage scaled microprocessor system," ISSCC Dig. Tech. Papers, pp. 294–295, Feb. 2000.

[8]   ITRS2003 Executive Summary Table 6a, p.57.

[9]   A. Agarwal, B. C. Paul and K. Roy, "Process variation in nano-scale memories: failure analysis and process tolerant architecture," Proc. CICC, pp. 353–356, Oct. 2004.

[10] A. Yamazaki, T. Fujino, K. Inoue, I. Hayashi, H. Noda, N. Watanabe, F. Morishita, J. Ootani, M. Kobayashi, K. Dosaka, Y. Morooka, H. Shimano, S. Soeda, A. Hachisuka, Y. Okumura, K. Arimoto, S. Wake and H. Ozaki, "A 56.8GB/s 0.18um embedded DRAM macro with dual port sense amplifier for 3D graphic controller," ISSCC Dig. Tech. Papers, pp. 394–395, Feb. 2000.

[11] O. Takahashi, S. H. Dhong, M. Ohkubo, S. Onishi, R. H. Dennard, R. Hannon, S. Crowder, S. S. Iyer, M. R. Wordeman, B. Davari, W. B. Weinberger and N. Aoki, "1-GHz fully pipelined 3.7-ns address access time 8 k x 1024 embedded synchronous DRAM macro," IEEE J. Solid-State Circuits, vol. 35, pp. 1673–1679, Nov. 2000.

[12] S. Hong, S. Kim, J.-K. Wee and S. Lee, "Low-voltage DRAM sensing scheme with offset-cancellation sense amplifier," IEEE J. Solid-State Circuits, vol. 37, pp. 1356–1360, Oct. 2002.

[13] J. Y. Sim, K. W. Kwon, J. H. Choi, S. H. Lee, D. M. Kim, H. R. Hwang, K. C. Chun, Y. H. Seo, H. S. Hwang, D. I. Seo, C. Kim and S. I. Cho, "A 1.0V 256Mb SDRAM with offset-compensated direct sensing and charge-recycled precharge schemes," ISSCC Dig. Tech. Papers, pp. 310–311, Feb. 2003.

[14] H. L. Kalter, C. H. Stapper, J. E. Barth Jr., J. DiLorenzo, C. E. Drake, J. A. Fifield, G. A. Kelley Jr., S. C. Lewis, W. B. van der Hoeven and J. A. Yankosky, "A 50-ns 16-Mb DRAM with a 10-ns data rate and on-chip ECC," IEEE J. Solid-State Circuits, vol. 25, pp. 1118–1128, Oct. 1990.

[15] K. Osada, Y. Saito, E. Ibe and K. Ishibashi, "16.7-fA/cell tunnel-leakage-suppressed 16-Mb SRAM for handling cosmic-ray-induced multierrors," IEEE J. Solid-State Circuits, vol. 38, pp. 1952–1957, Nov. 2003.

[16] M. Miyazaki, G. Ono, T. Hattori, K. Shiozawa, K. Uchiyama, and K. Ishibashi, "1000-MIPS/W microprocessor using speed-adaptive threshold-voltage CMOS with forward bias," ISSCC Dig. Tech. Papers, pp. 420–421, Feb. 2000.

[17] K. Hardee, F. Jones, D. Butler, M. Parris, M. Mound, H. Calendar, G. Jones, L. Aldrich, C. Gruenschlaeger, M. Miyabayashi, K. Taniguchi, and T. Arakawa, "A 0.6V 205MHz 19.5ns tRC 16Mb embedded DRAM," ISSCC Dig. Tech. Papers, pp. 200–201, Feb. 2004.

[18] R. Tsuchiya, M. Horiuchi, S. Kimura, M. Yamaoka, T. Kawahara, S. Maegawa, T. Ipposhi, Y. Ohji, and H. Matsuoka, "Silicon on thin BOX: a new paradigm of the CMOSFET for low-power and high-performance application featuring wide-range back-bias control," IEDM Dig. Tech. Papers, pp.631–634, Dec. 2004.

[19] C. H. Stapper and H.-S. Lee, "Synergistic fault-tolerance for memory chips," IEEE Trans. Computers, vol. 41, pp. 1078–1087, Sep. 1992.

[20] K. Itoh, M. Yamaoka, and T. Kawahara, "Impact of FD-SOI on deep-sub-100-nm CMOS LSIs–a view of memory designers–," IEEE Intl. SOI Conference Dig. Tech. Papers, pp.103–104, Oct. 2006.

[21] K. Itoh, M. Horiguchi, and T. Kawahara, "Ultra-low voltage nano-scale embedded RAMs," ISCAS Proceedings, pp.25–28, May 2006.

[22] A. Yamazaki, F. Horishita, N. Watanabe, T. Amano, M. Haraguchi, H. Noda, A. Hachisuka, K. Dosaka, K. Arimoto, S. Wake H. Ozaki and T. Yoshihara, "A study of sense-voltage margins in low-voltage-operating embedded DRAM Macros," IEICE Trans. Electron., vol. E88-C, pp. 2020–2027, Oct. 2005.

6
Reference Voltage Generators

6.1. Introduction

On-chip voltage converters are becoming increasingly important for ultra-low voltage nano-scale memories, as discussed in Chapter 1. They include reference generators, voltage down-converters, voltage up-converters and negative voltage generators using charge pump circuits, and level shifters. In particular, the reference voltage (V_{REF}) generator must create a well-regulated supply voltage for any variations of voltage, process, and temperature, because other converters operate based on the reference voltage. The V_{REF} generator is categorized as threshold-voltage (V_t)-referenced V_{REF} generator, V_t-difference (ΔV_t) V_{REF} generator, and the band-gap reference (BGR) generator. The characteristics of each circuit are summarized in Table 6.1. The V_t-referenced V_{REF} generator is easy to design. However, its output voltage has a large temperature coefficient and suffers from process fluctuation. Since MOST V_t has a negative temperature coefficient around $-2\,\text{mV}/°\text{C}$, the V_t shift is as large as 0.33 V in a junction-temperature range from $-40°\text{C}$ to $125°\text{C}$. Therefore, it is necessary to combine V_t with a physical parameter that has a positive temperature coefficient, as will be described in Section 6.2. In addition, chip-by-chip voltage trimming is indispensable to compensate for the process fluctuation. A $\Delta V_t\ V_{REF}$ generator operates by extracting the V_t difference of two MOSTs to allow the temperature coefficient to be almost cancelled out. However, the dependency on process fluctuation is large, and voltage trimming is still needed for mass production. Moreover, an additional process step to fabricate a low/high-V_t MOST is required. The BGR utilizes the base-emitter voltages V_{BE} of bipolar transistors (forward voltages of diodes). It features a small dependency on temperature or process fluctuation. Parasitic bipolar transistors fabricated with CMOS process, instead of a bipolar process, are usually used for process compatibility. A drawback of the BGR circuit is the limitation on the external supply voltage V_{DD}. Since the output voltage is inevitably the Si-bandgap voltage ($\sim 1.2\,\text{V}$), V_{DD} must be higher than 1.2 V. Some solutions to overcome this problem are proposed as will be described in Section 6.4.

This chapter describes their characteristics in detail from view point of memory circuit designers, including the burn-in test.

TABLE 6.1. Comparison of reference voltage generators.

	MOST V_t	MOST ΔV_t	Bandgap Ref.
Temperature Dependency	Large	Small	Small
Process Dependency	Large	Large	Medium - Small
Additional Process	None	Low/high-V_t MOST	None (Triple well)
Output Voltage	mV_t	$m\Delta V_t$	1.2 V
Minimum Operating Voltage	$mV_t + \alpha$	$V_{tn} + |V_{tp}| + \alpha$	1.2 V $+\alpha$

6.2. The V_t-Referenced V_{REF} Generator

Figure 6.1 shows a V_{REF} generator that consists of diode-connected nMOSTs for obtaining quite a high V_{REF}, such as 4 V. If body-bias effect is ignored, V_{REF} is equal to the product of the number of nMOSTs m and V_t. Although the generator is simple, it suffers from drawbacks such as V_{REF} setting inaccuracy: the variation in V_{REF} depends strongly on that in V_t, caused by variations in the fabrication process. The strong dependence on temperature is also serious. If V_{REF} is generated by a series connection of six nMOSTs, the variation in V_{REF} is six times that in V_t. For example, a temperature variation of 100 °C causes a variation in V_{REF} as large as 1.2V for a V_t temperature coefficient of about -2mV/°C.

Figure 6.2 shows another V_t-referenced generator [1] to obtain a low V_{REF}. The circuit configuration is similar to that of the voltage converter/trimming circuit that will be described in section 6.4. The output voltage V_{REF} is divided by the resistor R and by the on resistance of M_2, R_{C2}, and fed back to the differential amplifier A. The temperature-dependence of V_{REF} can be canceled if poly-Si is used for resistor R. The cancellation is explained by the following equation. If the open-loop gain of the two-stage amplifier composed of differential amplifier

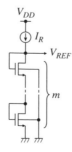

FIGURE 6.1. Reference voltage generator based on nMOST threshold voltage.

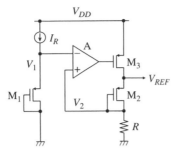

FIGURE 6.2. Reference voltage generator based on threshold voltage with improved temperature coefficient [1].

A and M_3 (common-source stage) is sufficiently large, the voltages V_1 and V_2 are almost equal. Therefore,

$$|V_{tp}| = V_2 = \frac{R}{R + R_{C2}} \cdot V_{REF} \qquad (6.1)$$

stands. The output voltage V_{REF} is expressed by

$$V_{REF} = |V_{tp}| \left(1 + \frac{R_{C2}}{R}\right). \qquad (6.2)$$

Here, the poly-Si resistance features no temperature dependence [2] if the sheet resistivity is around $50\Omega/\square$, while R_{C2} and $|V_{tp}|$ have positive and negative temperature coefficients, respectively. Thus, the temperature dependence is canceled. The whole voltage-down converter implementing the generator was designed for a 16-Mb DRAM [1]. The resultant performances were $\Delta V_{INT}/\Delta T =$ 0.9mV/°C, $\Delta V_{INT}/\Delta R = 18.6$mV/$\Omega$, $\Delta V_{INT}/\Delta V_{tp} = 1.5$, and $\Delta V_{INT}/\Delta V_{DD} =$ 0.07 for $V_{DD} = 5$V and $V_{INT} = 4$V. Standby current of the converter is 35μA, and the layout area of 400kΩ with 50-Ω/$\square$ resistivity is less than 0.1% of a 16-Mb chip area of 128mm². The drawback of the generator is a large variation of V_{REF} caused by variations of $|V_{tp}|$ and R.

Figure 6.3 shows a V_{REF} generator [3] based on the same principle without a differential amplifier. It is suitable for obtaining a low V_{REF} through automatically adjusting the gate-source bias voltage of a pMOST (M_1). If the parameters of R_1, M_1 and M_3 are set so that the voltage drop ($I_2 R_1$) is larger than the M_1 threshold voltage ($|V_{tp}|$), the current I separates into I_1 and I_2, as a result of M_1 being turned on. Any change in V_{REF} is eventually suppressed by the negative feedback of the circuit. For example, a positive change at V_{REF} is detected by M_2, enabling the M_1-gate voltage to drop and thus to increase I_1. I_2 is decreased, and the resulting reduced $I_2 R$-drop prevents M_2 from delivering more current. The generator can reduce the temperature dependence of V_{REF} by utilizing opposite

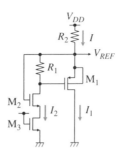

FIGURE 6.3. Reference voltage generator based on threshold voltage with improved temperature coefficient. Reproduced from [3] with permission; © 2006 IEEE.

temperature-coefficients of $|V_{tp}|$ and an equivalent resistance (R_{C23}) formed by M_2 and M_3, as explained previously. This is because V_{REF} is expressed as

$$V_{REF} = |V_{tp}| \left(1 + \frac{R_{C23}}{R_2}\right).$$ (6.3)

The V_{REF} generator provides excellent performance of an almost constant V_{REF} of 1.3V over a wide range of V_{DD} of 2–6V, $\Delta V_{REF}/\Delta T = 0.2\text{mV}/°\text{C}$, and current dissipation of $2\,\mu\text{A}$ at $V_{DD} = 2.5\text{V}$. The drawback, however, is a large V_{REF} variation caused by variations in $|V_{tp}|$ and R_{C23}/R.

Figure 6.4 is a current-mirror V_{REF} generator [4, 5]. A current mirror consisting of M_3 and M_4 allows the same bias-current (I) to flow to M_1 and R_1. If W/L of M_1 is large enough and I is small enough, the drop IR_1 is almost equal to $|V_{tp}|$. Therefore

$$I = \frac{|V_{tp}|}{R_1}.$$ (6.4)

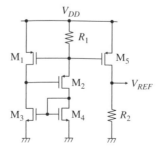

FIGURE 6.4. Reference voltage generator based on threshold voltage with improved temperature coefficient. Reproduced from [4] with permission; © 2006 IEEE.

Since the same current (I) flows to M_5, which is of the same size as M_1, V_{REF} is expressed as

$$V_{REF} = \frac{R_2}{R_1} \cdot |V_{tp}|. \qquad (6.5)$$

The variation in V_{REF} due to variation in V_{tp} is also suppressed by means of the laser trimming of R_1 and R_2. Therefore, the voltage conversion/trimming circuit described in Section 6.5 is not necessary. In addition, V_{REF} is less sensitive to temperature variation by selecting the optimum temperature coefficients of R_1 and R_2, so that the temperature dependence of V_{tp} is canceled by that of R_2/R_1. In fact, the V_{REF} generator for a 16-Mb DRAM [4] utilizes data-line polycide and doped-poly-Si as the materials for R_1 and R_2, respectively. The circuit revealed excellent experimental results: $V_{REF} = 1.935V$ at $V_{DD} = 5$ V and 25 °C, $\Delta V_{REF}/\Delta T = +0.15\text{mV}/°C$, and $\Delta V_{REF}/\Delta V_{DD} = +10\text{mV/V}$.

In summary, the V_t-referenced V_{REF} generator suffers from variation in V_{REF} caused by variation in V_t, though additional process steps are not needed.

6.3. The V_t-Difference (ΔV_t) V_{REF} Generator

6.3.1. Basic ΔV_t V_{REF} Generator

Figure 6.5 shows a V_{REF} generator that extracts the V_t difference (ΔV_t) between enhancement nMOSTs (EMOSTs) $M_{11} - M_{1m}$ and depletion nMOSTs (DMOSTs) $M_{21} - M_{2m}$ [6]. A small ΔV_t is favorable because a too large ΔV_t requires a large amount of channel implantation for DMOSTs and causes the increased leakage current. The circuit outputs m times of ΔV_t as output V_{REF}, that is,

$$V_{REF} = m(V_{te} - V_{td}) = m(V_{te} + |V_{td}|), \qquad (6.6)$$

where V_{te} (> 0) and V_{td} (< 0) are threshold voltages of EMOSTs and DMOSTs, respectively.

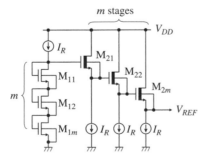

FIGURE 6.5. Reference voltage generator based on threshold-voltage difference between enhancement and depletion MOSTs [6].

(a)

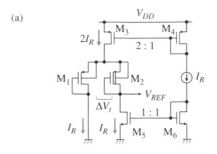

(b)

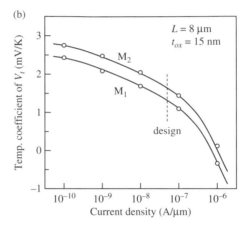

FIGURE 6.6. Reference voltage generator based on threshold-voltage difference between pMOSTs; principle (a), and measured temperature coefficient of threshold voltages (b). Reproduced from [7] with permission; © 2006 IEEE.

Figure 6.6(a) shows a V_{REF} generator [7] that utilizes the difference in V_t (ΔV_t) between two MOSTs (M_1, M_2). The ΔV_t is extracted by driving the MOSTs with differential current sources $2I_R$ and I_R. V_{REF} is less sensitive to variations in V_{DD} and temperature because it is determined only by ΔV_T. In addition, V_{REF} is immune to voltage fluctuations of the p-substrate and V_{DD} since pMOSTs in n-well are used, and their sources are not directly connected to V_{DD}. In the generator V_{REF} is equal to $\Delta V_t (= |V_{tp1}| - |V_{tp2}|)$ if two pMOSTs (M_1, M_2) are the same in size and the same current (I) flows. The output voltage and its temperature dependence are expressed as follows:

$$V_{REF} = |V_{tp1}| - |V_{tp2}|, \tag{6.7}$$

$$\frac{dV_{REF}}{dT} = \frac{d}{dT}\left(|V_{tp1}| - |V_{tp2}|\right) = \frac{k}{q}\ln\frac{N_1}{N_2} \tag{6.8}$$

where N_1 and N_2 are impurity concentrations of M_1 and M_2, respectively. To reduce the temperature dependence, a smaller N_1/N_2 (i.e. a smaller ΔV_t) is preferable, which inevitably causes a small V_{REF}. For example, V_{REF} is as small as 1.1V for $N_1/N_2 = 100$. Although the temperature-dependence of V_{tp} depends on the current density [8], the difference between M_1 and M_2 (i.e. temperature dependence of V_{REF}) is a constant as small as 0.4mV/°C, independent of the current density, as shown in Fig. 6.6(b).

The circuit shown in Fig. 6.7 [9] uses the above V_t-difference (ΔV_{tp}) scheme. It consists of a bias-current circuit, V_{REFN} generator for normal operation to keep V_{REFN} constant, and V_{REFB} generator for burn-in test to keep $V_{DD}-V_{REFB}$ constant as will be described in Section 6.5. The bias-current circuit composed of current-mirror circuits gives both V_{REFN} and V_{REFB} generators bias currents (I_R, $2I_R$), enabling the same current (I_R) to flow in M_1, M_2, and M_6-M_9. Here, $V_{DD}-V_{REFB}$ is equal to twice ΔV_{tp} because of the series connection of two same MOSTs. The doubled ΔV_{tp} is convenient for use in the trimmer. The V_{REFN} generator has only one enhanced-V_t (M_1) and one normal-V_t (M_2) MOST so that the circuit can operate in the low-V_{DD} region. This circuit uses a cascode-type circuit (M_3, M_4) for the current source $2I_R$ to reduce the voltage dependency on V_{DD} in normal mode. M_4 improves the constant-current characteristics of M_3 with suppressing dependence of the current on the drain-source voltage [5]. On the other hand, a simple circuit (M_5) is sufficient for the current source I_R, because the drain-source voltage of M_5 is almost constant.

Figure 6.8 shows a current-mirror V_{REF} generator [10] that utilizes a V_t-difference (ΔV_t). The circuit consists of a V_t-difference-to-current ($\Delta V_t - I$) converter, and a current-to-voltage (I-V) converter. It features a direct conversion from a small ΔV_t to a large V_{REF} without the voltage converter/trimming circuit described in Section 6.5. The MOSTs (M_1-M_4) and resistor R_R generate the constant current (I_R), and M_5 converts I_R to I_L. Then, resistor R_L convert I_L to V_{REF} that is a constant voltage determined only by the MOST-size ratio (W_5/W_1), resistance ratio (R_L/R_R) and ΔV_t. The detail is given in the following. If M_3 and M_4 have the same size and V_t, a current-mirror comprised by M_3 and M_4 allows the current (I_R) flowing from M_2 to M_4 to be the same as that flowing

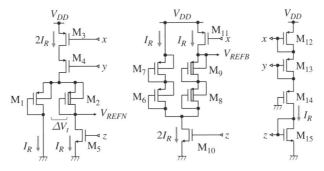

FIGURE 6.7. Reference voltage generator based on threshold-voltage difference between pMOSTs including burn-in voltage generator. Reproduced from [9] with permission; © 2006 IEEE.

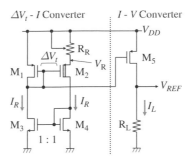

FIGURE 6.8. Reference voltage generator based on threshold-voltage difference between pMOSTs [10].

from M_1 to M_3. Since the saturation current of M_1 is the same as that of M_2,

$$\frac{\beta_1}{2}\left(V_{DD} - V_G - |V_{tp1}|\right)^2 = \frac{\beta_2}{2}\left(V_R - V_G - |V_{tp2}|\right)^2. \qquad (6.9)$$

Hence, by using $|V_{tp1}| - |V_{tp2}| = \Delta V_t$ and $\beta_1 = \beta_2$,

$$I_R = \frac{|V_{tp1}| - |V_{tp2}|}{R_R} = \frac{\Delta V_t}{R_R}. \qquad (6.10)$$

Another current mirror composed of M_1 and M_5 converts I_R to I_L $(= I_R W_5/W_1)$, which flows in R_L. Therefore,

$$V_{REF} = I_L R_L = \left(\frac{W_5}{W_1}\right)\left(\frac{R_L}{R_R}\right)|\Delta V_t|. \qquad (6.11)$$

If R_L and R_R are designed with the same width of poly-Si, the ratio R_L/R_R is almost independent of V_{DD}, temperature, and fabrication-process variations. W_5/W_1 can also be constant against parameter variations. In addition, the variation in ΔV_t can be compensated for by trimming R_R. Thus, a stabilized V_{REF} is obtained.

6.3.2. Application of ΔV_t V_{REF} Generator

The circuit shown in Fig. 6.9 [10] uses the above principle. The features are 1) dynamic operation for reducing operation current, and 2) generating both normal and burn-in voltages like the circuit in Fig. 6.7. The circuit consists of a V_t-difference-to-current $(\Delta V_t - I)$ converter, a current-to-voltage $(I\text{-}V)$ converter, a sample and hold (S/H) circuit, an inverter, and a normal/burn-in switch circuit. It operates as follows.

During a sampling period, all switches S_1-S_5 are closed. A reference current I_R is created in the $\Delta V_t - I$ converter, and it is converted to current I_L by a current mirror consisting of nMOST M_6. The current flows through R_L creating

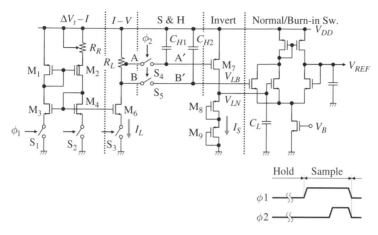

FIGURE 6.9. Reference voltage generator using sample/hold scheme. Reproduced from [10] with permission; © 2006 IEEE.

constant voltages relative to V_{DD} at nodes A and B. The voltages of nodes A and B are given by

$$V_A = V_{DD} - \left(\frac{W_6}{W_4}\right)\left(\frac{R_L}{R_R}\right)|\Delta V_t|, \qquad (6.12)$$

$$V_B = V_{DD} - \left(\frac{W_6}{W_4}\right)\left(\frac{R_L'}{R_R}\right)|\Delta V_t|. \qquad (6.13)$$

The $\Delta V_t - I$ converter and the I-V converter are enabled by clock φ_1, and the output voltages V_A and V_B are sampled on the hold capacitors C_{H2} and C_{H1}, respectively, by clock φ_2. Clock φ_2 is delayed to clock φ_1 to minimize the fluctuations of the output voltages at nodes A′ and B′. The voltage at node A′ provides a voltage V_{LB} for burn-in mode. On the other hand, in normal-operation mode, the voltage at node B′ relative to V_{DD} is converted to V_{LN} that is relative to the ground. Since M_7-M_9 in the inverter are in the saturation condition, their currents are expressed as

$$I_S = \frac{\beta_7}{2}\left(V_{EXT} - V_B' - |V_{tp7}|\right)^2 = \frac{\beta_8}{2}\left(V_{GS8} - |V_{tp8}|\right)^2 = \frac{\beta_9}{2}\left(V_{GS9} - |V_{tp9}|\right)^2. \qquad (6.14)$$

If M_7-M_9 are the same size, and the n-well of each MOST, which is separated from the others, is connected to the source to eliminate the body-bias effect and thus to obtain the same value of V_{tp}, (6.13) and (6.14) give the following equation:

$$V_{GS8} = V_{GS9} = I_L R_L'. \qquad (6.15)$$

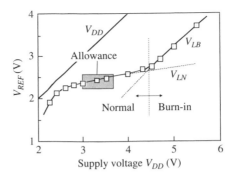

FIGURE 6.10. Supply-voltage dependency of reference voltage based on threshold-voltage difference. Reproduced from [10] with permission; © 2006 IEEE.

Thus, the same gate-source voltage of $I_L R_L'$ is supplied to M_7-M_9. Hence, V_{LN} is obtained as

$$V_{LN} = 2I_L R_L' = 2 \left(\frac{W_6}{W_4} \right) \left(\frac{R_L'}{R_R} |V_t| \right). \tag{6.16}$$

The voltages V_{LN} and V_{LB} are automatically switched with the switch circuit according to the operation mode and output as V_{REF}. The switch circuit consisting of an OR type current switch with a negative feedback loop outputs the larger voltage between V_{LN} and V_{LB}.

This V_{REF} generator was applied to a 64-Mbit DRAM [10]. The overall variation after trimming is ± 0.06 V ($\pm 2.5\%$) for $V_{DD} = 3.3 \pm 0.3$ V as shown in Fig. 6.10. The temperature coefficient is 370 ppm/°C for normal operation and -95 ppm/°C for burn-in as shown in Fig. 6.11. The average current is reduced to 0.3 μA when the activation interval is longer than 1 ms as shown in Fig. 6.12.

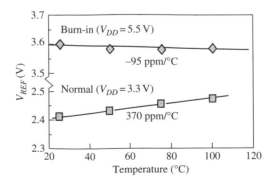

FIGURE 6.11. Temperature dependency of reference voltage based on threshold-voltage difference. Reproduced from [10] with permission; © 2006 IEEE.

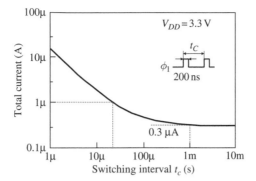

FIGURE 6.12. Power dissipation of reference voltage generator using sample/hold scheme. Reproduced from [10] with permission; © 2006 IEEE.

In summary, a quite stable V_{REF} is available by the V_t-difference V_{REF} generator. However, V_{REF} trimming might be still required, though the variation in V_{REF} is much smaller than in the V_t-referenced V_{REF} generator described in section 6.2. An additional process step to produce a V_t difference is also needed.

6.4. The Bandgap V_{REF} Generator

6.4.1. Principle

The bandgap reference (BGR) circuit utilizes the base-emitter voltages V_{BE} of bipolar transistors. Figure 6.13 shows the principle of the BGR [5, 11]. The voltage V_{BE} of a bipolar transistor is expressed as a function of collector current I_C:

$$V_{BE} = \frac{kT}{q} \ln \frac{I_C}{I_S}, \tag{6.17}$$

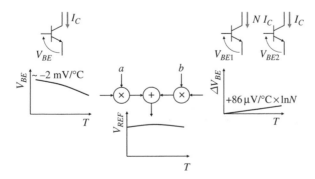

FIGURE 6.13. Operating principle of bandgap reference.

where k is Boltzmann constant, q is elementary charge, T is absolute temperature, I_S is the saturation current determined by device structure. The voltage V_{BE} has a negative temperature coefficient around $-2\text{mV}/^{\circ}\text{C}$. On the other hand, the V_{BE} difference of two bipolar transistors of the same I_S and different I_C is expressed as

$$\Delta V_{BE} = V_{BE1} - V_{BE2} = \frac{kT}{q} \ln N. \tag{6.18}$$

This is proportional to T, thus having a positive temperature coefficient. If we choose appropriate constants a and b for generating the reference voltage of

$$V_{REF} = aV_{BE} + b\Delta V_{BE} = aV_{BE} + b\ln N \cdot \frac{kT}{q}, \tag{6.19}$$

the temperature coefficients of V_{BE} and ΔV_{BE} can be cancelled by each other. Usually $a = 1$ and $b\ln N = 20\text{--}23$ are used and V_{REF} of 1.2–1.25V is obtained. The temperature dependency of ΔV_{BE} is linear as indicated by (6.18), while that of V_{BE} is a little concave down. The resultant V_{REF} is also a concave-down function of temperature. Although some techniques to compensate for this curvature are proposed [12], most of on-chip voltage converters do not require the compensation.

Now, the device structures of the bipolar transistors for BGR circuits are described. Figure 6.14 shows the cross sections of the parasitic bipolar transistors that are fabricated with a standard CMOS process. A parasitic pnp transistor has an emitter of p+ diffusion area, a base of n-well, and a collector of p-substrate. Since the p-substrate is usually connected to ground, this transistor must be used with collector grounded. A parasitic npn transistor has an emitter of n+ diffusion

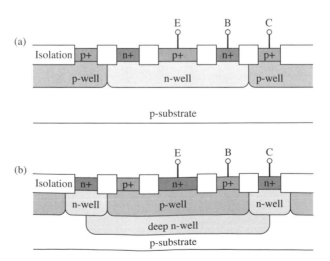

FIGURE 6.14. Device cross sections of parasitic bipolar transistors; a parasitic pnp transistor (a), and a parasitic npn transistor (b).

area, a base of p-well, and a collector of deep n-well. Although this device requires a triple-well structure, all three terminals can be at any potential. The common-emitter current gain of parasitic bipolar transistors is around 3–10. Therefore, the base currents cannot be neglected unlike in bipolar-circuit design.

6.4.2. Circuit Design

In this subsection BGR circuits to realize the above principle are described. The circuit shown in Fig. 6.15 utilizes parasitic pnp transistors Q_1 and Q_2 with diode connection (base and collector are connected together). The ratio between their emitter junctions is $1 : N$. The current mirror composed of pMOSTs M_1 and M_2 equalizes the collector currents of Q_1 and Q_2. Thus, the transistors have different sizes and an identical collector current, instead of an identical size and different currents as described above. Although both schemes can create different current density (I_C/I_S in equation (6.17)) and ΔV_{BE}, the former is usually used. Since the gates of nMOSTs M_3 and M_4 are connected together, their source voltages are equal. Accordingly, the ΔV_{BE} appears between both terminals of R_2, and the current

$$I_R = \frac{kT \ln N}{qR_2} \tag{6.20}$$

flows. The same current also flows through R_1 and through pnp transistor Q_3 due to the $1 : 1$ current mirror composed of pMOSTs M_2 and M_5. Therefore, the output voltage V_{REF} is given by

$$V_{REF} = V_{BE3} + I_R R_1 = V_{BE3} + \frac{kTR_1 \ln N}{qR_2}. \tag{6.21}$$

This BGR circuit requires a startup circuit that prevents the circuit from falling into an undesired operating point. If the circuit is at the normal operating point, the current I_R mirrored by M_1 and M_6 charges the gate of pMOST M_9. Therefore,

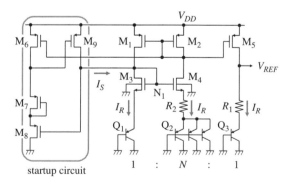

FIGURE 6.15. Bandgap reference circuit using parasitic pnp transistors.

the current I_S does not flow because M_9 is turned off. If the circuit is at the undesired operating point where all MOSTs M1-M4 are off, I_R does not flow. In this case M_9 is turned on because its gate is discharged through M_7 and M_8. The current I_S injected to node N_1 forces the current I_R to flow. The design method of the startup circuit is shown in Fig. 6.16. The loop composed of M_1-M_4 is opened, one end is connected to a voltage source V_{IN}, and the voltage V_{OUT} of the other end is observed. The operating points are indicated by the intersection points of V_{IN}-V_{OUT} characteristics and line $V_{IN} = V_{OUT}$. The circuit without startup circuit has two intersection points P_0 and P_1 as shown by the broken line. The former is the undesired operating point and the latter is the desired (normal) operating point. Since P_0 is a "metastable" point, the operating point drifts away from P_0 and toward P_1 after sufficient time. However, it takes a long time for the operating point to reach P_1, because the current I_R is extremely small when the operating point is around P_0. Therefore, I_R as well as V_{OUT} should be observed even if there is only one intersection point. The circuit with startup circuit has only one intersection point P_1 and a larger current I_R. Therefore, the operating point quickly reaches P_1. A dc current flows through M_6-M_8 in the startup circuit when the circuit is at the normal operating point. Although this current can be eliminated by starting up the circuit using a so-called "power-on reset" signal, such design should be avoided. This is because "power-on reset" is an unreliable signal which may not be generated in some startup conditions of V_{DD}.

The BGR circuit in Fig. 6.17 also utilizes parasitic pnp transistors Q_1 and Q_2 [5, 11, 13]. An operational amplifier with negative feedback equalizes the voltages of nodes N_1 and N_2. If $R_1 = R_2$, the collector currents of Q_1 and Q_2 are equal. Since the base-emitter-voltage difference ΔV_{BE} appears between both nodes of resistor R_3, the current I_R and the output voltage V_{REF} are expressed as

$$I_R = \frac{kT \ln N}{qR_3},$$
(6.22)

$$V_{REF} = V_{BE1} + I_R R_1 = V_{BE1} + \frac{kTR_1 \ln N}{qR_3}.$$
(6.23)

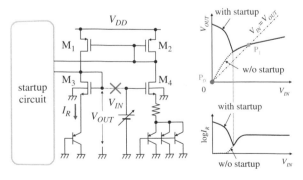

FIGURE 6.16. Design of startup circuit.

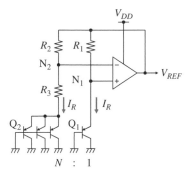

FIGURE 6.17. Bandgap reference circuit using parasitic pnp transistors.

The loop stability should be considered in the design of any circuit that utilizes negative feedback. The details of the loop stability and its improvement (compensation) are described in Chapter 7.

Figure 6.18 shows a BGR circuit using parasitic npn transistors [14]. This circuit also utilizes an operational amplifier with negative feedback to equalize the voltages of nodes N_1 and N_2. If $R_1 = R_2$, the collector currents of Q_1 and Q_2 are equal. Since the base-emitter-voltage difference ΔV_{BE} appears between both nodes of resistor R_3, the current I_R and the output voltage V_{REF} are expressed as

$$I_R = \frac{kT \ln N}{qR_3}, \tag{6.24}$$

$$V_{REF} = V_{BE1} + 2I_R R_4 = V_{BE1} + \frac{2kTR_4 \ln N}{qR_3}. \tag{6.25}$$

The loop stability should also be considered in this circuit. This circuit features a small voltage variation of V_{REF} as described below. The minimum operating voltage $V_{DD\,min}$, however, is higher by the voltage drop of resistors R_1, R_2 (several hundreds of mV-1 V).

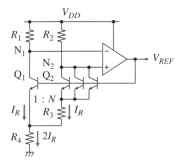

FIGURE 6.18. Bandgap reference circuit using parasitic npn transistors [14].

6.4.3. *Variation of Reference Voltage*

The voltage accuracy of a BGR circuit is mainly determined by the offset voltage of an operational amplifier, though variations in the resistor ratio and bipolar-transistor-size ratio also contribute to the voltage variation. If there is a threshold-voltage difference ΔV_t between the input pair MOSTs of an operational amplifier, as shown in Fig. 6.19(a), ΔV_t is equivalent to the offset voltage V_{OS} of the amplifier, as shown in (b). The V_{OS} of a CMOS amplifier is considerably larger than that of a bipolar amplifier mainly due to the intrinsic variation in V_t that is caused by the variation of dopant atoms in the channel region [15–17].

Let us consider the BGR circuit in Fig. 6.17. If the voltage gain of the operational amplifier is sufficiently large, the following equations holds:

$$I_2 R_3 = V_{BE1} - V_{BE2} + V_{OS} = \frac{kT}{q} \ln \frac{NI_1}{I_2} + V_{OS}, \tag{6.26}$$

$$I_1 R - V_{OS} = I_2 R, \tag{6.27}$$

where I_1 and I_2 are the emitter currents of Q_1 and Q_2, respectively, and $R = R_1 = R_2$. Differentiating these equations by V_{OS} results in

$$R_3 \frac{dI_2}{dV_{OS}} = \frac{kT}{q} \left(\frac{1}{I_1} \frac{dI_1}{dV_{OS}} - \frac{1}{I_2} \frac{dI_2}{dV_{OS}} \right) + 1, \tag{6.28}$$

$$R \frac{dI_1}{dV_{OS}} - 1 = R \frac{dI_2}{dV_{OS}}. \tag{6.29}$$

When $V_{OS} \to 0$, equation (6.28) becomes

$$R_3 \frac{dI_2}{dV_{OS}} = \frac{R_3}{\ln N} \left(\frac{dI_1}{dV_{OS}} - \frac{dI_2}{dV_{OS}} \right) + 1, \tag{6.30}$$

because I_1, $I_2 \to kT\ln N/qR_3$ (see equation (6.22)). From equations (6.29) and (6.30) we obtain

$$\left. \frac{dI_1}{dV_{OS}} \right|_{V_{OS} \to 0} = \frac{1}{R} + \frac{1}{R_3} + \frac{1}{R \ln N}. \tag{6.31}$$

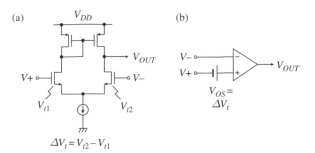

FIGURE 6.19. Offset voltage of differential amplifier.

From equation (6.31), the dependence of V_{REF} on V_{OS} is calculated as

$$\left.\frac{dV_{BE1}}{dV_{OS}}\right|_{V_{OS}\to0} = \frac{kT}{qI_1}\left.\frac{dI_1}{dV_{OS}}\right|_{V_{OS}\to0} = \frac{R_3}{R\ln N} + \frac{1}{\ln N} + \frac{R_3}{R(\ln N)^2}, \tag{6.32}$$

$$\left.\frac{dV_{REF}}{dV_{OS}}\right|_{V_{OS}\to0} = \frac{dV_{BE1}}{dV_{OS}} + R\left.\frac{dI_1}{dV_{OS}}\right|_{V_{OS}\to0} = 1 + \frac{2}{\ln N} + \frac{R}{R_3} + \frac{R_3}{R\ln N} + \frac{R_3}{R(\ln N)^2}. \tag{6.33}$$

Among the five terms on the right side of (6.33), the third term R/R_3 is dominant and is typically around ten. Therefore, the variation in V_{OS} is amplified by about one decade. For example, if the variation in V_{OS} is 10 mV, the variation in V_{REF} is as large as 100 mV (8%).

Next let us consider the circuit in Fig. 6.18 [25]. If the voltage gain of the operational amplifier is sufficiently large, the following equations hold:

$$I_2 R_3 = V_{BE1} - V_{BE2} = \frac{kT}{q}\ln\frac{NI_1}{I_2}, \tag{6.34}$$

$$\alpha I_2 R - \alpha I_1 R = V_{OS}, \tag{6.35}$$

where I_1 and I_2 are the emitter currents of Q_1 and Q_2, respectively, α is the common-base current gain of the npn transistors and $R = R_1 = R_2$. Differentiating (6.34) and (6.35) and similar calculations as above result in

$$\left.\frac{dV_{REF}}{dV_{OS}}\right|_{V_{OS}\to0} = \frac{R_3}{\alpha R\ln N}\left(1+\frac{1}{\ln N}\right) + \frac{R_4}{\alpha R}\left(1+\frac{2}{\ln N}\right). \tag{6.36}$$

This is usually less than unity, because $R_3 < R_4 < R$ in typical design. Therefore, the variation in V_{REF} is less than that in V_{OS}. This great difference can be qualitatively understood by comparing equations (6.26), (6.27), (6.34), and (6.35). In the circuit in Fig. 6.17, V_{OS} becomes an error in ΔV_{BE} directly, as in equation (6.26). On the other hand, V_{OS} has no direct effect on ΔV_{BE} of the circuit in Fig. 6.18, as shown in (6.34), but only causes emitter-current error as shown in (6.35).

The simulated dependencies of V_{REF} on V_{OS}, shown in Fig. 6.20, are very close to the calculations from equations (6.33) and (6.36). Figure 6.21 shows the measured results of the voltage variations. The standard deviation σ of the circuit in Fig. 6.18 (lower histogram) is as small as 3 mV, which is one tenth that of the circuit in Fig. 6.17 (upper histogram).

6.4.4. The Bandgap V_{REF} Generators for Low Supply Voltage

The output voltage of a BGR circuit, V_{REF}, is inevitably Si-bandgap voltage (~ 1.2 V), which is derived by substituting $a = 1$ into equation (6.19). Since V_{REF} determines the lower limit, V_{DDmin}, of the external supply voltage, V_{REF} must

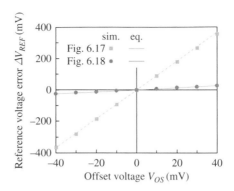

FIGURE 6.20. Relationship between offset voltage and output-voltage error.

be lowered for low-voltage operation. A lower V_{REF} is generated if a smaller value of $a(< 1)$ is adopted, maintaining the ratio between a and b. The circuits to realize this principle are described below.

Figure 6.22 shows the low-voltage BGR circuit proposed in reference [18]. The operational amplifier with negative feedback equalizes the voltages of nodes N_1 and N_2. Three pMOSTs M_1, M_2 and M_3 have the same size and the same gate voltage. Since the current of two pnp transistors Q_1 and Q_2 are equal, ΔV_{BE} appears between both ends of R_2. The current I_R is expressed as

$$I_R = \frac{kT \ln N}{q R_2}. \tag{6.37}$$

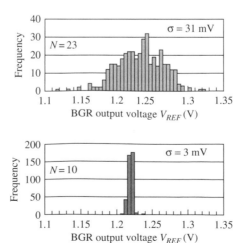

FIGURE 6.21. Measured voltage variations.

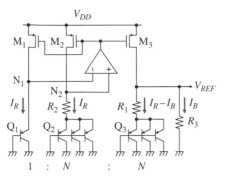

FIGURE 6.22. Bandgap reference circuit for low supply voltage. Reproduced from [18] with permission; © 2006 IEEE.

The current of M_3 is also I_R, and the current of R_1 is less than I_R by the current of R_3. Therefore, the output voltage V_{REF} is expressed as

$$V_{REF} = V_{BE3} + (I_R - I_B) R_1 = I_B R_3. \tag{6.38}$$

Equations (6.37) and (6.38) yield

$$V_{REF} = \frac{R_3}{R_1 + R_3} \left(V_{BE3} + \frac{R_1}{R_2} \cdot \frac{kT \ln N}{q} \right). \tag{6.39}$$

Comparing this equation and (6.19) shows that $a = R_3/(R_1 + R_3) < 1$. Thus, a lower V_{REF} is obtained (e. g. 0.67 V in [18]).

Figure 6.23 shows the circuit proposed in reference [19]. The operational amplifier with negative feedback equalizes the voltages of nodes N_1 and N_2. Three pMOSTs M_1, M_2 and M_3 have the same size and the same gate voltage. If $R_1 = R_2 = R$, the current of pnp transistors Q_1 and Q_2 are equal. Thus, ΔV_{BE} appears between both ends of R_3. The currents I_A and I_B are expressed as

$$I_A = \frac{kT \ln N}{qR_3}, \tag{6.40}$$

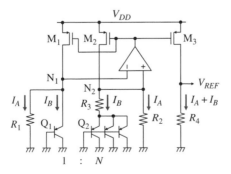

FIGURE 6.23. Bandgap reference circuit for low supply voltage [19].

$$I_B = \frac{V_{BE1}}{R}.$$

(6.41)

Since the sum of I_A and I_B flows through R_4, the output voltage V_{BGR} is given by

$$V_{REF} = (I_A + I_B)\, R_4 = \frac{R_4}{R} \cdot V_{BE1} + \frac{R_4}{R_3} \cdot \frac{kT \ln N}{q}.$$

(6.42)

If we design $R_4 < R$, a lower V_{REF} is obtained (e. g. 0.515 V in [19]).

Both the circuits in Figs. 6.22 and 6.23 require that the amplifiers operate at low V_{DD}. Reference [19] utilizes low-V_t nMOSTs as input devices of the amplifier.

6.5. The Reference Voltage Converter/Trimming Circuit

The output voltage V_{REF} of a reference voltage generator is not necessarily equal to the voltage V_{INT} that is supplied to internal circuit. In addition the voltage level of V_{REF} might vary with process fluctuations. Thus, it is necessary to convert the voltage level of V_{REF}. The voltage variation can be suppressed by adjusting the conversion ratio (voltage trimming). In this section the circuit for converting the voltage level of V_{REF} is described.

6.5.1. Basic Design

A reference voltage converter/trimming circuit is shown in Fig. 6.24. The input voltage V_{REF} is converted into V_{REF}', which is divided by resistors and fed back to the input of the operational amplifier. The conversion ratio, which is given by

$$\frac{V'_{REF}}{V_{REF}} = \frac{R_1 + R_2}{R_2},$$

(6.43)

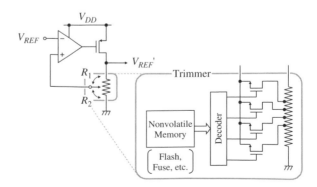

FIGURE 6.24. Reference voltage converter/trimming circuit.

can be adjusted by changing the ratio $R_1 : R_2$. The decoder decodes the trimming data and turns on one of the trimming MOSTs, selecting the feedback voltage. Although the trimmer in the figure has only four trimming taps for simplicity, an actual trimmer usually has 8–32 taps according to the required voltage accuracy. Nonvolatile programmable elements are necessary to memorize the trimming data. Fuses, which are also used for redundancy, are usually used in memory LSIs, while flash memory cells are used in logic LSIs with flash memories. The internal supply voltage V_{INT} is measured in probe test and appropriate trimming data are programmed so that V_{INT} is within the target voltage range.

One design issue of this circuit is the loop stability because it utilizes an operational amplifier with negative feedback. Another design issue is the threshold voltage V_{tn} of the trimming MOSTs. The MOSTs must be at non-saturation so that the feedback voltage does not drop due to V_{tn}., which is increased due to the body effect. The best way is to use the parallel connection of a pMOST and an nMOST as a trimming device, though the area is increased.

6.5.2. Design for Burn-In Test

(1) Burn-In Test Condition

Burn-in test is a kind of reliability test for LSIs to quickly get rid of potential defects. As a result of acceleration through applying a high stress voltage and high temperature, potential defects show up quickly. The test also must be applied to chips that incorporate an on-chip voltage down-converter, although special care should be taken so that the converter does not prevent the application of a stress voltage to the internal circuit during the burn-in test. Figure 6.25 shows the concept of the burn-in test for LSIs using an on-chip voltage down-converter. The core circuit (L_1) operates at V_{INT} while I/O circuit (L_2) and the voltage down-converter (VDC) operate at V_{DD}. The voltages during normal operation (point N) are $V_{INT} = V_{INTN}$, and $V_{DD} = V_{DDN}$, while $V_{INT} = V_{INTB}$ and $V_{DD} = V_{DDB}$ during the burn-in test.

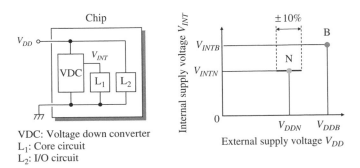

VDC: Voltage down converter
L_1: Core circuit
L_2: I/O circuit

FIGURE 6.25. The concept of burn-in test for LSIs using on-chip voltage down converter.

The burn-in voltages V_{INTB} and V_{DDB} are determined as follows. Generally, the lifetime of gate oxide t_{BD} is expressed as

$$t_{BD} \propto \exp\left(\frac{G}{E_{OX}}\right) = \exp\left(\frac{Gt_{OX}}{V_{OX}}\right), \qquad (6.44)$$

where E_{OX} is the electric field across the oxide, V_{OX} the applied voltage, t_{OX} the oxide thickness and G is a constant of around $320\,\mathrm{MV/cm}$ [20]. Therefore, the acceleration factor of L_1 is given by

$$k_1 = \exp\left(\frac{Gt_{OX1}}{V_{INTB}}\right) \Big/ \exp\left(\frac{Gt_{OX1}}{V_{INTN}}\right) = \exp\left\{Gt_{OX1}\left(\frac{1}{V_{INTB}} - \frac{1}{V_{INTN}}\right)\right\}, \qquad (6.45)$$

and that of L_2 is given by

$$k_2 = \exp\left\{Gt_{OX2}\left(\frac{1}{V_{DDB}} - \frac{1}{V_{DDN}}\right)\right\}. \qquad (6.46)$$

where t_{OX1} and t_{OX2} are the oxide thicknesses of MOSTs in L_1 and L_2, respectively.

To successfully perform the burn-in test, the following conditions must be satisfied.

Uniform Acceleration Throughout Chip: The stress condition during normal operation should be uniformly accelerated during the burn-in test for each circuit. Thus the acceleration factors k_1 and k_2 must be equal. Equations (6.45) and (6.46) result in

$$t_{OX1}\left(\frac{1}{V_{INTB}} - \frac{1}{V_{INTN}}\right) = t_{OX2}\left(\frac{1}{V_{DDB}} - \frac{1}{V_{DDN}}\right). \qquad (6.47)$$

Constant V_{DL} at around Normal-Operation V_{DD}: To obtain a stable operation V_{DL} must be almost constant even for the maximum V_{DD}-variation (at least $\pm 10\%$) that is guaranteed at the normal-operation V_{DD}. Therefore,

$$V_{INT} \equiv V_{INTN} \quad (0.9V_{DDN} \leqslant V_{DD} \leqslant 1.1V_{DDN}) \qquad (6.48)$$

(2) Burn-In Voltage Generation

Figure 6.26(b) illustrates a reference voltage converter [9], which meets the above requirements. The converter features switching of a line V_{RN} via a flat line-V_{RN} to a line V_{RB} at a certain V_{DD}. The line V_{RN} governs normal-operation characteristics, while the line V_{RB} governs burn-in (i.e. V_{INTB}) characteristics. When V_{DD} is low enough, $V_{INT} = V_{DD}$ because M_1 is always turned on. When V_{DD} is higher than $r_1 V_{REFN}$, however, the output V_{INT} is fixed to V_{INTN}. On the other hand, V_{RB}-characteristics are obtained by using two comparators (CP$_2$, and CP$_3$), a pMOST (M$_2$) and an nMOST (M$_3$). The reference voltage, $V_{DD} - V_{REFB}$, is generated based on V_{DD}. When V_{DD} is low enough, the resultant high voltage at the CP$_3$ output makes the CP$_2$ input (V_{RB}) low. Since another CP$_2$ input

(a)

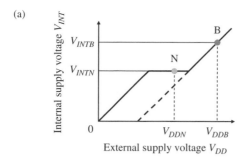

(b)

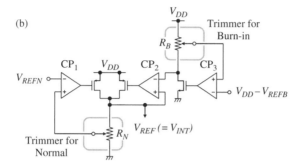

FIGURE 6.26. A burn-in test circuit (a) and its characteristics (b) [9].

voltage is V_{DD} as a result of CP_1 and M_1 operation, the CP_2 output voltage is high enough to cut off M_2. When V_{DD} is higher than $V_{DD} - V_{REFB}$, however, V_{RB} becomes $V_{DD} - r_2 V_{REFB}$. Here, $r_2 = (R_{21} + R_{22})/R_{22}$. If the resultant V_{RB} voltage is higher than V_{INT}, M_2 is turned on while M_1 is turned off. Thus, V_{INT} is in turn determined by V_{RB}, eventually causing the characteristics shown in Fig. 6.26(a). Here, the condition to ensure the flatness of V_{INT} against V_{DD}-variations at $V_{DDN} \pm 10\%$ is expressed as

$$V_{DDB} - V_{INTB} > 1.1 V_{DDN} - V_{INTN}. \qquad (6.49)$$

For example, if $V_{DDN} = 5\,\text{V}$, $V_{INTN} = 3.3\,\text{V}$, $V_{DDB} = 8\,\text{V}$, and $V_{INTB} = 5.3\,\text{V}$, (6.49) is satisfied, and the V_{INT} flatness is ensured.

Figure 6.27 shows another burn-in test circuit (Fig. 6.27(b)) and V_{INT}-characteristics (Fig. 6.27(a)) [21]. It features an excellent V_{INT} flatness against wide variations in V_{DD} at around V_{DDN}, and a flexible V_{DDB}-setting to meet user's requirements. In order to more quickly perform burn-in test a high V_{DDB} is better. In addition, the flatness of V_{INT} must be ensured with a wider V_{DD} margin at V_{DDN}. Both requirements are satisfied with the thick solid line in Fig. 6.27(a), in which V_{DDB} is assumed to be higher than 6.3 V when $V_{DDN} = 5\,\text{V}$ and $V_{INTN} = 3.3\,\text{V}$ (i.e. the slope of V_{RB} line $= V_{INTB}/V_{DDB} = 2/3$). The characteristics are realized by using the circuit shown in Fig. 6.27(b). When V_{DD} is low enough M_2 is off, enabling $V_{RB} = 0$. Thus, V_{INT} is under the control of a fixed V_{REF1} (1.65 V), so that V_{INTN} is fixed to 3.3 V at $V_{DD} > 2V_{REF1}$ (3.3 V) because of

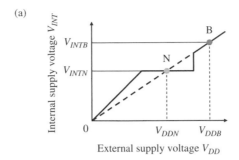

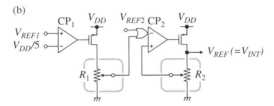

FIGURE 6.27. A burn-in test circuit (a) and its characteristics (b) [21].

a voltage-division ratio of 1/2 at R_2. When V_{DD} is increased further to the extent that $V_{DD}/5$ is higher than V_{REF2} (1.26 V) - that is, $V_{DD} > 5V_{REF2}$ (6.3 V)-V_{RB} is $V_{DD}/3$ as a result of turning on M_2. The V_{RB} is higher than V_{REF1}, allowing the V_{INT} characteristics to be governed by V_{RB} at V_{DD} higher than 6.3 V. Eventually, V_{INT} changes at a rate of $(2/3)V_{DD}$, lying on the V_{RB} line.

Figure 6.28 shows another example for burn-in test [4]. In this scheme the reduced voltage, V_{INT}, during normal operation is switched to V_{DD} during the

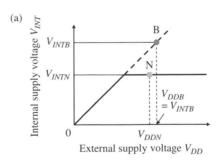

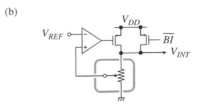

FIGURE 6.28. A burn-in test circuit (a) and its characteristics (b) [4].

burn-in test with $V_{DDB} = V_{INTB}$. The drawback is that during the burn-in test the appropriate stress voltage for the large devices in the L_2 block in Fig. 6.25 may destroy the small devices in the L_1 block. In other words, the appropriate stress voltage for the L_1 block becomes an insufficient stress voltage for the L_2 block. Thus, the scheme is acceptable only when the difference between V_{DDN} and V_{INTN} is small, or the scale of integration of the L_1 block is much larger than that of the L_2 block, so that the reliability of the whole chip is determined by that of the L_1 block. The burn-in test starts with a burn-in test control signal, $\overline{BI}$.

Figure 6.29 shows another example for burn-in test. In this scheme the voltage V_{INT} during burn-in test is raised from its nominal value. The circuit features that a trimmer and a test circuit share a resistor string. The highest tap is selected during burn-in so that the highest voltage is output as V_{REF}. Selecting a higher or lower test tap than the default one changes the level of $V_{REF} (= V_{INT})$, thus enabling a voltage-margin test of the core circuit.

Figure 6.30 shows the measured V_{INT} of an on-chip voltage down-converter implemented in a microcontroller before and after trimming [22]. Most of the samples are out of the target voltage range (1.45–1.6V) before trimming, but are within the range after trimming.

(a)

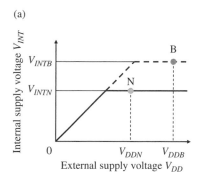

(b)

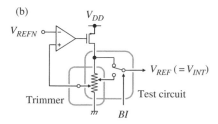

FIGURE 6.29. A burn-in test circuit (a) and its characteristics (b).

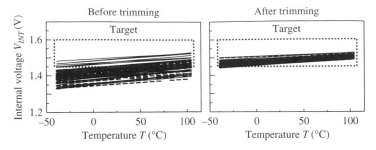

FIGURE 6.30. Measured voltage variation before and after trimming. Reproduced from [22] with permission; © 2006 IEEE.

6.6. Layout Design of V_{REF} Generator

Layout design as well as circuit design is a critical issue for high-precision analog circuits like reference voltage generators [26]. In this section, layout design methods for matching devices and for suppressing the effect of parasitic resistance will be described.

The characteristics (V_t, gain factor, etc.) of MOSTs for a current-mirror circuit and for input devices of a differential amplifier should be matched. To match the MOSTs, their layout patterns must be geometrically congruent. In addition, the directions of currents should be the same as shown in Fig. 6.31(c), not (a) or (b). This is because the source and the drain might not be symmetrical due to the source/drain formation, where impurity ions are implanted with a small angle from the perpendicular line. A better matching is obtained by the common-centroid layout [23], where two MOSTs share a common "centroid". In Fig. 6.32, M_1 is composed of left-upper and right-lower devices, where M_2 is composed of left-lower and right-upper devices. Even if there exists a tendency of characteristics from left to right, for example, M1 and M2 can be matched. The common-centroid layout is effective not only for MOSTs but also for other devices [24]. Figure 6.33(a) shows a layout pattern of bipolar transistors for a bandgap reference circuit described in Section 6.4. Nine devices are arranged in 3×3 matrix. The center one is for one transistor Q_1, while the others are

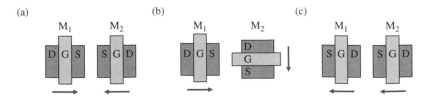

FIGURE 6.31. Layout of paired MOSTs; unfavorable layout (a), (b), and favorable layout (c).

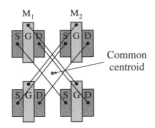

FIGURE 6.32. Common-centroid layout of paired MOSTs [23].

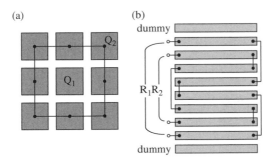

FIGURE 6.33. Common-centroid layout of bipolar transistors (a), and resistors (b). Reproduced from [24] with permission; © 2006 IEEE.

for Q_2. Thus, Q_1 and Q_2 have a common centroid and a size ratio of 1:8. Figure 6.33(b) shows resistors with common centroid. In this layout pattern, two dummy resistors that are not used are arranged at both edges. This is because edge elements might have different characteristics due to the irregularity of arrangement.

Next, the layout design for devices with an accurate ratio of $1:n$ is discussed. Generally connecting n identical devices is more favorable than a device with a constant multiplied by n. Figure 6.34 shows the layout patterns of a current-mirror circuit with a ratio of 1:2. The layout in Fig. 6.34(a), where the gate width

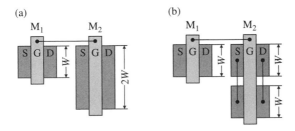

FIGURE 6.34. Layout of current-mirror circuit with mirror ratio of 1:2; unfavorable layout (a) and favorable layout (b).

of M_2 is twice that of M_1, is not favorable because of the following reason. If the effective channel width W_{eff} is smaller than the designed gate width W by ΔW, the effective mirror ratio is $W_{eff1}: W_{eff2} = (W - \Delta W) : (2W - \Delta W)$. This is not equal to 1:2 unless $\Delta W = 0$. On the other hand, the effective mirror ratio of Fig. 6.34(b) is $W_{eff1} : W_{eff2} = (W - \Delta W) : (2W - 2\Delta W) = 1:2$, irrespective of ΔW. The layout of Fig. 6.35(b) is more favorable than that of (a) for two resistors with a resistance ratio of 1:2, because of the similar reason. Figure 6.35(c) shows the most favorable layout pattern using the common-centroid technique.

Next, the layout design to suppress the effect of parasitic resistance is described. Designers should pay attention to a wiring through which a large current flows. Figure 6.36 shows the layout patterns of a current-mirror circuit. The layout of (a), where the sources of M_1 and M_2 are respectively connected to the main power line, is not favorable. This is because the voltage drop due to the parasitic resistance r_1 and a large current I causes the difference in source potentials and the error of mirror ratio. The layout of (b), where the sources are connected together to the main line, can avoid the source-potential difference. This layout is acceptable if the current I does not change. If I varies, however, the parasitic resistance r_2 and the variation of I generate a noise on the sources. The best layout design is to completely separate the source wiring from the main power line as shown in (c). Figure 6.37 shows the layout of an entire on-chip voltage down converter. The power-supply terminals (V_{DD1}, V_{SS1}) for the reference voltage generator and reference voltage converter should be separated

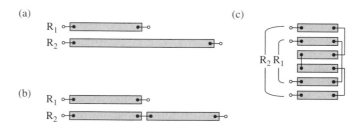

FIGURE 6.35. Layout of resistors with resistance ratio of 1:2; unfavorable layout (a), favorable layout (b) and the most favorable (common-centroid) layout (c).

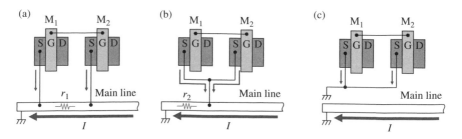

FIGURE 6.36. Layout of current-mirror circuit; unfavorable layout (a) (b), and favorable layout (c).

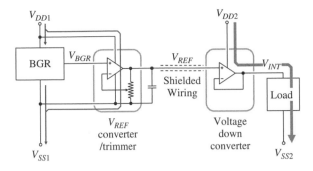

FIGURE 6.37. Layout of on-chip voltage down converter.

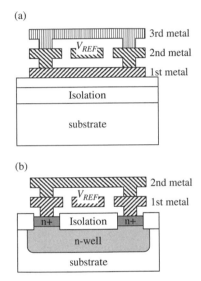

FIGURE 6.38. Cross sections of shielded wirings using three metal layers (a), and two metal layers and n-well (b).

from the terminals (V_{DD2}, V_{SS2}) for a large load current. The long wiring of the reference voltage V_{REF} should be shielded, because it has high impedance. Figure 6.38 shows the cross sections of shielded wirings; (a) utilizes three metal layers, while (b) utilizes two metal layers and a well.

References

[1] D.-S. Min, S. Cho, D. S. Jun, D.-J. Lee, Y. Seok and D. Chin, "Temperature-compensation circuit techniques for high-density CMOS DRAM's," IEEE J. Solid-State Circuits, vol. 27, pp. 626–631, Apr. 1992.
[2] K. Itoh, *VLSI Memory Design* (Baifukan, Tokyo 1994) (in Japanese).

[3] S.-M. Yoo, E. Haq, S.-H. Lee, Y.-H. Choi, S.-I.-Cho, N.-S. Kang and D. Chin, "Variable Vcc design techniques for battery-operated DRAM's," IEEE J. Solid-State Circuits, vol. 28, pp. 499–503, Apr. 1993.

[4] H. Hidaka, K. Arimoto, K. Hirayama, M. Hayashikoshi, M. Asakura, M. Tsukude, T. Oishi, S. Kawai, K. Suma, Y. Konishi, K. Tanaka, W. Wakamiya, Y. Ohno and K. Fujishima, "A 34-ns 16-Mb DRAM with controllable voltage down-converter," IEEE J. Solid-State Circuits, vol. 27, pp. 1020–1027, July 1992.

[5] P. R. Gray, P. J. Hurst, S. H. Lewis and R. G. Meyer, *Analysis and design of analog integrated circuits*, 4th ed. (John Wiley, New York 2001), Chap. 4.

[6] K. Ishibashi, K. Sasaki and H. Toyoshima, "A voltage down converter with submicroampere standby current for low-power static RAM's," IEEE J. Solid-State Circuits, vol. 27, pp. 920–926, June 1992.

[7] M. Horiguchi, M. Aoki, J. Etoh, H. Tanaka, S. Ikenaga, K. Itoh, K. Kajigaya, H. Kotani, K. Ohshima and T. Matsumoto, "A tunable CMOS-DRAM voltage limiter with stabilized feedback amplifier," IEEE J. Solid-State Circuits, vol. 25, pp. 1129–1135, Oct. 1990.

[8] R. A. Blauschild, P. A. Tucci, R. S. Muller and R. G. Meyer, "A new NMOS temperature-stable voltage reference," IEEE J. Solid-State Circuits, vol. SC-13, pp. 767–773, Dec. 1978.

[9] M. Horiguchi, M. Aoki, J. Etoh, K. Itoh, K. Kajigaya, A. Nozoe and T. Matsumoto, "Dual-regulator dual-decoding-trimmer DRAM voltage limiter for burn-in test," IEEE J. Solid-State Circuits, vol. 26, pp. 1544–1549, Nov. 1991.

[10] H. Tanaka, Y. Nakagome, J. Etoh, E. Yamasaki, M. Aoki and K. Miyazawa, "Sub-1-μA dynamic reference voltage generator for battery-operated DRAM's," IEEE J. Solid-State Circuits, vol. 29, pp. 448–453, Apr. 1994.

[11] B. Razavi, *Design of Analog CMOS Integrated Circuits*, (McGraw-Hill, 2001), Chap. 11.

[12] B.-S. Song and P. R. Gray, "A precision curvature-compensated CMOS bandgap reference," IEEE J. Solid-State Circuits, vol. SC-18, pp. 634–643, Dec. 1983.

[13] K. E. Kuijk, "A precision reference voltage source," IEEE J. Solid-State Circuits, vol. SC-8, pp. 222–226, June 1973.

[14] A. P. Brokaw, "A simple tree-terminal IC bandgap reference," IEEE J. Solid-State Circuits, vol. SC-9, pp. 388–393, Dec. 1974.

[15] D. Burnett, K. Erington, C. Subramanian and K. Baker, "Implications of fundamental threshold voltage variations for high-density SRAM and logic circuits," in Symp. VLSI Technology Dig. Tech. Papers, June 1994, pp. 15–16.

[16] Y. Taur, D. A. Buchanan, W. Chen, D. J. Frank, K. E. Ismail, S.-H. Lo, G. A. Sai-Halasz, R. G. Viswanathan, H.-J. C. Wann, S. J. Wind and H.-S. Wong, "CMOS scaling into the nanometer regime," Proc. IEEE, vol. 85, pp. 486–504, Apr. 1997.

[17] X. Tang, V. De and J. Meindl, "Intrinsic MOSFET parameter fluctuations due to random dopant placement," IEEE Trans. VLSI Systems, vol. 5, pp. 369–376, Dec. 1997.

[18] H. Neuteboom, B. M. J. Kup and M. Janssens, "A DSP-based hearing instrument IC," IEEE J. Solid-State Circuits, vol. 32, pp. 1790–1806, Nov. 1997.

[19] H. Banba, H. Shiga, A. Umezawa, T. Miyaba, T. Tanzawa, S. Atsumi and K. Sakui, "A CMOS bandgap reference circuit with sub-1-V operation," IEEE J. Solid-State Circuits, vol. 34, pp. 670–674, May 1999.

[20] J. C. Lee, I.-C. Chen and C. Hu, "Modeling and characterization of gate oxide reliability," IEEE Trans. Electron Devices, vol. 35, pp. 2268–2278, Dec. 1988.

[21] R. S. Mao, H. H. Chao, Y. C. Chi, P. W. Chung, C. H. Hsieh, C. M. Lin and N. C. C. Lu, "A new on-chip voltage regulator for high density CMOS DRAMs," in Symp. VLSI Circuits Dig. Tech. Papers, June 1992, pp. 108–109.

[22] M. Hiraki, K. Fukui and T. Ito, "A low-power microcontroller having a 0.5-μA standby current on-chip regulator with dual-reference scheme," IEEE J. Solid-State Circuits, vol. 39, pp. 661–666, Apr. 2004.

[23] K. Ishibashi, K. Komiyaji, S. Morita, T. Aoto, S. Ikeda, K. Asayama, A. Koike, T. Yamanaka, N. Hashimoto, H. Iida, F. Kojima, K. Motohashi and K. Sasaki, "A 12.5-ns 16-Mb CMOS SRAM with common-centroid-geometry-layout sense amplifiers," IEEE J. Solid-State Circuits, vol. 29, pp. 411–418, Apr. 1994.

[24] P. K. T. Mok and K. N. Leung, "Design considerations of recent advanced low-voltage low-temperature-coefficient CMOS bandgap voltage reference," in Proc. CICC, Oct. 2004, pp. 635–642.

[25] K. Fukuda, M. Hiraki, M. Horiguchi, T. Akiba, S. Ichiki and H. Tsunoda, "Voltage generation circuit and semiconductor integrated circuit device," PCT Publication No. WO/2005/062150, July 2005.

[26] B. Razavi, Design of Analog CMOS Integrated Circuits, (McGraw-Hill, 2001), Chap. 18.

7
Voltage Down-Converters

7.1. Introduction

On-chip voltage converters are becoming increasingly important for ultra-low voltage nano-scale memories, as discussed in Chapter 1. In particular, the on-chip voltage down-converter used to generate an internal supply voltage V_{INT}, which is lower than the external supply voltage V_{DD}, is indispensable for modern memories as well as low-voltage logic LSIs. The converter allows a single external power supply that can remain constant according to standards even when the internal supply changes. Thus, the converter has made it possible to quadruple memory capacity with the same V_{DD} despite the ever-lower device breakdown voltage. Moreover, it has allowed the successive chip-shrinking with scaled devices under a fixed memory capacity and V_{DD}, enabling reduced bit-cost. In addition, the converter has many advantages including a well-fixed internal voltage regardless of unregulated supply-voltages from various batteries, protection of internal core-circuits against high voltages within a wide range of voltage variations, and an adjustable internal voltage (V_{DL}) in accordance with variations in supply voltage (V_{DD}), temperature, and device parameters to compensate for speed and leakage variations. The approach using the voltage down-converter has also been applied to low-end micro-controllers (MCUs). Key design issues for the converter are the efficiency of voltage conversion, degree of voltage setting accuracy and stability in the output voltage, load-current delivering capability, power of the converter itself (especially during the standby period), speed and recovery time, and cost of implementation.

The voltage down-converter is categorized as the series regulator, the switching regulator, and the switched-capacitor regulator. The characteristics of each circuit are summarized in Table 7.1. The series regulator does not require off-chip components, and is widely used as on-chip voltage down-converters for memories and MCUs. However, the power-conversion efficiency is low, because V_{DD} is reduced to V_{INT} by the on-resistance of the output transistor. The difference between V_{DD} and V_{INT} corresponds to the power loss because the input current I_{DD} equals to the output current I_{INT} except for the idling current, as shown in Fig. 7.1. To achieve higher power efficiency, a reactance element that stores electric energy is required. A switching regulator, which utilizes an inductor and a capacitor, can realize an efficiency of over 90%, because current ratio I_{INT}/I_{DD} is equal to the reciprocal of the voltage ratio V_{INT}/V_{DD} except for the idling current. It suffers from, however, the increase in

TABLE 7.1. Comparison of voltage down-converters. T_C is the cycle time of clock.

	Series	Switching	Switched C.
Votage-conversion ratio	Any ratio $\left(\begin{array}{c}\text{except}\\ V_{DD} \sim V_{INT}\end{array}\right)$	Any ratio	Integer ratio
Power efficiency	$< \frac{V_{INT}}{V_{DD}}$	$> 90\%$	$> 80\%$
Off-chip components	None (1 C)	1 L, 1 C, 1 Diode, (2 MOSTs)	n C
Additional terminals	0–1	≥ 2	$2n - 1$
Transient Response	< 10ns	$> \frac{T_C}{2}$	$> \frac{T_C}{2}$
EMI	Low	High	Medium

board area because off-chip components: an inductor, a capacitor and a diode (and switching MOSTs) are required. The switching regulator also suffers from electromagnetic interference (EMI) caused by switching the inductor current. A switched-capacitor regulator, which utilizes only capacitors, also realizes a high efficiency. However, it also requires off-chip capacitors. Another drawback of the switched-capacitor regulator is that the voltage-conversion ratio realized is fixed to an integer ratio determined by the circuit topology, as will be explained in Section 7.4. If the conversion ratio is, for example, 1/2, $V_{DD}/2 - V_{INT}$ corresponds the power loss because $I_{DD} = I_{INT}/2$ except for the idling current, as shown in Fig. 7.1. A series regulator can quickly respond to the fluctuation of load current at a speed of the error amplifier. On the other hand, a

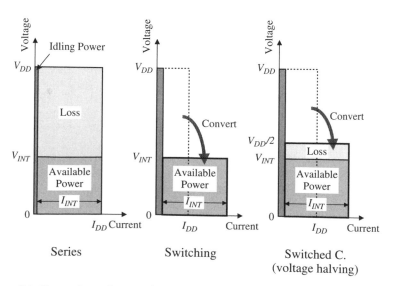

FIGURE 7.1. Comparison of conversion efficiency. Horizontal and vertical axes are current and voltage, respectively. The area of each rectangle, therefore, corresponds to power.

switching regulator and a switched-capacitor regulator cannot respond until the next switching. Therefore, a large off-chip capacitor is needed to suppress the output-voltage droop.

This chapter describes circuit designs of on-chip voltage down-converters in detail from the view point of memory designers.

7.2. The Series Regulator

In the past, voltage down-converters without negative feedback [1, 2, 3] utilizing the V_t of a MOST as a reference have been proposed. Eventually, however, the negative-feedback circuit consisting of an output MOST and a comparator (error amplifier) [4, 5, 6, 7] shown in Fig. 7.2(a) has become standard for commercial

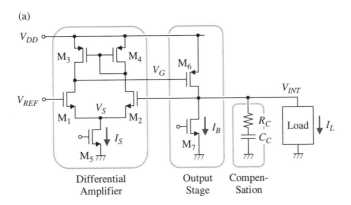

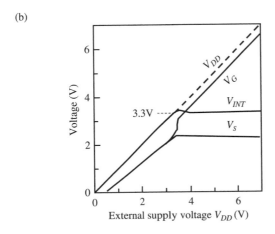

FIGURE 7.2. A typical voltage down converter; circuit diagram (a), and DC characteristics (b). Reproduced from [5] with permission; © 2006 IEEE.

LSI designs. The error amplifier compares the output voltage (V_{INT}) with the reference voltage (V_{REF}). The resulting output signal controls the gate of the output MOST (M_6), so that V_{INT} is stabilized. In other words, when the load current (I_L) flows to ground, the V_{INT} starts to drop, since M_6 works as an impedance. As soon as the resultant V_{INT} becomes lower than V_{REF}, M_6 starts to turn on, so that it charges up the load. When the resultant V_{INT} exceeds V_{REF}, however, the M_6 gate voltage V_G is raised, so as to stop the charge up. As a result, the circuit compensates for the drop in V_{INT}. A larger drop in V_{INT} allows the output to be charged up more quickly, because the feedback loop responds more quickly with a larger amplified voltage V_G. A large size of M_6 (i.e. a larger W/L) also charges up the load more quickly. The output voltage is stabilized in this manner. The compensation circuit is for stabilizing the feedback loop as will be discussed later. The DC characteristics of the regulator [5] are shown in Fig. 7.2(b).

The voltage down-converter must offer a well-regulated output voltage that is immune to any variation of the operating conditions. V_{INT} must be regulated well even under a dynamic load, where the heavy load capacitance is dynamically changed and various current pulses flow as a result of the load (i.e. internal core circuit) operation. The voltage regularity of V_{INT} is determined by the circuit characteristics, such as the response time, the current-driving capability, and the feedback-loop stability, and by the load characteristics. Thus, to design the voltage down converter, a deep understanding of both the circuit and the load characteristics is required. Otherwise, V_{INT} may result in oscillation or ringing waveforms.

7.2.1. DC Characteristics

The practical voltage down converters proposed so far can be categorized as the nMOST-output circuit [8, 9, 10, 11, 12] and the pMOST-output circuit, each of which has a negative-feedback loop, as shown in Figs. 7.3(a) and 7.4(a),

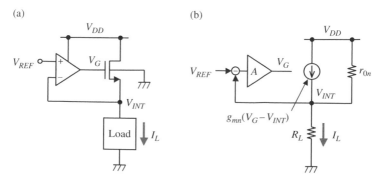

FIGURE 7.3. An nMOST-output voltage down converter; circuit diagram (a), and equivalent circuit (b).

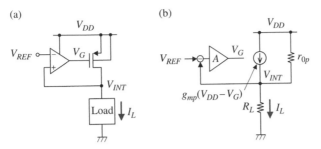

FIGURE 7.4. A pMOST-output voltage down converter; circuit diagram (a), and equivalent circuit (b).

respectively. Note that the output voltage is fed back to the non-inverting input terminal of the differential amplifier in Fig. 7.4(a), because the pMOST common-source output stage inverts the signal phase. The load regulation (the voltage drop due to load current) of each circuit is calculated as follows. The equivalent circuit of each circuit is shown in Figs. 7.3(b) and 7.4(b). In the case of the nMOST-output circuit, the following equation stands:

$$g_{mn}(V_G - V_{INT}) + \frac{V_{DD} - V_{INT}}{r_{0n}} = g_{mn}\{A(V_{REF} - V_{INT}) - V_{INT}\}$$

$$+ \frac{V_{DD} - V_{INT}}{r_{0n}} = I_L = \frac{V_{INT}}{R_L}, \qquad (7.1)$$

where g_{mn} and r_{0n} are the transconductance and drain output resistance of the output nMOST, respectively, and A is the voltage gain of the differential amplifier. From this equation, the effective output resistance is calculated as

$$R_{OUT} = -\frac{dV_{INT}}{dI_L} = \frac{r_{0n}}{1 + (A+1)g_{mn}r_{0n}} \cong \frac{1}{(A+1)g_{mn}}. \qquad (7.2)$$

In the case of the pMOST-output circuit,

$$g_{mp}(V_{DD} - V_G) + \frac{V_{DD} - V_{INT}}{r_{0p}} = g_{mp}\{V_{DD} - A(V_{INT} - V_{REF})\}$$

$$+ \frac{V_{DD} - V_{INT}}{r_{0p}} = I_L = \frac{V_{INT}}{R_L}, \qquad (7.3)$$

$$R_{OUT} = -\frac{dV_{INT}}{dI_L} = \frac{r_{0p}}{1 + Ag_{mp}r_{0p}} \cong \frac{1}{Ag_{mp}}. \qquad (7.4)$$

The nMOST-output circuit has fewer stability problems due to a smaller loop gain, because the source-follower output stage has no voltage gain. Therefore, it does not require a large compensation capacitor, unlike the pMOST-output circuit. Despite the smaller loop gain, the load regulation is comparable to that of the pMOST-output circuit as suggested by (7.2) and (7.4), as long as V_{DD}

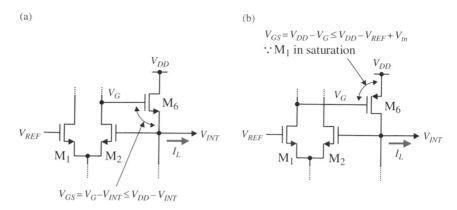

FIGURE 7.5. Current driving capability of voltage down converters; an nMOST-output circuit (a), and a pMOST-output circuit (b).

is sufficiently high. For lower V_{DD}, however, the maximum driving current of the nMOST drastically decreases. This is because the maximum effective gate-source voltage, expressed as

$$V_{GS\,max} - V_{tn} = V_{DD} - V_{INT} - V_{tn} \qquad (7.5)$$

becomes small (see Fig. 7.5(a)). In particular, it is prominent for a low dropout voltage (i.e. small difference between V_{DD} and V_{INT}). Note an increased threshold voltage (V_{tn}) due to the substrate-bias effect (i.e. body effect). For example, the resultant V_{tn} is as large as 1 V, which is not suitable for low dropout applications such as $V_{DD} = 1.8\,\text{V}$ and $V_{INT} = 1.5\,\text{V}$. On the other hand, the pMOST-output circuit (Fig. 7.4(a)) allows the pMOST to turn on more strongly with a higher gate-source voltage and a smaller threshold voltage, due to the absence of any body effect. In this case the maximum effective gate-source voltage is expressed as

$$V_{GS\,max} - |V_{tp}| = V_{DD} - V_{REF} + V_{tn} - |V_{tp}|, \qquad (7.6)$$

because the lowest pMOST gate voltage, i.e. the lowest drain voltages of paired nMOSTs in the saturation condition, must be $V_{REF} - V_{tn}$ (see Fig. 7.5(b)). Thus, lower V_{DD} and lower dropout applications are acceptable as shown in Fig. 7.6 [13]. However, the circuit needs a compensation circuit for feedback-loop stability, because it is a negative feedback circuit with a large loop gain produced by a high-gain error amplifier and a high-gain pMOST common-source stage. Moreover, the pMOST size tends to be large due to low transconductance of the pMOST.

Several attempts have been made for both circuits to improve the current driving capability. Figure 7.7(a) shows a boosted gate nMOST-output voltage down converter [10, 12]. Since the gate voltage can exceed V_{DD}, the converter can have a low dropout voltage. However, the voltage up converter (VUC) and

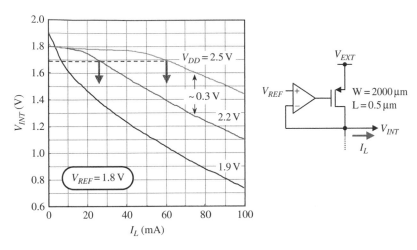

FIGURE 7.6. Simulated current driving capability of a pMOST-output voltage down converter [13].

the amplifier circuit need to be designed with care so as not to destroy the low-voltage devices. If a voltage doubler is used as the pump, devices could see a stress as high as $2V_{DD}$. So, longer channel devices, device stacking, and other circuit techniques are necessary to reduce the stress on each device. The same applies to the amplifier which typically has devices stacked. In addition, the pump and amplifier consume large area and power, as a result of the boosted gate voltage. The depletion nMOST-output circuit shown in Fig. 7.7(b) also improves the current driving capability because $V_{tn} < 0$ in equation (7.5). However, an additional process step to produce a depletion nMOST is needed. The output voltage division [7, 14] shown in Fig. 7.8(a) has been widely used to improve

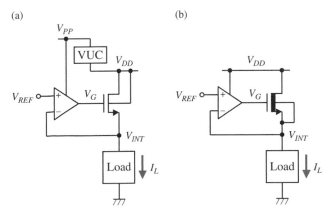

FIGURE 7.7. Improvement of the current driving capability of an nMOST-output voltage down converter; gate boosting (a) [10], and depletion-mode output transistor (b) [9].

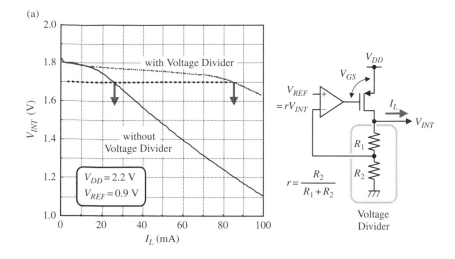

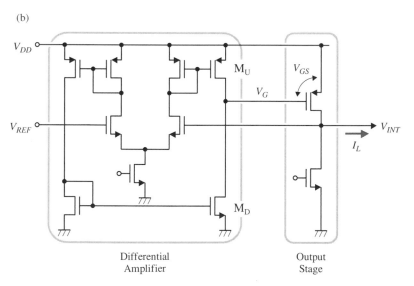

FIGURE 7.8. Improvement of the current driving capability of a pMOST-output voltage down converter, using a voltage divider (a), and a push-pull differential amplifier (b).

the current driving capability of the pMOST-output circuit. In this circuit, a high V_{INT} is divided using resistors (or MOSTs), R_1 and R_2, so that the resultant lower V_{REF} increases the voltage swing at the pMOST gate and the current driving capability as shown in Fig. 7.8(a). However, the circuit suffers from the following drawbacks. First, the lower V_{REF} poses a difficulty in setting the common-mode level of the differential amplifier, especially for low V_{INT}. Second, the response time of the feedback loop is slow, which causes a possible loop instability. A large RC delay caused by the input capacitance of the differential amplifier and the high resistivity of R_1 and R_2 is responsible for the slow response. Third, the resistors for V_{INT} division need a large layout area to achieve an extremely high resistivity of $M\Omega$ using a poly-Si layer, whose sheet resistivity is around $100\ \Omega/\square$. One solution to the latter two problems is using two pMOSTs with identical W/L instead of the resistors [14] (therefore the division ratio $r = 1/2$). The resistivity of $M\Omega$ can be easily realized by a long-channel pMOST, and the gate capacitances of the pMOSTs compensate for the RC delay. Another technique to improve the current driving capability of the pMOST-output circuit is shown in Fig. 7.8(b). This circuit features a push-pull-type differential amplifier. The pull-up pMOST M_U and the pull-down nMOST M_D realize a large voltage swing of the pMOST gate, V_G. Since M_D can operate in a triode region, the minimum V_G is as low as almost $0\ V$. The maximum effective gate-source voltage of the output pMOST is expressed as

$$V_{GS\max} - |V_{tp}| \cong V_{DD} - |V_{tp}|, \tag{7.7}$$

thus increasing the current driving capability. A buffer circuit inserted between the amplifier and the pMOST [5, 15] also delivers a large voltage swing to the pMOST gate in spite of a small swing at the amplifier output. However, an additional pole produced by the buffer tends to make the feedback loop unstable. Eventually, however, pMOST-output circuit has become widely used in the commercial LSI's after the stability problems are solved. Phase compensation using an on-chip capacitor and resistor is the key to stabilization, as discussed later.

7.2.2. Transient Characteristics

In this section the guiding design-principles are given, based on the transient analysis for the typical voltage-down converter shown in Fig. 7.2(a). The converter is a negative feedback amplifier consisting of two-stage amplifiers; a current-mirror differential amplifier for the first stage, and an output pMOST (M_6) for the second stage, which also works as a common-source amplifier. The output voltage, that is, the down converted voltage (V_{INT}) is the internal supply voltage, and thus the common source voltage of pMOSTs in the CMOS core circuit. The total current of the core circuit is the load current of the converter. Although V_{INT} becomes lower than V_{REF} when the load current I_L starts to flow, it can be quickly corrected by the negative feedback. However, once V_{INT} is higher than V_{REF}, it is no longer corrected under a light-load condition, that is $I_L \sim 0$. The voltage is then held at the level for a long time, because the increased M_6 gate voltage cuts off Q_6. The condition of V_{INT} being higher than V_{REF} could be developed when a positive coupling from other wires to the V_{INT} line occurs. Thus, a small I_B is needed for discharging the V_{INT} down to V_{REF}. Here is an analysis [16] to obtain the optimum design parameters of the converter, assuming the same channel length L for all of the MOSTs and the same channel width for the paired MOSTs.

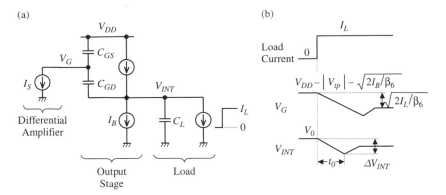

FIGURE 7.9. Large-signal equivalent circuit (a), and step response (b) [16].

When the load current (I_L) is at a maximum, the drain-source voltage (V_{DS6}) of M_6 in the saturation region is given by

$$V_{DS6} = V_{DD} - V_{INT} \geq V_{GS6} - V_{tp} = \sqrt{2I_L \Big/ \Big(\frac{W_6}{L}\beta_{p0}\Big)}; \qquad (7.8)$$

$$\therefore \frac{W_6}{L} \geq \frac{2I_L}{V_{DS6}^2 \beta_{p0}}, \qquad (7.9)$$

where W_6 is the channel width of M_6, V_{tp}, and β_{p0} are the threshold voltage and the gain factor of pMOSTs, and a small $I_B(<< I_L)$ is assumed. Here, voltages are all expressed using absolute values. On the other hand, when I_L is zero, a small I_B makes V_{INT} equal to V_{REF}, allowing a half I_S to flow through M_3 in the saturation region. Hence,

$$V_{DS3} = V_{GS6} = \sqrt{2I_B \Big/ \Big(\frac{W_6}{L}\beta_{p0}\Big)} + V_{tp}, \qquad (7.10)$$

$$V_{DS3} \geq V_{GS3} - V_{tp} = \sqrt{I_S \Big/ \Big(\frac{W_3}{L}\beta_{p0}\Big)}; \qquad (7.11)$$

$$\therefore \frac{W_3}{L} = \frac{W_4}{L} \geq \frac{I_S}{V_{DS3}^2 \beta_{p0}}. \qquad (7.12)$$

Here, the constant current (I_S) is given by using a large signal equivalent circuit in Fig. 7.9(a), as follows (see Fig. 7.9(b)). When the load current steps up from 0 to I_L, V_{INT} instantaneously falls, so that a more current flows through M_1 although an equal current of a half I_S flowed into M_1 and M_2. Instead, the current flowing through M_2 and M_4 is reduced so as to also reduce the current of M_3 in the current-mirror circuit. Eventually, the current difference between M_1 and M_3 discharges the M_6 gate. Therefore, when V_{INT} falls to the extent that I_S

fully flows through M_1, I_S discharges the M_6 gate. Since M_6 is in the saturation region and the gate-drain capacitance (C_{gd}) is thus negligible,

$$C_{gs}\frac{dV_G}{dt}+I_S=0 \tag{7.13}$$

is obtained. Solving this equation by using the initial V_{GS6} expressed by equations (7.10),

$$\int_{V_{DD}-V_{tp}-\sqrt{2I_B/\beta_6}}^{V_G} dV_G = -\int_0^t \frac{I_S}{C_{gs}}dt, \tag{7.14}$$

$$\therefore V_G = -V_{tp}+V_{DD}-\sqrt{\frac{2I_B}{\beta_6}}-\frac{I_S}{C_{gs}}t \tag{7.15}$$

are given, where $\beta_6=\beta_{p0}W_6/L$. Moreover, V_{INT} is the solution is the differential equation

$$C_L\frac{dV_{INT}}{dt}+I_L+I_B=\frac{\beta_6}{2}\left(V_{DD}-V_G-V_{tp}\right)^2. \tag{7.16}$$

By substituting equation (7.15) into this equation, we can obtain V_{INT} as

$$V_{INT}=\frac{I_S^2\beta_6}{6C_LC_{gs}^2}t^3+\frac{\sqrt{2I_B\beta_6}I_S}{2C_LC_{gs}}t^2-\frac{I_L}{C_L}t+V_0, \tag{7.17}$$

where V_0 is the initial value of V_{INT}. The time (t_0) when V_{INT} falls to the lowest voltage is obtained by $dV_{INT}/dt=0$, as

$$t_0=\frac{\sqrt{2}C_{gs}}{I_S\sqrt{\beta_6}}\left(\sqrt{I_L+I_B}-\sqrt{I_B}\right). \tag{7.18}$$

V_{INT} at this time is derived from substituting (7.18) into (7.17) and assuming that $I_L\gg I_B$, as follows:

$$V_{DL}=V_0-\frac{2\sqrt{2}C_{gs}I_L^{1.5}}{3C_LI_S\sqrt{\beta_6}}. \tag{7.19}$$

The second term of the right side denotes the output-voltage variation (ΔV_{INT}). Assuming that C_{gs} is two-thirds of the total gate capacitance, I_S is given as

$$V_{INT}=V_0+\Delta V_{INT}, \tag{7.20}$$

$$\Delta V_{INT}=-\frac{4\sqrt{2}\,I_L^{1.5}C_{OX}L^{1.5}\sqrt{W_6}}{9C_LI_S\sqrt{\beta_{p0}}}, \tag{7.21}$$

$$\therefore I_S=-\frac{4\sqrt{2}\,I_L^{1.5}C_{OX}L^{1.5}\sqrt{W_6}}{9C_LI_S\sqrt{\beta_{p0}}\Delta V_{INT}}, \tag{7.22}$$

where C_{OX} is the gate capacitance per unit area. It is obvious that a smaller I_L, a larger β_6, and a larger I_S are effective to reduce ΔV_{INT}.

The sizes of input nMOSTs (M_1, M_2) are determined by the requirement for the gain A of the differential amplifier, as follows. The output voltage of the differential amplifier, when the input voltages are equal (i.e. $V_{INT} = V_{REF}$), is $V_{DD} - V_{GS4}$ because $V_{GS3} = V_{GS4}$. When V_{INT} differs from V_{REF}, a variation, $A(V_{DL} - V_{REF})$, is added to the above output voltage. Since the resultant voltage is V_G, the following steady state condition is established:

$$(V_{DD} - V_{GS4}) + A(V_{INT} - V_{REF}) = V_{DD} - G_{GS6}. \tag{7.23}$$

Hence, by using (7.8) and the equation:

$$V_{GS4} = V_{tp} + \sqrt{\frac{I_S}{\beta_4}}, \tag{7.24}$$

we obtain

$$A = \frac{\sqrt{I_S/\beta_4} - \sqrt{2I_L/\beta_6}}{\varepsilon V_{REF}}, \tag{7.25}$$

$$\varepsilon = \frac{V_{INT} - V_{REF}}{V_{REF}}, \tag{7.26}$$

where ε is the setting accuracy of V_{INT} for V_{REF}. As discussed in Chapter 1 the gain of the differential amplifier is expressed as follows:

$$A = r_1 g_{m1}, \tag{7.27}$$

$$r_1 = \frac{1}{(\lambda_p + \lambda_n) I_S/2}, \tag{7.28}$$

$$g_{m1} = \sqrt{\frac{I_S W_1 \beta_{n0}}{L}}, \tag{7.29}$$

where β_{n0} is the gain factor of nMOSTs, λ_p and λ_n are channel-length modulation constants for pMOSTs (M_3, M_4) and nMOSTs (M_1, M_2), respectively. Consequently, the sizes of M_1 and M_2 are obtained as

$$\frac{W_1}{L} = \frac{W_2}{L} = \left\{ \frac{A(\lambda_p + \lambda_n)}{2} \right\}^2 \frac{I_S}{\beta_{n0}}, \tag{7.30}$$

Both M_5 and M_7, which must always operate in the saturation region to give a constant current, have wide voltage margins. Therefore, their sizes are easily obtained, if their gate voltages are determined so as to achieve the necessary I_S and I_B (usually µA).

Here is an example of design [16]. MOST parameters such as $W_6 \geq 2{,}000\,\mu\text{m}$, $W_3 = W_4 \geq 250\,\mu\text{m}$, $W_1 = W_2 \geq 290\,\mu\text{m}$, $I_S = 1.6\,\text{mA}$, and

$r_1 = 8.9 \, \text{k}\Omega$ were obtained for the design conditions of $V_{DD} = 5 \, \text{V}$, $V_{INT} = 3 \, \text{V}$, $V_{DS6} = 2 \, \text{V}$, $V_{REF} = 3 \, \text{V}$, $I_L \leq 100 \, \text{mA}$, $I_B = 2.5 \, \mu\text{A}$, $\Delta V_{INT}/V_{INT} \leq 10\%$, $(V_{INT} - V_{REF})/V_{REF} \leq 1\%$, $C_L = 1 \, \text{nF}$, $L = 1.2 \, \mu\text{m}$, $\beta_{p0} = 3 \times 10^{-5} \text{S/V}$, $\beta_{n0} = 8 \times 10^{-5} \text{S/V}$, $g_{m1} = 4.9 \, \text{mS}$, $C_{OX} = 2.3 \, \text{fF}/\mu\text{m}^2$, $V_{tp} = 0.5 \, \text{V}$, $\lambda_p = 0.1/\text{V}$, and $\lambda_n = 0.04/\text{V}$. In fact, an actual 0.5-μm 16-Mb DRAM with a voltage down converter [16] that used the similar parameters yielded experimental results of $\Delta V_{INT}/V_{INT} = 6\%$ and $(V_{INT} - V_{REF})/V_{REF} = 2\%$.

7.2.3. AC Characteristics and Phase Compensation

Stable conditions. The key to designing a negative feedback amplifier is to cope with its inherent instabilities, such as ringing or oscillation. The design requires the investigation of the frequency response of the amplifier. Let us consider a multi-stage amplifier in Fig. 7.10. Assuming that each stage has a low-frequency gain of a_i, output resistance of r_i, and load capacitance of $C_i (i = 1, 2, 3)$, the transfer function is expressed as

$$a(s) = \frac{v_{out}(s)}{v_{in}(s)} = \frac{a_0}{\left(1 + s/\omega_{P1}\right)\left(1 + s/\omega_{P2}\right)\left(1 + s/\omega_{P3}\right)}. \tag{7.31}$$

Here $a_0 = a_1 a_2 a_3$ is total gain at low frequency and ω_{P1}, ω_{P2}, ω_{P3} are "poles", which is given by

$$\omega_{P1} = \frac{1}{C_1 r_1}, \quad \omega_{P2} = \frac{1}{C_2 r_2}, \quad \omega_{P3} = \frac{1}{C_3 r_3} \tag{7.32}$$

[17, 18]. The response to a sinusoidal input of angular frequency ω is given by substituting $s = j\omega$ into equation (7.31). The AC characteristics of an amplifier, whose transfer function is expressed by (7.31), are analyzed by examining both the gain and the phase characteristics of the function. The gain characteristics are examined on the plane of $\log \omega$ (rad/s) and $20 \log_{10} |a(j\omega)|$ (decibels, dB), while the phase characteristics are examined on the plane of $\log \omega$ and phase arg $a(j\omega)$. Before going into detail, we investigate here how the gain and phase behave on the planes, citing the example of typical circuits.

Let us begin with a single-stage amplifier, whose transfer function is expressed as

$$a(s) = \frac{a_0}{\left(1 + s/\omega_{P1}\right)}. \tag{7.33}$$

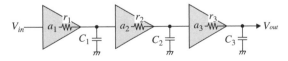

FIGURE 7.10. A multi-stage amplifier

The gain and the phase are expressed as follows:

$$|a\,(j\omega)| = \frac{a_0}{\sqrt{\left(1 + \omega/\omega_{P1}\right)^2}}, \qquad (7.34)$$

$$20\log|a\,(j\omega)| = 20\log a_0 - 10\log\left\{1 + \left(\frac{\omega}{\omega_{P1}}\right)^2\right\}, \qquad (7.35)$$

$$\arg a\,(j\omega) = -\arctan\frac{\omega}{\omega_{P1}}. \qquad (7.36)$$

Figure 7.11 shows the gain and phase characteristics. The gain and phase are about $20\log a_0$ (dB) and $0°$ at $\omega \ll \omega_{p1}$, respectively. However, they are $20\log a_0 - 3$ (dB) and $-45°$ at $\omega = \omega_{p1}$, and $20\log a_0 + 20\log\omega_{p1} - 20\log\omega$ (dB) and $-90°$ at $\omega \gg \omega_{p1}$. In other words, the gain curve is flat at $\omega < \omega_{p1}$, but it decreases by 3 dB at $\omega = \omega_{p1}$, and then decreases by 20 dB per decade in ω. As for phase, it is $0°$ at $\omega = 0$. However, it is delayed with increasing ω, reaching $-45°$ at $\omega = \omega_{p1}$, and $-90°$ at $\omega = \infty$.

The gain and phase characteristics of a multi-stage amplifier is shown in Fig. 7.12. Here, poles are assumed to be sufficiently (more than one decade) separated from each other. The gain is flat at $\omega < \omega_{p1}$, decreases by 20 dB per decade for $\omega_{P1} < \omega < \omega_{P2}$, decreases by 40 dB per decade for $\omega_{P2} < \omega < \omega_{P3}$, and decreases by 60 dB per decade for $\omega_{P3} < \omega$. As for phase, it is $0°$ at $\omega = 0$, $-45°$ at $\omega = \omega_{p1}$, $-135°$ at $\omega = \omega_{p2}$, and $-225°$ at $\omega = \omega_{p3}$. The frequency at which the gain is 0 dB is an important point for AC characteristics called the unity-gain frequency.

The transfer functions of some circuits have an s-term in the numerator, such as

$$a\,(s) = \frac{a_0\left(1 + s/\omega_Z\right)}{\left(1 + s/\omega_{P1}\right)\left(1 + s/\omega_{P2}\right)\left(1 + s/\omega_{P3}\right)}, \qquad (7.37)$$

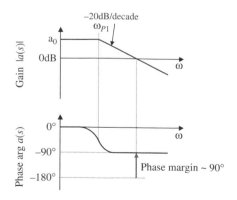

FIGURE 7.11. Gain and phase characteristics of a single-stage amplifier.

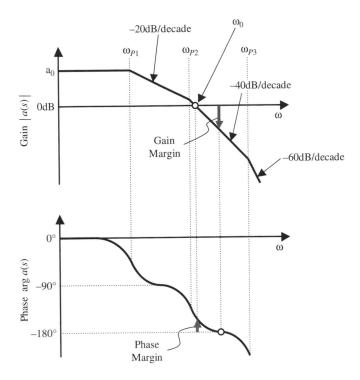

FIGURE 7.12. Gain and phase characteristics of a multi-stage amplifier.

where, ω_Z is a zero. The role of a zero is opposite that of a pole, as analyzed in the same manner as above. The gain increases by 3 dB at $\omega = \omega_Z$, and increases at a rate of 20 dB/decade for $\omega > \omega_Z$. The phase advances to the positive with an increaseing ω, reaching $+45°$ at $\omega = \omega_Z$, and $+90°$ at $\omega = \infty$. If the zero coincides one of the poles, $\omega_Z = \omega_{Pi}$, ω_{Pi} is cancelled out, as described later.

Let us consider the voltage follower shown in Fig. 7.13 using the amplifier with negative feedback. The transfer function of this circuit is expressed as

$$\{V_{REF}(s) - \beta V_{INT}(s)\} a(s) = V_{INT}(s), \tag{7.38}$$

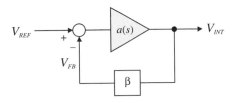

FIGURE 7.13. A voltage follower using negative feedback.

$$\therefore A\left(s\right) = \frac{V_{INT}\left(s\right)}{V_{REF}\left(s\right)} = \frac{a\left(s\right)}{1+\beta \cdot a\left(s\right)}. \tag{7.39}$$

The function $A(s)$ is the transfer function with feedback and is called "closed-loop" transfer function, while $a(s)$ is a "open-loop" transfer function. Here, β is a feedback ratio between feedback voltage V_{FB} and output voltage V_{INT}. In the case of voltage down converters, β is usually unity. However, the circuit in Fig. 7.8(a) and the reference voltage converters described in Section 6.5 have smaller β. The following analysis assumes $\beta = 1$, because it is the severest condition for loop stability.

When $\beta = 1$, the equation (7.39) becomes

$$A\left(s\right) = \frac{V_{INT}\left(s\right)}{V_{REF}\left(s\right)} = \frac{a\left(s\right)}{1+a\left(s\right)}. \tag{7.40}$$

Since the phase shift of $a(s)$ is almost zero and the absolute value (gain) is sufficiently large at DC or low frequency, $A(s) \approx 1$ and $V_{INT}(s) \approx V_{REF}(s)$ are derived from (7.40). This means that the voltage V_{INT} follows V_{REF}, and that the circuit functions as a voltage follower. At high frequency, however, the behavior of the circuit differs from that of a voltage follower. The frequency characteristics of $a(s)$ must be investigated to judge the loop stability.

To stabilize the closed-loop transfer function $A(s)$, the gain of $a(s)$ must be lower than unity (0 dB) at the frequency where the phase of $a(s)$ reaches $-180°$ because of the following reason. Since negative feedback itself causes a phase shift of $-180°$, the total phase shift of the entire loop becomes $-360°$ (positive feedback). If the gain is larger than 0 dB at this frequency, a signal with the same phase and larger amplitude than the original signal is fed again to the amplifier. Thus, the amplitude of the signal increases with time, and oscillation occurs. Therefore, the gain at this frequency must be lower than 0 dB. In other words (see Fig. 7.12) the phase shift must be within $-180°$ at the unity-gain frequency, where the gain reaches 0 dB. The margin φ_M between the phase at the unity-gain angular frequency ω_0 and $-180°$ is expressed as

$$\varphi_M = \arg a\left(j\omega_0\right) + 180°. \tag{7.41}$$

This is called "phase margin", an important parameter to judge the loop stability.

Oscillation does not occur as long as the phase margin is positive. Then, is an extremely small positive phase margin acceptable? The answer is "no" because of the following two reasons. First, the circuit must be designed with a sufficient margin against PVT (process, voltage, temperature) variations. Second, a small phase margin causes unstable operation of the circuit. Figure 7.14 shows the frequency characteristics of closed-loop gain $|A(j\omega)|$ (the absolute value of closed-loop transfer function) with the phase margin φ_M as a parameter (see Section A7.1). Here, ω_0 is the unity-gain angular frequency. The closed-loop gain $|A(j\omega)|$ is almost 0 dB at low frequency ($\omega << \omega_0$) as describe above. If the phase margin is large, $|A(j\omega)|$ has no peaks and gradually decreases with

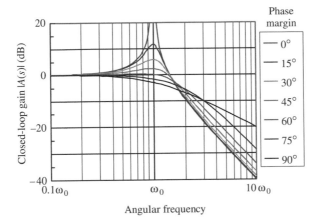

FIGURE 7.14. Normalized closed-loop gain for various phase margin, where ω_0 is the unity-gain angular frequency.

frequency. If the phase margin is small, however, $|A(j\omega)|$ has a peak at $\omega \approx \omega_0$. The peak gain is expressed as

$$|A(j\omega)|_{\max} \approx \frac{1}{2 \sin \dfrac{\varphi_M}{2}}. \tag{7.42}$$

This is 5.7 dB for a phase margin of 30° and 11.7 dB for 15°. Circuits under such conditions are not suitable for voltage down converters because a large ringing of the output voltage might be generated by the fluctuations of V_{DD} and load current. The operating waveform of a circuit with a phase margin of almost 0° is shown in Fig. 7.15. The output (V_{INT}) starts to ring when a large load current of

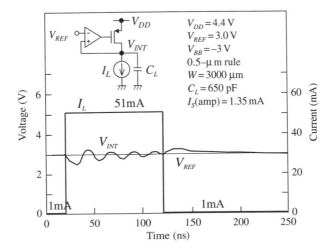

FIGURE 7.15. A simulated oscillatory V_{INT} waveform without phase compensation.

51 mA is applied, though such a ringing never happens for a small load current of 1 mA. The ideal phase margin is 60° [18]. In the design of actual circuits, however, a phase margin of 60° may result in a large layout area. Accordingly a phase margin of 45° is a general design criterion.

The design policies to ensure a sufficient phase margin are as follows. First, the poles, which cause phase shift, should be as few as possible. This means the number of stages should be as small as possible because each stage has at least one pole as shown in Fig. 7.10. A single-stage amplifier causes no loop-stability problem because a phase margin of at least 90° is ensured as shown in Fig. 7.11. A voltage down converter using a single-stage amplifier is reported [9]. However, it is difficult to obtain good DC characteristics with a single-stage amplifier because of an insufficient voltage gain. Therefore, most voltage down converters utilize two-stage amplifiers. Second, the poles should be separated from each other in frequency. If the poles are close to each other, the phase delays before the gain sufficiently decreases. The ratio between the lowest-frequency pole (dominant pole) ω_1 and the second-lowest-frequency pole ω_2 is important. The third lowest and the following poles do not have a large effect on the phase margin. Third, the open-loop gain of the amplifier should not be too large. Although a large open-loop gain is favorable for DC characteristics as suggested by equations (7.2) and (7.4), a too much open-loop gain causes unstable AC characteristics. Figure 7.16 shows the relationship between pole-frequency ratio ω_2/ω_1 and the low-frequency gain a_0 necessary to ensure a phase margin of 45° or 60° (See Section A7.1). Roughly speaking the pole-frequency ratio must be at least a_0. If $a_0 = 1000$ (60dB), for example, ω_1 and ω_2 should be separated from each other by at least three decades.

The voltage down-converter and phase margin. In this section, a voltage down-converter shown in Figure 7.17(a) is analyzed with its AC small-signal equivalent circuit, and then the stable condition is given.

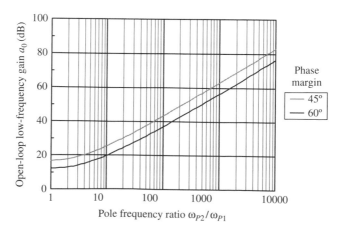

FIGURE 7.16. Relationship between pole-frequency ratio and open-loop low-frequency gain.

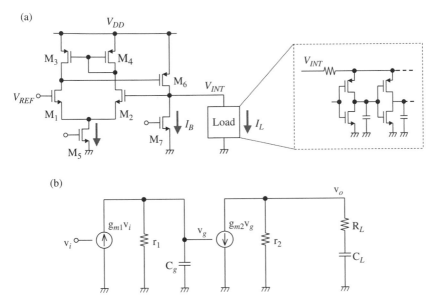

FIGURE 7.17. A voltage down converter (a) and its open-loop small-signal equivalent circuit (b) [16].

When a small signal is inputted, the circuit shown in Fig. 7.17(a) is expressed as a superposition of a DC components circuit (i.e. a fixed DC bias circuit) without any input signal and a small signal AC component circuit. Therefore, AC analysis can be carried out by extracting only the small signal AC component circuit shown in Fig. 7.17(b). Here, to simplify the analysis of feedback loop stability, the loop is made open. Through investigating the whole gain and phase characteristics of the open-loop circuit, we can estimate the characteristics of the actual closed-loop circuit. In the following, we will study how to obtain the equivalent circuit shown in Fig. 7.17(b).

Figure 7.18(a) and (b) show the differential amplifier and its equivalent circuit. Capital letters are used for DC biasings, while small letters are used for small signals. The sizes of the paired MOSTs (M_3 and M_4, M_1 and M_2) are the same and the output resistance and the transconductances of the nMOST and the pMOST are r_n and r_p, and g_{mn} and g_{mp}, respectively. All of the MOSTs are in saturation mode, and the r_p of M4 is negligible because of the forward biased diode connection. Thus, the current equations at nodes A, B, and C in Fig. 7.18(b) are as follows:

$$g_{mp}v_1 + \frac{v_1 - v_3}{r_n} + g_{mn}(v_1 - v_3) = 0, \qquad (7.43)$$

$$g_{mp}v_1 + \frac{v_2}{r_p} - g_{mn}v_3 + \frac{v_2 - v_3}{r_n} = 0, \qquad (7.44)$$

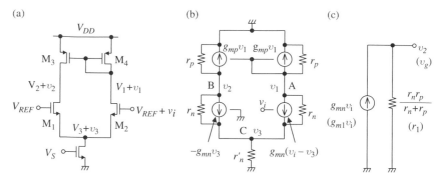

(a)

(b)

(c)

FIGURE 7.18. A differential amplifier (a), its small-signal equivalent circuit (b), and the simplified equivalent circuit (c). Reproduced from [37] with permission of Springer.

$$-\frac{v_1 - v_3}{r_n} - g_{mn}\left(v_i - v_3\right) - \frac{v_2 - v_3}{r_n} + g_{mn}v_3 + \frac{v_3}{r'_n} = 0. \qquad (7.45)$$

Assuming that $g_{mn}r_n \gg 1$, $gm_p r_p \gg 1$, and $g_{mn}r'_n \gg 1$, v_2 is obtained as follows:

$$v_2 = \frac{r_n r_p}{r_n + r_p} g_{mn} v_i. \qquad (7.46)$$

Thus, the amplifier is eventually simplified to the circuit shown in Fig. 7.18(c). Therefore, the whole voltage down converter is expressed with the equivalent circuit shown in Fig. 7.17(b), where g_{m2} and r_2 are the transconductance and output resistance of the output pMOST (M_6), respectively, C_g is the sum of the gate-source capacitance and the gate-drain capacitance of M_6. The load is modeled by the series connection of R_L and C_L. R_L corresponds to the sum of the wiring resistance of the V_{INT} line and the on resistances of the pMOSTs in the CMOS core circuit, and C_L corresponds to the sum of the wiring capacitance of the V_{INT} line and the capacitances seen through pMOSTs in the core circuit.

The circuit shown in Fig. 7.17(b) is analyzed with Laplace transformation as

$$-g_{m1}v_i + \frac{v_g}{r_1} + sC_g v_g = 0, \qquad (7.47)$$

$$g_{m2}v_g + \frac{v_o}{r_2} + \frac{v_o}{R_L + 1/sC_L} = 0, \qquad (7.48)$$

$$\therefore a\left(s\right) = \frac{v_o}{v_i} = \frac{-g_{m1}r_1}{sC_g r_1 + 1} \cdot \frac{g_{m2}r_2\left(sC_L R_L + 1\right)}{sC_L\left(R_L + r_2\right) + 1}. \qquad (7.49)$$

By transforming into the complex-frequency plane ($s = j\omega$), we obtain

$$a\left(j\omega\right) = \frac{-g_{m1}r_1 g_{m2}r_2\left(1 + j\omega/\omega_z\right)}{\left(1 + j\omega/\omega_{P1}\right)\left(1 + j\omega/\omega_{P2}\right)}, \qquad (7.50)$$

$$\omega_{P1} = \frac{1}{C_g r_1}, \quad \omega_{P2} = \frac{1}{C_L(R_L + r_2)}, \quad \omega_Z = \frac{1}{C_L R_L}, \quad (7.51)$$

which has two poles (ω_{P1}, ω_{P2}) and one zero (ω_Z). Usually, ω_Z is much higher than ω_{P1} and ω_{P2} because R_L (several Ω) $<< r_1$ and r_2, and it is located at the frequency where the gain is sufficiently reduced on the ω-gain plane. Therefore, ω_Z has no substantial effect on the gain and phase characteristics, making (7.48) almost equal to

$$a(j\omega) = \frac{-g_{m1} r_1 g_{m2} r_2}{(1 + j\omega/\omega_{P1})(1 + j\omega/\omega_{P2})}. \quad (7.52)$$

The phase margin can be calculated from this equation in the same manner as Fig. 7.12.

Above described is the analytical method to evaluate the phase margin. Two evaluation methods through circuit simulation are next explained. One uses an open-loop circuit, while the other uses a closed-loop circuit.

Figure 7.19 shows the test circuit to calculate an open-loop transfer function. The feedback loop is opened by cutting off a path (usually the feedback path). An AC voltage source v_{in} is connected to one end, and the gain and phase of the other end are observed. The following two modifications are also needed. The first modification is to insert a low-pass filter (LPF) at the opened path so that DC bias conditions are not changed. The cut-off frequency of this LPF must be sufficiently lower than the minimum simulation frequency f_{min}, that is,

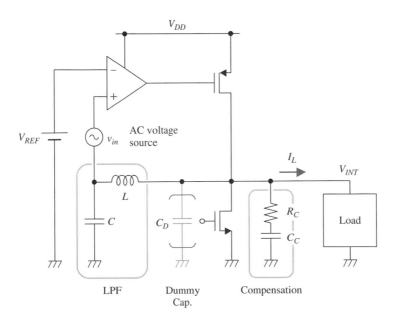

FIGURE 7.19. Simulation method for phase margin (open loop).

$$\frac{1}{2\pi\sqrt{LC}} \ll f_{min}. \tag{7.53}$$

In addition, the impedance $2\Pi f_{min}L$ of the inductor at f_{min} must be sufficiently larger than that of V_{INT} so that C does not affect the capacitance of the node V_{INT}. Since this LPF is only for the sake of simulation convenience, unrealistic values (e. g. $L = 1$ GH, $C = 1$ F) are allowed. The second modification is to add a dummy capacitance C_D to compensate for the input capacitance of the differential amplifier, which is separated from the node V_{INT} by the LPF. The capacitance C_D, however, can be omitted if the capacitance of node V_{INT} is sufficiently large.

Figure 7.20 shows the test circuit to calculate a closed-loop transfer function. This method is based on the dependence of the peak gain on the phase margin shown in Fig. 7.14. An ac voltage source v_{in} is connected in series with V_{REF} and the peak gain at the output is observed. The phase margin φ_M is calculated using equation (7.42). This method is effective especially in the case that two or more voltage down converters share an output node. One drawback of this method is that the error of (7.42) becomes larger for $\varphi_M > 45°$. (A more accurate equation is (A7.12), which also assumes that the circuit has only two poles sufficiently separated from each other.) However, $\varphi_M > 45°$ means a sufficient phase margin and a small evaluation error may be allowed.

An accurate model of the load (internal circuit) is critical for either method, because the frequency characteristics of the output stage are greatly dependent on the load capacitance and output current. For example the output current greatly changes with the mode (active/standby) of the internal circuit. In memory

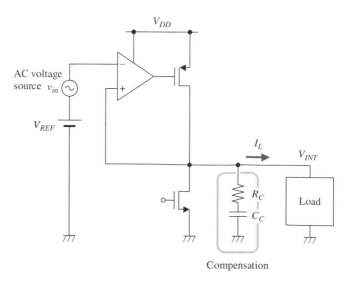

FIGURE 7.20. Simulation method for phase margin (closed loop)

LSIs the load capacitance is greatly dependent on whether the memory array is connected to V_{INT} or not [19]. In these cases, we must evaluate the phase margins in both conditions. The method in Fig. 7.20 is used to experimentally evaluate the phase margin as described later, because the method in Fig. 7.19, which requires cutting a wire and inserting a low-pass filter, is hard to apply.

Compensation. To stabilize the voltage down converter, the phase at the unity-gain frequency should be larger than $-135°$ (i.e. phase margin larger than $45°$), as explained previously. Improving the phase margin by modifying the original transfer function is called compensation. This is realized by shifting ω_{P1} to a sufficiently higher frequency or to a frequency that is sufficiently lower than ω_{P2} for a given ω_{P2}. This is also achieved by shifting ω_{P2} for a fixed ω_{P1}. In any case, the key to stabilization is a sufficient separation of two poles.

Three kinds of compensation (dominant pole compensation, pole-zero compensation, Miller compensation) have been proposed for application to on-chip voltage down converters, as shown in Table 7.2. Our major concerns are not only the stability, but also the size of the compensation devices, because the devices must be incorporated in a chip.

Dominant pole compensation adds a compensation capacitor (C_C) at the output V_{INT} as shown in Fig. 7.21(a). Figure 7.21(b) shows the small-signal equivalent circuit. Here, C_g is the gate-source capacitance of M_6, that is, $(2/3)C_{OX}LW_6$

TABLE 7.2. Comparison between phase compensation methods. Reproduced from [37] with permission of Springer.

Compensation	Gain/phase characteristics	Poles and zeros of open-loop gain function	Stabilizing condition
Dominant pole	Fig. 7.21(c)	$\omega_{P1} = \frac{1}{C_g r_1}$, $\omega_{P2} \cong \frac{1}{r_2(C_C+C_L)}$	(a) $R_L = 0$ $\quad C_C \geq \frac{a_0 C_g r_1}{\sqrt{2} r_2} - C_L$
		$\omega_{P3} \cong \frac{C_C+C_L}{C_C C_L R_L}$, $\omega_{Z1} \cong \frac{1}{C_L R_L}$	(b) $R_L \geq \frac{C_g r_1}{(3+2\sqrt{2})C_L}$
		$a_0 = g_{m1} r_1 g_{m2} r_2$	$C_C = 0$
Pole-zero	Fig. 7.22(c)(d)	$\omega_{P1} = \frac{1}{C_g r_1}$, $\omega_{P2} \cong \frac{1}{r_2(C_C+C_L)}$	(a) $R_L = 0$ $\quad C_C R_C = C_g r_1$
		$\omega_{P3} \cong \frac{C_C+C_L}{C_C C_L (R_C+R_L)}$	$C_C \geq \sqrt{\frac{a_0 C_L C_g r_1}{\sqrt{2} r_2}} - C_L$
		$\omega_{Z1} = \frac{1}{C_L R_L}$, $\omega_{Z2} = \frac{1}{C_C R_C}$	(b) $R_L > 0$ $\quad C_C R_C = C_g r_1$
		$a_0 = g_{m1} r_1 g_{m2} r_2$	$C_C \geq \frac{1}{k-1}\left(\frac{r_1 C_g}{R_L} - kC_L\right)$ $(k = 3+2\sqrt{2})$
Miller	Fig. 7.24(c)	$\omega_{P1} \cong \frac{1}{C_C g_{m2} r_1 r_2}$, $\omega_{P2} \cong \frac{g_{m2}}{C_L}$	(a) $C_g << C_C$, C_L $\quad C_C \geq \frac{C_L g_{m1}}{\sqrt{2} g_{m2}}$
		$\omega_{Z1} = \frac{1}{R_L C_L}$, $\omega_{Z2} \cong -\frac{g_{m2}}{C_C}$	(b) $C_g \sim C_C$, C_L $\quad C_C \geq \frac{1}{2}\sqrt{\frac{A^2}{2} + B} + \frac{\sqrt{2}}{4}A$
		$a_0 = g_{m1} r_1 g_{m2} r_2$	$A = \frac{(C_L+C_g)g_{m1}}{g_{m2}}$
			$B = \frac{2\sqrt{2}C_L C_g g_{m1}}{g_{m2}}$

(a)

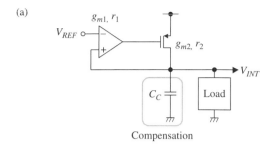

(b)

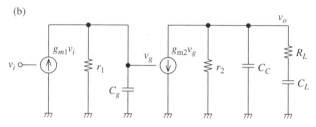

(c)

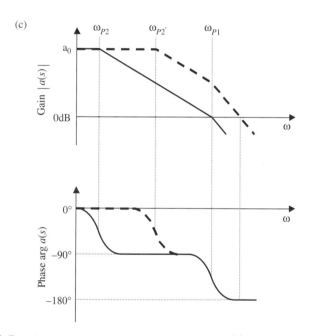

FIGURE 7.21. Dominant pole compensation; circuit diagram (a), its small-signal equivalent circuit (b), and gain and phase characteristics (c) [16].

(C_{OX} is the gate capacitance per unit area), because M_6 is in the saturation region. The dotted curves in Fig. 7.21(c) represent the gain and phase curves before compensation, while the solid curves are after compensation. The resulting open-loop transfer function has three poles (ω_{P1}, ω_{P2}, and ω_{P3}) and one zero (ω_{Z1}). A small R_L and large $C_L(>> C_g)$ allow ω_{P3}, $\omega_{Z1} >> \omega_{P1}$, ω_{P2}, and $\omega_{P1} >> \omega_{P2}$. Thus ω_{P2} shifts from $\omega_{P2}' = 1/[C_L(R_L + r_2)]$ expressed in (7.51) to a lower frequency of $\omega_{P2} = 1/[(C_C + C_L)r_2]$, enabling a wider phase margin of $45°$. However, the compensation requires a quite large C_C, which is expressed as

$$C_C \geq \frac{a_0 C_g r_1}{\sqrt{2} r_2} - C_L \tag{7.54}$$

where $r_2 = 1/(\lambda_p I_L)$ [16]. The large gain a_0 (typically 60 dB) is responsible for the large C_C. For example, the same parameters as in subsection 7.2.2 result in $C_C \geq 100$ nF. If C_C is fabricated with a 15-nm oxide sandwiched between the n-well and the poly-Si, it occupies as much as 50 mm^2. Dominant-pole compensation is therefore not suitable for on-chip integration. However, if an off-chip capacitor is available, it is a simple and effective solution both for compensation and for V_{INT} stabilization.

Pole-zero compensation is the most promising for on-chip voltage down converters. In this compensation, the resistor R_C and the capacitor C_C are added at the output of the circuit shown in Fig. 7.22(a). Figure 7.22(b) shows the small-signal equivalent circuit. The current equations at the nodes, expressed using the Laplace transformation, are as follows:

$$-g_{m1}v_i + \frac{v_g}{r_1} + sC_g v_g = 0, \tag{7.55}$$

$$g_{m2}v_g + \frac{v_o}{r_2} + \frac{v_o}{R_C + \dfrac{1}{sC_C}} + \frac{v_o}{R_L + \dfrac{1}{sC_L}} = 0. \tag{7.56}$$

Hence, the open-loop transfer function is expressed as

$$a(j\omega) = \frac{v_o}{v_i} = \frac{a_0\left(1 + j\omega/\omega_{Z1}\right)\left(1 + j\omega/\omega_{Z2}\right)}{\left(1 + j\omega/\omega_{P1}\right)\left(1 + j\omega/\omega_{P2}\right)\left(1 + j\omega/\omega_{P3}\right)}, \tag{7.57}$$

where $s = j\omega$, and the poles, zeros, and low-frequency gain are defined as follows:

$$\omega_{P1} = \frac{1}{C_g r_1}, \tag{7.58}$$

$$\omega_{P2} = \frac{1}{C_C(R_C + r_2) + C_L(R_L + r_2)} \approx \frac{1}{(C_C + C_L)r_2}, \quad (R_C, R_L << r_2) \tag{7.59}$$

$$\omega_{P3} = \frac{C_C(R_C + r_2) + C_L(R_L + r_2)}{C_C C_L(R_C R_L + r_2 R_C + r_2 R_L)} \approx \frac{C_C + C_L}{C_C C_L(R_C + R_L)}, \tag{7.60}$$

$$\omega_{Z1} = \frac{1}{C_L R_L}, \tag{7.61}$$

$$\omega_{Z2} = \frac{1}{C_C R_C}, \tag{7.62}$$

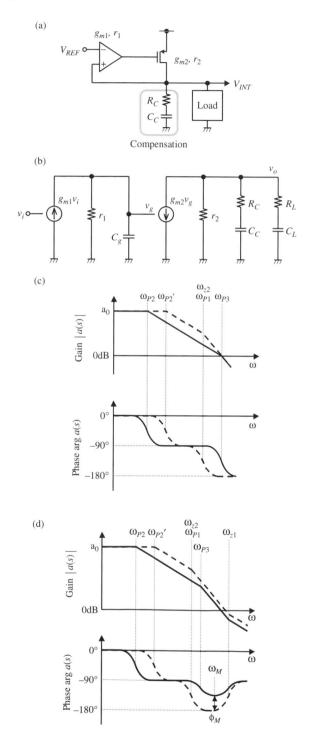

FIGURE 7.22. Pole-zero cancellation compensation; circuit diagram (a), its small-signal equivalent circuit (b), and gain and phase characteristics for $R_L = 0$ (c), and $R_L > 0$ (d) [16].

$$a_0 = g_{m1} r_1 g_{m2} r_2. \tag{7.63}$$

Therefore the transfer function has three poles (ω_{P1}, ω_{P2}, and ω_{P3}) and two zeros (ω_{Z1}, ω_{Z2}). When comparing with (7.51) it is obvious that ω_{P3} and ω_{Z2} are newly generated, and ω_{P2} shifts to a lower frequency because of an additional component of $C_C(R_C + r_2)$. Note that ω_{Z2} plays an important role, because ω_{P1} can be canceled by ω_{Z2} if $\omega_{Z2} = \omega_{P1}$, that is,

$$C_C R_C = C_g r_1. \tag{7.64}$$

This implies that a decrease in gain of $-20\,\text{dB/decade}$ and a phase delay of over $-45°$ at $\omega > \omega_{P1}$ are canceled by an increase in gain of $+20\,\text{dB/decade}$ and a phase advance of over $+45°$ at $\omega > \omega_{Z2}$. Under these conditions, stabilizing conditions for the converter are obtained for two cases of $R_L = 0$ and $R_L > 0$, as follows.

Case of $R_L = 0$. In this case only two poles (ω_{P2} and ω_{P3}) remain in $a(j\omega)$ because $\omega_{Z1} = \infty$, and thus

$$a(j\omega) = \frac{-a_0}{(1 + j\omega/\omega_{P2})(1 + j\omega/\omega_{P3})}. \tag{7.65}$$

This is the same as (7.52) because $\omega_{P3} \gg \omega_{P2}$, if we regard ω_{P3} as ω_{P1} in (7.52). The stabilizing condition, where the gain at $\omega = \omega_{P3}$ is smaller than $0\,\text{dB}$ and the phase is $-135°$ (see Fig. 7.22(c)), is obtained as

$$|a(j\omega_{P3})| = \frac{a_0}{\sqrt{2} \dfrac{\omega_{P3}}{\omega_{P2}}} \leq 1. \tag{7.66}$$

Substituting (7.59), (7.60) and (7.64) into this equation and solving for C_C and R_C results in the following design equations: [16],

$$C_C \geq \sqrt{\frac{a_0 C_L C_g r_1}{\sqrt{2} r_2}} - C_L, \tag{7.67}$$

$$R_C \leq \frac{C_g r_1}{\sqrt{\dfrac{a_0 C_L C_g r_1}{\sqrt{2} r_2}} - C_L}. \tag{7.68}$$

The necessary C_C for the phase margin of $45°$ is rather smaller than that for dominant-pole compensation. This is because it is proportional to the square root of a_0, instead of a_0 for dominant-pole compensation.

Case of $R_L > 0$. The transfer function is expressed as

$$a(j\omega) = \frac{-a_0(1 + j\omega/\omega_{Z1})}{(1 + j\omega/\omega_{P2})(1 + j\omega/\omega_{P3})}. \tag{7.69}$$

The gain and phase characteristics are shown in Fig. 7.22(d). The zero ω_{Z1} increases the gain and advances the phase. Therefore, the decreasing rate of the gain is -20dB/decade in $\omega > \omega_{P2}$, is -40dB/decade in $\omega > \omega_{P3}$, and is again -20dB/decade in $\omega > \omega_{Z1}$. The phase is $-45°$ at $\omega = \omega_{P2}$ and exceeds $-90°$ due to the pole ω_{P3}, but comes back toward $-90°$ due to ω_{Z1}. That is, the phase has a minimum value at a frequency between ω_{P3} and ω_{Z1}. This is explained by the following analysis. From (7.69) the phase is expressed as

$$\arg a\,(j\omega) = -\arctan\frac{\omega}{\omega_2} - \arctan\frac{\omega}{\omega_3} + \arctan\frac{\omega}{\omega_{Z1}}$$

$$\approx -90° - \arctan\frac{\omega}{\omega_3} + \arctan\frac{\omega}{\omega_{Z1}} \qquad (7.70)$$

if we deal with $\omega \gg \omega_{P2}$. Hence, by using the addition theorem of tangent,

$$\tan\,(\arg a\,(j\omega) + 90°) = \frac{-1/\omega_{P3} + 1/\omega_{Z1}}{1/\omega + \omega/\omega_{P3}\omega_{Z1}}. \qquad (7.71)$$

The right-hand side of the equation is a minimum at

$$\omega = \omega_M = \sqrt{\omega_{P3}\omega_{Z1}}. \qquad (7.72)$$

Since the stabilizing condition is $\arg a(j\omega_M) \geq -135°$, and thus $\tan(\arg a(j\omega_M) + 90°) \geq -1$, (7.71) is expressed as

$$4\omega_{P3}\omega_{Z1} \geq (\omega_{Z1} - \omega_{P3})^2. \qquad (7.73)$$

Thus, by using (7.64), the limitations for ω_{Z1}, C_C and R_C are obtained as follows:

$$\frac{\omega_{Z1}}{\omega_{P3}} \leq 3 + 2\sqrt{2}, \qquad (7.74)$$

$$C_C \geq \frac{C_g r_1/R_L - \left(3 + 2\sqrt{2}\right)C_L}{2 + 2\sqrt{2}}, \qquad (7.75)$$

$$R_C \leq \frac{\left(2 + 2\sqrt{2}\right)C_g r_1}{C_g r_1/R_L - \left(3 + 2\sqrt{2}\right)C_L}. \qquad (7.76)$$

On the other hand, the gain at $\omega = \omega_M$ is derived using (7.72) as follows:

$$|a\,(j\omega_M)| = \frac{a_0\sqrt{1 + \left(\dfrac{\omega_M}{\omega_{Z1}}\right)^2}}{\sqrt{1 + \left(\dfrac{\omega_M}{\omega_{P2}}\right)^2}\sqrt{1 + \left(\dfrac{\omega_M}{\omega_{P3}}\right)^2}} \approx a_0\left(\frac{\omega_{P2}}{\omega_{Z1}}\right). \qquad (7.77)$$

Note that as long as there is a zero and the minimum phase is larger than $-135°$ (i.e. the phase margin $\varphi_M \geq 45°$), even a gain of larger than unity realizes a stable circuit. The condition $|a(j\omega_M)| < 1$ would result in an over margin, as follows.

In the case shown in Fig. 7.22(c), where the circuit is substantially characterized only by the poles, the phase delays from $-135°$ to $-180°$ when ω is increased beyond ω_{P3}. The phase delay increases the degree of instability toward positive feedback at $-180°$. Thus, to ensure stability, the gain must be strictly limited to less than unity. In the case shown in Fig. 7.22(d), however, where there is the zero influential to the AC characteristics, a phase margin φ_M larger than $45°$ is ensured at any ω as long as the minimum φ_M at $\omega = \omega_M$ is larger than $45°$. The limitation on the gain is not necessarily needed, because the gain at $\omega = \omega_M$ is low enough as (7.77). Therefore, (7.75) and (7.76) are stable conditions.

The voltage down converter must be stable against wide changes of load parameters (R_L, C_L, I_L) caused by internal circuit operation. Figure 7.23(a) shows the relationship between the parameter changes and the phase margin for

(a)

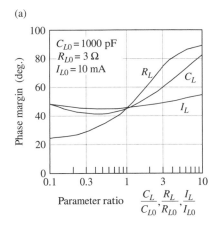

(b)

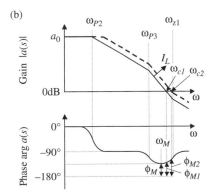

FIGURE 7.23. A sensitivity analysis for pole-zero cancellation; phase margin versus load parameters (a), and gain and phase characteristics (b). Reproduced from [16] with permission; © 2006 IEICE.

pole-zero compensation [16]. Here, the phase margin is defined as the margin at $\omega = \omega_{C1}$ where the gain is $0\,\text{dB}$, as shown by φ_{M1} in Fig. 7.23(b). φ_{M1} is derived from analyzing (7.67), assuming $C_C = 1.92\,\text{nF}$ and $R_C = 23.4\,\Omega$, which are obtained by substituting normal values of $R_{L0} = 3\,\Omega$, $C_{L0} = 1\,\text{nF}$, $I_{L0} = 10\,\text{mA}$, $C_g = 7.2\,\text{pF}$, $r_1 = 6.25\,\text{k}\Omega$, $\lambda_p = 0.1/\text{V}$, $\lambda_n = 0.04/\text{V}$, in (7.75) and (7.76). The parameters are all for a $0.5 - \mu\text{m}$ 16-Mbit DRAM. φ_{M1} increases with increasing R_L or C_L because ω_{Z1} approaches ω_{Z3}. However, it is almost a minimum at the normal value of I_{L0} when I_L is changed. This can be explained using Fig. 7.23(b). The pole ω_{P2} increases with I_L because $r_2 = (\lambda_p I_L)^{-1}$, which makes the gain curve shifted upwards with a fixed location of ω_M that is independent of I_L (i.e. a thick dashed line in the figure). Thus, the unity-gain angular frequency is shifted from ω_{C1} to ω_{C2} and the phase margin is increased from φ_{M1} to φ_{M2}. On the contrary, when I_L is decreased, φ_M continues to decrease until it reaches $45°$, and then it

(a)

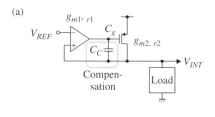

(b)

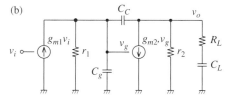

(c)

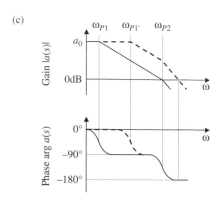

FIGURE 7.24. Miller compensation; circuit diagram (a), its small-signal equivalent circuit (b), and gain and phase characteristics (c) [20].

increases. In any case, the phase margin is not sensitive to variations in I_L. There is an important implication here, that the voltage down converter is stable against drasitic changes in the load current. Careful attention should be paid to the reduction of R_L because φ_M decreases with the reduction. In this case, C_C and R_C must again be obtained by substituting the minimum R_L for (7.75) and (7.76). Thus, the voltage down converter is stable because a phase margin larger than $45°$ is always ensured for R_L larger than the minimum value.

Miller compensation adds a compensation capacitor (C_C) between the input and output of the pMOST output stage as shown in Fig. 7.24. The necessary C_C can be reduced by utilizing the Miller effect of the output stage. The open-loop transfer function has two poles and two zeros as shown in Table 7.2 [20]. Here, ω_{P1} is generated by the r_1 and the C_C magnified through the Miller effect, ω_{P2} is generated by the output resistance of the pMOST output stage, $1/g_{m2}$, and C_L, ω_{Z1} is generated by R_L and C_L, and ω_{Z2} is the zero in the right-half plane generated by the feed-forward path through C_C [17]. In usual design both R_L and $1/g_{m2}$ are small, and thus ω_{Z1} and ω_{Z2} are negligible, because they are sufficiently higher than ω_{P1} and ω_{P2}. Note that the pole without compensation, ω_{P1}', shifts drastically to a lower frequency, ω_{P1}. This is because the load capacitance of the first stage is increased from C_g to $C_C g_{m2} r_2$ by the Miller effect ($g_{m2} r_2$ is the gain of the output stage). The stabilizing condition is expressed as

$$C_C \geq \frac{1}{2} \sqrt{\frac{1}{2} \left\{ \frac{(C_L + C_g) g_{m1}}{g_{m2}} \right\}^2 + \frac{2\sqrt{2} C_L C_g g_{m1}}{g_{m2}} + \frac{\sqrt{2} (C_L + C_g) g_{m1}}{4 g_{m2}}}. \quad (7.78)$$

This is smaller than that for dominant-pole or pole-zero compensations. For example, a small C_C of 50 pF is enough for stable operation of a 0.5-μm 16-Mbit DRAM with $C_L = 500$ pF, $0 \leq R_L \leq 10\ \Omega$, and $I_L = 100$ mA [20]. The drawback of this compensation, however, is a poor power-supply rejection ratio (PSRR) as discussed in the next subsection.

MOS capacitors are suitable for compensation capacitors because of their large capacitance per unit area. In the case of the pole-zero compensation, the device structure shown in Fig. 7.25(a) is widely used [21]. The n+ diffusion area in n-well is one electrode C, and the poly-Si gate is the other electrode A. The device is in the accumulation state (electrons, the major carriers, are accumulated under the gate), as long as the potential of A is higher than that of C. The device can be used as a compensation capacitor despite the voltage dependence of capacitance for pole-zero compensation, because the applied voltage is almost constant, V_{INT}. On the other hand, the polarity of the voltage might change in the Miller compensation. Therefore, the parallel connection of a pair of MOS capacitors with opposite directions shown in Fig. 7.25(b) or a voltage-independent capacitor, such as an MIM (metal insulator metal) capacitor is used. In the former case, the parasitic junction capacitances between n-well and p-substrate should be included in the simulation.

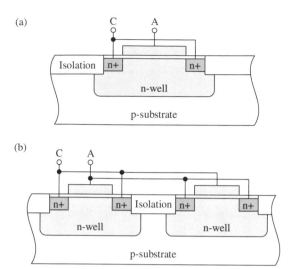

FIGURE 7.25. Device cross sections of compensation capacitor [21].

7.2.4. PSRR

The stability of the internal supply voltage V_{INT} against V_{DD} fluctuation is important for high-density LSIs. If the fluctuation of V_{DD} supplied to a voltage converter propagates to its output voltage V_{INT}, the fluctuation of V_{INT} might cause a malfunction of the internal circuit. The ratio between V_{DD} fluctuation and V_{INT} fluctuation is called the PSRR (power supply rejection ratio) and is defined by the following equation (see Fig. 7.26):

$$PSRR = 20 \log_{10} \frac{|v_{int}|}{|v_{dd}|} \ [dB],\qquad(7.79)$$

where v_{dd} and v_{int} are the AC components of V_{DD} and V_{INT}, respectively. The PSRR is generally a function of frequency and is evaluated by an AC circuit simulation of the circuit in Fig. 7.26. The target for PSRR should be lower than $-20\,dB$ (V_{INT} fluctuation is smaller than one tenth V_{DD} fluctuation) in the design of on-chip voltage converters.

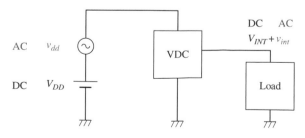

FIGURE 7.26. Simulation method for PSRR.

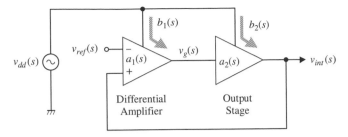

FIGURE 7.27. Small-signal equivalent circuit for PSRR analysis.

The small-signal equivalent circuit to analyze the PSRR of a voltage down converter is shown in Fig. 7.27. Here, v_{ref} denotes the AC component of reference voltage V_{REF}, $a_i(s)$ is the transfer function from input to output (without V_{DD} fluctuation) of each stage, $b_i(s)$ is the transfer function from V_{DD} to output (without input fluctuation) of each stage ($i = 1, 2$). The following equations stand from the principle of superposition

$$v_{int}(s) = \left\{ \left(v_{int}(s) - v_{ref}(s) \right) a_1(s) + v_{dd}(s) b_1(s) \right\} a_2(s) + v_{dd}(s) b_2(s),$$

$$(7.80)$$

$$\frac{v_{int}(s)}{v_{dd}(s)} = -\frac{a_1(s) a_2(s) v_{ref}(s)}{(1 - a_1(s) a_2(s)) v_{dd}(s)} + \frac{b_1(s) a_2(s) + b_2(s)}{1 - a_1(s) a_2(s)}$$

$$\cong \frac{v_{ref}(s)}{v_{dd}(s)} + \frac{b_1(s) a_2(s) + b_2(s)}{1 - a_1(s) a_2(s)}. \qquad (7.81)$$

The first term of (7.81) is the component due to V_{REF} fluctuation, which translates almost directly into V_{INT} variation. This is reduced by improving the PSRR of the V_{REF} generator. The second term indicates the stability of the converter circuit itself against V_{DD} fluctuation. This is reduced by designing so that $b_1(s)a_2(s)$ and $b_2(s)$ are cancelled out by each other as described below.

The transfer functions b_1 and b_2 at low frequency are calculated using the equivalent circuits in Fig. 7.28. First, analyzing the equivalent circuit of the differential amplifier in Fig. 7.28(a),

$$g_{mp}(v_1 - v_{dd}) + \frac{v_1 - v_{dd}}{r_p} - g_{mn}v_3 + \frac{v_1 - v_3}{r_n} = 0, \qquad (7.82)$$

$$g_{mp}(v_1 - v_{dd}) + \frac{v_g - v_{dd}}{r_p} - g_{mn}v_3 + \frac{v_g - v_3}{r_n} = 0, \qquad (7.83)$$

$$-\frac{v_1 - v_3}{r_n} + g_{mn}v_3 - \frac{v_g - v_3}{r_n} + g_{mn}v_3 + \frac{v_3}{r_n'} = 0 \qquad (7.84)$$

Solving these equations assuming that $g_{mn}r_n' \gg 1$, similar to the case of Fig. 7.18,

$$b_1 = \frac{v_g}{v_{dd}} \cong 1 \qquad (7.85)$$

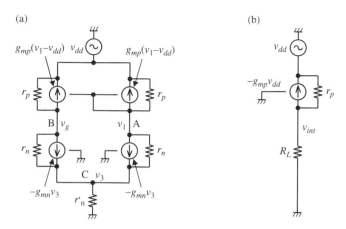

FIGURE 7.28. Small-signal equivalent circuits of the differential amplifier (a), and the output stage (b) for PSRR analysis.

is obtained. This equation means that the output voltage V_G of the differential amplifier almost follows V_{DD}. Next, from the equivalent circuit of the output stage in Fig. 7.28(b),

$$- g_{mp}v_{dd} + \frac{v_{int} - v_{dd}}{r_{0p}} + \frac{v_{int}}{R_L} = 0, \tag{7.86}$$

$$b_2 = \frac{v_{int}}{v_{dd}} = \frac{1 + g_{mp}r_{0p}}{1 + r_{0p}/R_L} \cong g_{mp}\left(r_{0p}//R_L\right) \tag{7.87}$$

Since $a_2 = -g_{mp}(r_{0p}//R_L)$, substituting (7.85), (7.87) and $v_{ref} = 0$ into (7.81) yields

$$\frac{v_{int}(s)}{v_{dd}(s)} \cong 0. \tag{7.88}$$

Thus, the voltage down converter has a good PSRR at low frequency because $b_1 a_2$ and b_2 are cancelled out by each other. In other words, the gate voltage V_G and the source voltage V_{DD} of the output pMOST synchronously fluctuate, and the V_{DD} fluctuation does not propagate to the drain, V_{INT}.

At higher frequency, however, the PSRR deteriorates due to parasitic capacitances that are neglected in Fig. 7.28, especially in the case of Miller compensation shown in Fig. 7.24. The drain and gate of the output pMOST are short-circuited by the Miller capacitor C_C at high frequency as shown by the equivalent circuit in Fig. 7.29(a). Since the current source in this circuit is controlled by the voltage difference between its both terminals, it can be replaced by a resistor as shown in Fig. 7.29(b). The PSRR is calculated from this circuit as

$$\frac{v_{int}(s)}{v_{dd}(s)} = \frac{R_L}{(1/g_{mp})//r_{0p} + R_L} \cong \frac{g_{mp}R_L}{1 + g_{mp}R_L}. \tag{7.89}$$

The PSRR is poor especially for large R_L, that is, small load current.

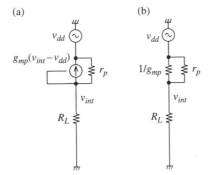

FIGURE 7.29. Small-signal equivalent circuits of a voltage down converter with Miller compensation.

7.2.5. Low-Power Design

An on-chip voltage down converter itself consumes an idling current as shown by I_S and I_B in Fig. 7.2(a). This idling current should be minimized while the chip is in a low-power mode (standby or sleep). Fortunately, neither a large current driving capability nor quick response is required during low-power mode, while both are required during active mode to suppress the voltage dip (see ΔV_{INT} in Fig. 7.9). Thus, switching the idling current is widely used for low-power design.

A typical example is shown in Fig. 7.30, which is similar to the basic voltage down-converter in Fig. 7.2(a) if pMOSTs and nMOSTs in the differential amplifier are exchanged. The constant-current source composed of M_{5A} and M_{5S} is controlled by a clock φ [4, 22]. For example, the current, that is, I_S in equation (7.22), is increased by turning on M_{5A} synchronously with the chip operation. During low-power mode, M_{5A} is turned off to save the power. M_{5S} is always turned on with a small current to maintain the V_{INT} level while the load current is quite small. Power dissipations were 15mW and 2mW at active mode and standby mode, respectively.

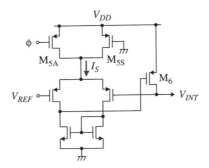

FIGURE 7.30. A current switching voltage down converter. Reproduced from [4] with permission; © 2006 IEEE.

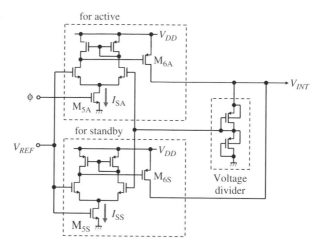

FIGURE 7.31. Parallel connection of voltage down converters [14, 23]. Reproduced from [14] with permission; © 2006 IEEE.

Figure 7.31 shows parallel connection of two voltage down-converters (VDCs) [14, 23]: a low-power circuit for supplying a small current (0.2 mA at $V_{DD} = 5$ V) during the standby mode, and a high-power circuit for supplying a large enough current during the active mode with φ activation.

Figure 7.32 shows parallel connection of not only two VDCs but also two reference-voltage (V_{REF}) generators [24]. During standby mode, the V_{REF} generator for active, which is composed of a bandgap reference (BGR) and a trimmer,

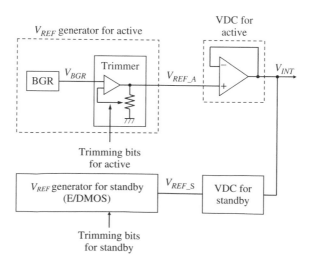

FIGURE 7.32. Parallel connection of voltage down converters. Reproduced from [24] with permission; © 2006 IEEE.

is turned off. The V_{REF} generator for standby, which generates its reference based on the V_t difference of enhancement MOSTs and depletion MOSTs, is instead turned on. The idling current during standby is as small as $0.5\,\mu A$ at the sacrifice of voltage accuracy. However, the degraded accuracy still fits in the required voltage window because the window for standby mode is fairly wide.

7.2.6. Applications

Figure 7.33 shows an on-chip VDC circuit using the pole-zero compensation implemented in a 16-Mbit DRAM [16]. To prevent the zero frequency from deviating from its designed value, the compensation circuit is located close to the VDC. The voltage bounce of the internal supply voltage V_{INT} during operation is as small as 0.2 V as shown in Fig. 7.34. Figure 7.35 shows the stability of V_{INT} when 1.5-Vp-p noise is imposed on the external supply voltage V_{DD}. The noise appearing in V_{INT} is as small as 50 mV. Figure 7.36 shows the closed-loop gain of the VDC circuit for various load currents. The peak gain is almost independent of load current and is 2 dB, from which we can calculate a phase margin of 47° from equation (7.42).

Figure 7.37 shows a microcontroller having an on-chip VDC[24]. To reduce the voltage drop due to parasitic wiring resistance, the VDC circuit for active mode is divided into 19 equivalent pieces and inserted into 19 places of I/O area. Since these places were originally unused, the VDC for active mode does not increase the chip size.

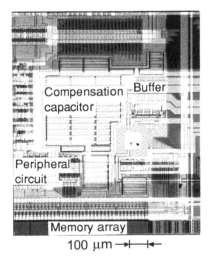

FIGURE 7.33. A voltage down converter applied to a DRAM. Reproduced from [16] with permission; © 2006 IEICE.

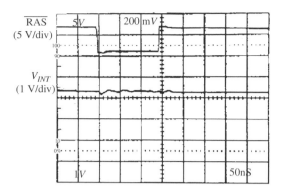

FIGURE 7.34. Waveform of V_{INT} for load operation. Reproduced from [16] with permission; © 2006 IEICE.

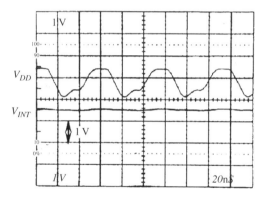

FIGURE 7.35. Waveforms of V_{EXT} and V_{INT}. Reproduced from [16] with permission; © 2006 IEICE.

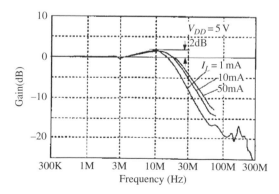

FIGURE 7.36. Measured closed-loop gain. Reproduced from [16] with permission; © 2006 IEICE.

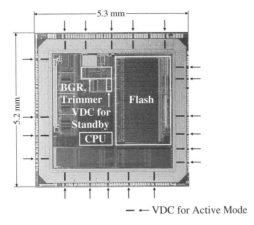

FIGURE 7.37. A voltage down converter applied to a microcontroller. Reproduced from [24] with permission; © 2006 IEEE.

7.3. The Switching Regulator

Since a series regulator reduces the power supply voltage V_{DD} to V_{INT} by the on resistance of the output transistor, the difference between V_{DD} and V_{INT} corresponds to the power loss. Therefore, the power conversion efficiency is lower than V_{INT}/V_{DD}. To achieve higher power efficiency, a reactance element that stores electric energy is required. A switching regulator utilizes an inductor and a capacitor as the reactance elements, while a switched-capacitor regulator described in Section **7.4** utilizes only capacitors.

Figure 7.38 shows the circuit diagram of a typical switching regulator. The pulse generator alternately turns on the two power MOSTs M_1 and M_2 and generates a pulse with a prescribed duty cycle at node N_1. The low-pass filter composed of an inductor L and a capacitor C smoothes this pulse, generating

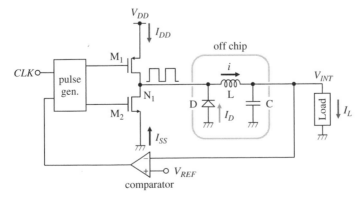

FIGURE 7.38. Switching voltage down converter.

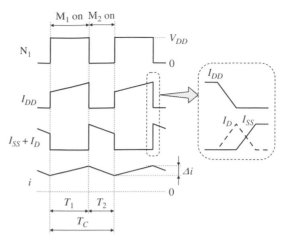

FIGURE 7.39. Operating waveforms of switching voltage down converter.

an output voltage V_{INT}. The operating waveforms are shown in Fig. 7.39. Here, T_1 and T_2 denote the periods in which M_1 and M_2 are turned on, respectively. We assume that C is sufficiently large and that the fluctuation of the voltage V_{INT} can be ignored. We also assume that the on resistances of M_1 and M_2 are negligible for simplicity. A current flows from V_{DD} through M_1 and L during T_1. Since the voltage $V_{DD} - V_{INT}$ is applied to L, the increase of inductor current i during this period is expressed as

$$L \cdot \Delta i = T_1 \left(V_{DD} - V_{INT} \right). \tag{7.90}$$

Since the voltage $-V_{INT}$ is applied to L during T_2, the decrease of i during this period is expressed as

$$L \cdot (-\Delta i) = T_2 \left(-V_{INT} \right). \tag{7.91}$$

From the equilibrium condition of i,

$$V_{INT} = \frac{T_1}{T_1 + T_2} \cdot V_{DD} \tag{7.92}$$

is derived. Thus, the output voltage V_{INT} is proportional to the duty cycle of the pulse. Therefore, monitoring the level of V_{INT} by the comparator and adjusting the duty cycle of the pulse are required to generate the desired voltage V_{INT}. The output current I_L is equal to the average of i. The current I_{DD} supplied from V_{DD} is equal to i during T_1, while it is zero during T_2. Therefore, the average current is derived as

$$\overline{I_{DD}} = \frac{T_1}{T_1 + T_2} \cdot \overline{i} = \frac{T_1}{T_1 + T_2} \cdot I_L. \tag{7.93}$$

From equations (7.92) and (7.93),

$$V_{INT} I_L = V_{DD} \overline{I_{DD}} \tag{7.94}$$

is derived. Thus, the circuit achieves a power efficiency of 100% under the ideal conditions of no parasitic resistances.

The usual design of a switching regulator requires an inductor of several μH and a capacitor of several μF. Such devices are difficult to integrate in a monolithic chip, and are usually placed off chip along with a diode. The power MOSTs are also off chip according to the design. However, it is not impossible that a capacitor and even an inductor are integrated on the chip, if the switching frequency is extremely high [25, 26].

The diode D prevents the sudden change of inductor current i during the switching of the power MOSTs. Turning off M_1 occurs a little earlier than turning on M_2 to prevent both MOSTs from being simultaneously turned on. Therefore, the rise of current I_{SS} through M_2 occurs slightly after the fall of current I_{DD} through M_1. During the switching, the diode supplies the inductor with current I_D, thus preventing the sudden change of $i = I_{DD} + I_{SS} + I_D$. The diode plays the same role during the switching from M_2 to M_1.

The design of a switching regulator requires attention to EMI (electro-magnetic interference) because switching a current causes noise. Figure 7.40 shows a technique [27] proposed to reduce the switching noise. The switching MOST M_1 (M_2) is divided into M_{11} (M_{21}) with a large channel width and M_{12} (M_{22}) with a small channel width. When turning on, M_{12} (M_{22}) is firstly turned on followed by M_{11} (M_{21}). When turning off, M_{11} (M_{21}) is firstly turned off followed by M_{12} (M_{22}). The sudden change of current I_{DD} (I_{SS}) is thereby avoided and the switching noise is reduced. It is reported that a regulator using this divided-switch technique achieved a power efficiency of over 90% as shown in Fig. 7.41.

The idling (quiescent) current of a switching regulator is larger than that of a series regulator because of the charging and discharging currents to the gates of power MOSTs. This degrades the power efficiency of a switching regulator relative to that of a series regulator when the load current I_L is small (light load). Figure 7.42 shows a solution to this problem that uses both regulators [28]. The

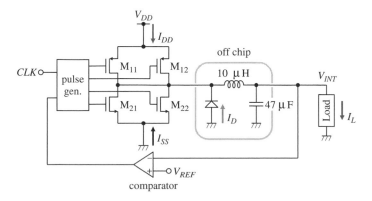

FIGURE 7.40. Switching-noise reducing scheme [27].

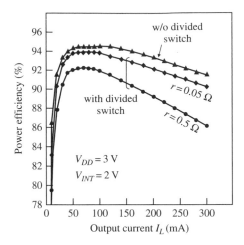

FIGURE 7.41. Power efficiency of switching voltage down converters with and without divided-switch technique; r denotes the parasitic resistance of inductor. Reproduced from [27] with permission; © 2006 IEEE.

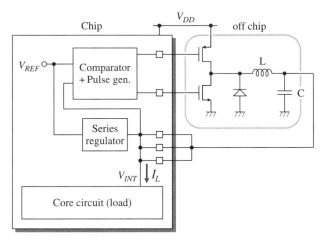

FIGURE 7.42. An on-chip hybrid regulator. [28].

switching regulator is for a heavy load, and the series regulator is for a light load, thus improving the power efficiency.

7.4. The Switched-Capacitor Regulator

A switched-capacitor regulator realizes voltage up/down conversion by periodically switching the connection among the external supply voltage and capacitors. Only voltage down conversion is described in this section, and voltage up conversion will be described in Chapter 8.

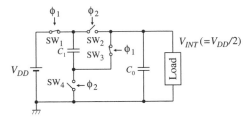

FIGURE 7.43. Switched-capacitor voltage down converter (voltage halving).

Figure 7.43 shows a switched-capacitor voltage halving circuit [29, 30, 31] using two capacitors. The capacitor C_0 holds the output voltage V_{INT}, while C_1 is the fly capacitor to transfer charge from V_{DD} to C_0. Two-phase non-overlap clock signals φ_1 and φ_2 alternately turn on/off four switches SW_1-SW_4. The equivalent circuits and the operating waveforms are shown in Figs. 7.44 and 7.45, respectively. Here, R_1-R_4 denote the on resistances of SW_1-SW_4, respectively, V_1 is the voltage between the both terminals of C_1. The fly capacitor C_1 is charged during period T_1 through SW_1 and SW_3 and is discharged during period T_2 through SW_2 and SW_4.

Let us consider the equilibrium state, assuming that $R_1 = R_2 = R_3 = R_4 = R$, and $T_1 = T_2 = T = d\,T_C/2$ (d is the ratio of the period during which either switch is on), for simplicity. We also assume that C_0 is sufficiently large and that the fluctuation of the voltage V_{INT} can be ignored. During period T_1,

$$i_{DD} = C_1 \frac{dV_1}{dt} = \frac{V_{DD} - V_{INT} - V_1}{2R} \tag{7.95}$$

(a)

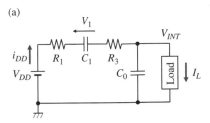

(b)

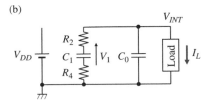

FIGURE 7.44. Equivalent circuits of switched-capacitor voltage down converter; period T_1 (a), and period T_2 (b).

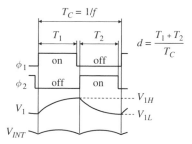

FIGURE 7.45. Operating waveforms of switched-capacitor voltage down converter.

The voltage V_1 increases from V_{1L} to V_{1H} during this period, and V_{1H} is expressed as

$$V_{1H} = V_{DD} - V_{INT} - (V_{DD} - V_{INT} - V_{1L}) \exp\left(-\frac{T}{2C_1 R}\right). \qquad (7.96)$$

During period T_2,

$$C_1 \frac{dV_1}{dt} = \frac{V_{INT} - V_1}{2R} \qquad (7.97)$$

stands. The voltage V_1 decreases from V_{1H} to V_{1L} during this period, and V_{1L} is expressed as

$$V_{1L} = V_{INT} + (V_{1H} - V_{INT}) \exp\left(-\frac{T}{2C_1 R}\right). \qquad (7.98)$$

From equations (7.96) and (7.98) we obtain

$$V_{1H} - V_{1L} = \frac{1 - e}{1 + e} \cdot (V_{DD} - 2V_{INT}), \qquad (7.99)$$

where

$$e = \exp\left(-\frac{T}{2C_1 R}\right). \qquad (7.100)$$

From the equilibrium condition of C_1, the charge increase during T_1 is equal to the charge decrease during T_2, and is expressed by the following equation:

$$\int_{T_1} \frac{V_{DD} - V_{INT} - V_1}{2R} dt = \int_{T_2} \frac{V_{INT} - V_1}{2R} dt = C_1 (V_{1H} - V_{1L}). \qquad (7.101)$$

On the other hand, the integral of the charge flowing into C_0 during a cycle is equal to zero, that is,

$$\int_{T_1} \frac{V_{DD} - V_{INT} - V_1}{2R} dt + \int_{T_2} \frac{V_{INT} - V_1}{2R} dt - I_L T_C = 0. \qquad (7.102)$$

From (7.99), (7.101) and (7.102) we obtain

$$I_L T_C = 2C_1 (V_{1H} - V_{1L}) = \frac{2(1-e)}{1+e} \cdot (V_{DD} - 2V_{INT}), \tag{7.103}$$

$$V_{INT} = \frac{V_{DD}}{2} - \frac{T_C(1+e)}{4C_1(1-e)} \cdot I_L = \frac{V_{DD}}{2} - \frac{T_C}{4C_1 \tanh \dfrac{d\, T_C}{8C_1 R}} \cdot I_L. \tag{7.104}$$

Thus, the circuit is equivalent to a voltage source with electromotive force of $E = V_{DD}/2$ and output resistance of

$$R_{OUT} = \frac{T_C}{4C_1 \tanh \dfrac{d\, T_C}{8C_1 R}}, \tag{7.105}$$

as shown in Fig. 7.46(a). When the on resistance R can be ignored, the equation (7.105) becomes

$$R_{OUT} = \frac{T_C}{4C_1}. \tag{7.106}$$

Therefore a higher switching frequency (smaller T_C) or a larger C_1 is required to reduce R_{OUT}. On the contrary, when T_C/C_1 is sufficiently small, (7.105) becomes

$$R_{OUT} = \frac{2R}{d}. \tag{7.107}$$

The relationship between R_{OUT} and switching frequency $f = 1/T_C$ is shown in Fig. 7.46(b). At low frequency R_{OUT} decreases according to equation (7.106), eventually reaching a lower limit of (7.107).

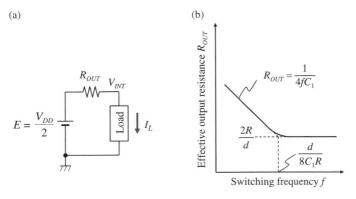

(a)   (b)

FIGURE 7.46. Equivalent circuit of switched-capacitor voltage down converter (a) and effective output resistance as a funtion of switching frequency (b).

The current i_{DD} supplied from V_{DD} is expressed by equation (7.95) during T_1, while it is zero during T_2. Therefore the average current is derived using (7.99) and (7.104) as

$$I_{DD} = \frac{1}{T} \int_{T_1} C_1 \frac{dV_1}{dt} dt = \frac{C_1(V_{1H} - V_{1L})}{T} = \frac{C_1}{T} \cdot \frac{1-e}{1+e} (V_{DD} - 2V_{INT}) = \frac{I_L}{2}.$$
(7.108)

Thus, the supply current is a half the load current. The power efficiency η is calculated as

$$\eta = \frac{V_{INT} I_L}{V_{DD} I_{DD}} = \frac{V_{DD} - 2I_L R_{OUT}}{V_{DD}} = 1 - \frac{2I_L R_{OUT}}{V_{DD}}.$$
(7.109)

The efficiency is 100% in the ideal case $(R_{OUT} = 0)$. It is lowered, however, by the power loss, which corresponds to the voltage drop due to R_{OUT} (see Fig. 7.1).

It should be noted that E in Fig. 7.46(a) is independent of C_1 or C_0. The voltage-conversion ratio (in this case, 1/2) is determined by the circuit topology, not by the capacitor ratio. Therefore a different conversion ratio requires a different circuit topology. Figure 7.47 shows the equivalent circuits of switched-capacitor voltage down converters with conversion ratios of 1/3 and 2/3. Both utilize three capacitors. Generally, the more complex the conversion ratio (rational number) is, the more capacitors are needed [32]. Hence, a practical way to generate a desired V_{INT} is to adopt a circuit with a simple rational-number conversion ratio that is a little higher than the desired conversion ratio, and to drop the output voltage down to V_{INT} by R_{OUT}. This voltage drop, however, causes a power loss.

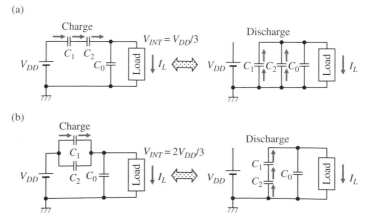

FIGURE 7.47. Equivalent circuits of switched-capacitor voltage down converters; 1/3 voltage generator (a) and 2/3 voltage generator (b).

7.5. The Half -V_{DD} Generator

The voltage down-converters described above generate V_{INT} that is constant independent of V_{DD}. However, some applications require voltage converters that generate a voltage proportional to V_{DD}. A typical example is the half-V_{DD} generator for a DRAM. It generates a data line (bit line) precharge voltage and a memory-cell plate voltage, both of which are usually half the data-line high level because of the following reason. The former voltage is a reference level for amplifying read signal from memory cells, while the latter voltage reduces the electric field across the dielectric of memory-cell capacitors. In particular, the accuracy of the former voltage is essential for the high-S/N design of the sense amplifiers as described in Chapter 2.

Figure 7.48 shows a half voltage generator [33] consisting of a bias generator and push-pull output stage. If R_1 and R_2 are high enough resistances, the output voltage is expressed by

$$V_{OUT} = \frac{V_{DD} - V_{tn} + |V_{tp}|}{2}, \qquad (7.110)$$

where V_{tn} and V_{tp} are threshold voltages of nMOST and pMOST, respectively. If $V_{tn} = |V_{tp}| = V_t$, the voltages at nodes N_0, N_1, N_2 are $V_{DD}/2$, $(V_{DD}/2) + V_t$, and $(V_{DD}/2) - V_t$, respectively. Thus, the output is stabilized at $V_{DD}/2$. Since both gate-source voltage of M_3 and M_4 are V_t, they are weakly turned on, allowing a small subthreshold current to flow. Any change in the output voltage is eventually suppressed, because either of the two is strongly turned on, providing several mA. No DC current flows through the output stage because of the small difference in V_{tp} between M_2 and M_4 due to body effect. In addition, if high-resistance ($\sim M\Omega$) poly-Si strings are used as R_1 and R_2, the resultant standby current is μA-level. The variation of V_{OUT} caused by V_{tn} and V_{tp} variations is reduced by adjusting R_1 and/or R_2. However, this generator has the following drawbacks when V_{DD} is

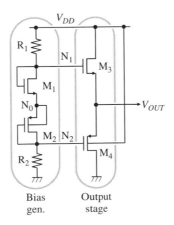

FIGURE 7.48. Half-V_{DD} generator [33].

lowered. The voltage setting accuracy gets worse for a fixed variation in V_{tn} and V_{tp} as suggested by (7.110). In addition, a high-speed response is more difficult to achieve for the ever-increasing load capacitance with increasing memory capacity. For example, for a load (i. e. a cell-capacitor plate) capacitance of a 64-Mbit DRAM as heavy as 115 nF, the rise time of the $V_{DD}/2$ waveform during power-on is as long as 160 μs at $V_{DD} = 1.5$ V [34]. Such a slow response might be detrimental to normal operation.

The circuit [34] shown in Fig. 7.49 solves the possible problems mentioned above. One feature is the separation of the voltage divider and the bias generator, to improve the accuracy by about one order of magnitude. Only the difference in threshold voltages between the MOSTs with the same conduction type (M_1 and M_4, or M_2 and M_5), which is usually small, is related to the accuracy. Another feature is the addition of a push-pull current-mirror amplifier and a tri-state output driver to speed up operation. The amplifier outputs $V_{DD}/2$ at steady state, because the gate voltages of M_4 and M_5 are $V_{DD}/2 + V_{tn}$ and $V_{DD}/2 - |V_{tp}|$, respectively. When the output voltage falls from $V_{DD}/2$ during a load operation, the charging current i flows as a result of more forward biasing of M_4. Thus, the mirrored current ni (n is the ratio between M_7 and M_3) charges up the M_{11} gate. The resulting large current I charges up the load, so that the load recovers to the original $V_{DD}/2$. On the contrary, when the output voltage increases, the lower part of the amplifier suppresses the voltage variation in the same manner. The tri-state driver enhances the driving capability. This circuit has achieved a 30-fold speed increase at $V_{DD} = 1.5$ V, compared with that in Fig. 7.48.

The circuit [35, 36] shown in Fig. 7.50(a) has the same feature. The amplifier A_1 and the output MOST M_1 suppress the decrease in the output voltage V_{OUT}, while A_2 and M_2 suppress the increase in V_{OUT}. This circuit has the "dead band" around $V_{OUT} = V_{DD}/2$, where both M_1 and M_2 are turned off as shown in Fig. 7.50(b). This prevents a large current flowing from V_{DD} to ground through M_1 and M_2. The

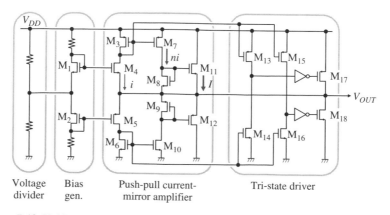

FIGURE 7.49. Half-V_{DD} generator for low-voltage operation [34].

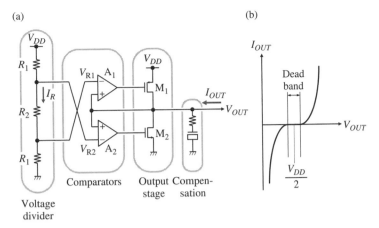

FIGURE 7.50. Half-V_{DD} generator for low-voltage operation using a differential amplifier; circuit diagram (a), and DC characteristics (b) [35].

width of the dead band is determined by the difference between V_{R1} and V_{R2} that is equal to $I_R R_2$. The compensation circuit stabilizes the feedback loops as described in Section 7.2.

A7.1. Relationship between Phase Margin and Loop Gain

Let us consider an open-loop transfer function with two poles, ω_{P1} and ω_{P2}:

$$a(j\omega) = \frac{a_0}{\left(1 + j\dfrac{\omega}{\omega_{P1}}\right)\left(1 + j\dfrac{\omega}{\omega_{P2}}\right)} = \frac{a_0}{(1 + jpx)(1 + jx)}, \tag{A7.1}$$

where, $x = \omega/\omega_{P2}$, and $p = \omega_{P2}/\omega_{P1}$ (pole-frequency ratio). When we define ω_0 as the unity-gain angular frequency (i.e. $|a(j\omega_0)| = 1$) and $x_0 = \omega_0/\omega_{P2}$, the equation

$$a_0^2 = \left(1 + p^2 x_0^2\right)\left(1 + x_0^2\right) \tag{A7.2}$$

stands. The phase margin is expressed as

$$\varphi_M = \arg a(j\omega_0) + 180° = -\arctan \frac{x_0(p+1)}{1 - px_0^2}. \tag{A7.3}$$

From this equation, we obtain

$$x_0 = \frac{p + 1 + \sqrt{(p+1)^2 + 4p\tan\varphi_M}}{2p\tan\varphi_M}. \tag{A7.4}$$

Assuming that the two poles are sufficiently separated from each other ($p \gg 1$), equations (A7.1), (A7.2), and (A7.4) are approximated by

$$a\,(j\omega) = \frac{a_0/p}{-x^2 + jx},\tag{A7.5}$$

$$a_0^2 = p^2\left(x_0^4 + x_0^2\right),\tag{A7.6}$$

$$x_0 = \frac{1}{\tan \varphi_M},\tag{A7.7}$$

respectively. From (A7.6) and (A7.7), we obtain

$$\frac{a_0}{p} = \frac{1}{\sin \varphi_M \tan \varphi_M}.\tag{A7.8}$$

The closed-loop transfer function is calculated from (A7.5):

$$A\,(j\omega) = \frac{a\,(j\omega)}{1 + a\,(j\omega)} = \frac{a_0/p}{a_0/p - x^2 + jx} = \frac{a_0/p}{a_0/p - X^2 x_0^2 + jX x_0},\tag{A7.9}$$

where, $X = \omega/\omega_0$. Substituting (A7.7) and (A7.8) into (A7.9) yields

$$A\,(j\omega) = \frac{\dfrac{1}{\sin \varphi_M \tan \varphi_M}}{\dfrac{1}{\sin \varphi_M \tan \varphi_M} - \dfrac{X^2}{\tan^2 \varphi_M} + j\dfrac{X}{\tan \varphi_M}} = \frac{1}{1 - X^2 \cos \varphi_M + jX \sin \varphi_M}.\tag{A7.10}$$

The closed-loop gain is given by

$$|A\,(j\omega)| = \frac{1}{\sqrt{(1 - X^2 \cos \varphi_M)^2 + X^2 \sin^2 \varphi_M}}$$

$$= \frac{1}{\sqrt{\left\{X^2 \cos \varphi_M + \left(\dfrac{\sin \varphi_M \tan \varphi_M}{2} - 1\right)\right\}^2 + \sin \varphi_M \tan \varphi_M - \dfrac{\sin^2 \varphi_M \tan^2 \varphi_M}{4}}}.\tag{A7.11}$$

Plotting this equation results in Fig. 7.14. From this equation, we can see that if $\sin \varphi_M \tan \varphi_M > 2$ ($\varphi_M > 65.5°$), the closed-loop gain has no peaks. On the other hand, if $\sin \varphi_M \tan \varphi_M < 2$, it has a peak of

$$|A\,(j\omega)|_{\max} = \frac{1}{\sqrt{\sin \varphi_M \tan \varphi_M - \dfrac{\sin^2 \varphi_M \tan^2 \varphi_M}{4}}} = \frac{2}{\tan \varphi_M \sqrt{4 \cos \varphi_M - \sin^2 \varphi_M}}.\tag{A7.12}$$

The peak gain is also calculated as follows. If φ_M is small, the angular frequency at which the peak exists is approximately ω_0. Since $|a(j\omega_0)| = 1$ and $\arg a(j\omega_0) = \varphi_M - 180°$,

$$a(j\omega_0) = -\cos\varphi_M - j\sin\varphi_M, \tag{A7.13}$$

stands. Therefore, the peak gain is expressed as

$$|A(j\omega)|_{max} \approx |A(j\omega_0)| = \left| \frac{-\cos\varphi_M - j\sin\varphi_M}{1 - \cos\varphi_M - j\sin\varphi_M} \right| = \frac{1}{2\sin\dfrac{\varphi_M}{2}}. \tag{A7.14}$$

Thus, equation (7.42) is derived, which is approximately equal to (A7.12) for small φ_M.

Substituting (A7.4) into (A7.2) yields

$$a_0 = \frac{(p+1)\sqrt{(p+1)^2 + 2p\tan^2\varphi_M} + (p+1)\sqrt{(p+1)^2 + 4p\tan^2\varphi_M}}{\sqrt{2}p\sin\varphi_M\tan\varphi_M}. \tag{A7.15}$$

For example, if $\varphi_M = 45°$,

$$a_0 = \frac{(p+1)\sqrt{p^2 + 4p + 1} + (p+1)\sqrt{p^2 + 6p + 1}}{p}, \tag{A7.16}$$

and if $\varphi_M = 60°$,

$$a_0 = \frac{\sqrt{2}(p+1)\sqrt{p^2 + 8p + 1} + (p+1)\sqrt{p^2 + 14p + 1}}{3p}. \tag{A7.17}$$

Plotting (A7.16) and (A7.17) results in Fig. 7.16.

References

[1] T. Mano, J. Yamada, J. Inoue and S. Nakajima, "Submicron VLSI memory circuits," in ISSCC Dig. Tech. Papers, Feb. 1983, pp. 234–235.
[2] K. Itoh, R. Hori, J. Etoh, S. Asai, N. Hashimoto, K. Yagi and H. Sunami, "An experimental 1Mb DRAM with on-chip voltage limiter," in ISSCC Dig. Tech. Papers, Feb. 1984, pp. 282–283.
[3] M. Takada, T. Takeshima, M. Sakamoto, T. Shimizu, H. Abiko, T. Katoh, M. Kikuchi, S. Takahashi, Y. Sato and Y. Inoue, "A 4Mb DRAM with half internal-voltage bitline precharge," in ISSCC Dig. Tech. Papers, Feb. 1986, pp. 270–271.
[4] T. Furuyama, T. Ohsawa, Y. Watanabe, H. Ishiuchi, T. Watanabe, T. Tanaka, K. Natori and O. Ozawa, "An experimental 4-Mbit CMOS DRAM," IEEE J. Solid-State Circuits, vol. SC-21, pp. 605–611, Oct. 1986.

[5] M. Horiguchi, M. Aoki, H. Tanaka, J. Etoh, Y. Nakagome, S. Ikenaga, Y. Kawamoto and K. Itoh, "Dual-operating-voltage scheme for a single 5-V 16-Mbit DRAM," in IEEE J. Solid-State Circuits, Oct. 1988, pp. 1128–1132.

[6] S. Hayakawa, K. Sato, A. Aono, T. Yoshida, T. Ohtani and K. Ochii, "A process-insensitivity voltage down converter suitable for half-micron SRAM's," in Symp. VLSI Circuits Dig. Tech. Papers, Aug. 1988, pp. 53–54.

[7] H. Hidaka, K. Arimoto, K. Hirayama, M. Hayashikoshi, M. Asakura, M. Tsukude, T. Oishi, S. Kawai, K. Suma, Y. Konishi, K. Tanaka, W. Wakamiya, Y. Ohno and K. Fujishima, "A 34-ns 16-Mb DRAM with controllable voltage down-converter," IEEE J. Solid-State Circuits, vol. 27, pp. 1020–1027, July 1992.

[8] A. L. Roberts, J. H. Dreibelbis, G. M. Braceras, J. A. Gabric, L. E. Gilbert, R. B. Goodwin, E. L. Hedberg, T. M. Maffitt, L. G. Meunier, D. S. Moran, P. K. Nguyen, D. E. Reed, D. R. Reismiller and R. A. Sasaki, "A 256K SRAM with on-chip power supply conversion," in ISSCC Dig. Tech. Papers, Feb. 1987, pp. 252–253.

[9] K. Ishibashi, K. Sasaki and H. Toyoshima, "A voltage down converter with submicroampere standby current for low-power static RAM's," IEEE J. Solid-State Circuits, vol. 27, pp. 920–926, June 1992.

[10] H. J. Shin, S. K. Reynolds, K. R. Wrenner, T. Rajeevakumar, S. Gowda and D. J. Pearson, "Low-dropout on-chip voltage regulator for low-power circuits," in Symp. Low Power Electronics Dig. Tech. Papers, Oct. 1994, pp. 76–77.

[11] S.-J. Jou and T.-L. Chen, "On-chip voltage down converter for low-power digital system," IEEE Trans. Circuits and Systems-II, vol. 45, pp. 617–625, May 1998.

[12] G. W. den Besten and B. Nauta, "Embedded 5 V-to-3.3 V voltage regulator for supplying digital IC's in 3.3 V CMOS technology," IEEE J. Solid-State Circuits, vol. 33, pp. 956–962, July 1998.

[13] Y. Nakagome, "Voltage regulator design for low voltage DRAMs", Symp. VLSI Circuits, Memory Design Short Course, June 1998.

[14] D. Chin, C. Kim, Y. Choi, D.-S. Min, H. S. Hwang, H. Choi, S. Cho, T. Y. Chung, C. J. Park, Y. Shin, K. Suh and Y. E. Park, "An experimental 16-Mbit DRAM with reduced peak-current noise," IEEE J. Solid-State Circuits, vol. 24, pp. 1191–1197, Oct. 1989.

[15] T. Ooishi, Y. Komiya, K. Hamade, M. Asakura, K. Yasuda, K. Furutani, T. Kato, H. Hidaka and H. Ozaki, "A mixed-mode voltage down converter with impedance adjustment circuitry for low-voltage high-frequency memories," IEEE J. Solid-State Circuits, vol. 31, pp. 575–585, Apr. 1996.

[16] H. Tanaka, M. Aoki, J. Etoh, M. Horiguchi, K. Itoh, K. Kajigaya and T. Matsumoto, "Stabilization of voltage limiter circuit for high-density DRAM's using pole-zero compensation," IEICE Trans. Electron., vol. E75-C, pp. 1333–1343, Nov. 1992.

[17] P. R. Gray, P. J. Hurst, S. H. Lewis and R. G. Meyer, *Analysis and design of analog integrated circuits*, 4th ed. (John Wiley, New York 2001), Chap. 9.

[18] B. Razavi, *Design of Analog CMOS Integrated Circuits*, (McGraw-Hill, 2001), Chap. 6, 10.

[19] M. Horiguchi, M. Aoki, J. Etoh, H. Tanaka, S. Ikenaga, K. Itoh, K. Kajigaya, H. Kotani, K. Ohshima and T. Matsumoto, "A tunable CMOS-DRAM voltage limiter with stabilized feedback amplifier," IEEE J. Solid-State Circuits, vol. 25, pp. 1129–1135, Oct. 1990.

[20] H. Tanaka, M. Aoki, J. Etoh, M. Horiguchi, K. Itoh, K. Kajigaya and T. Matsumoto, "Stabilization of voltage limiter circuit for high-density DRAM's using Miller compensation," IEICE Trans., vol. J75-C-II, pp. 425–433, Aug. 1992.

[21] M. Horiguchi, M. Aoki, K. Itoh, Y. Nakagome, N. Miyake, T. Noda, J. Etoh, H. Tanaka and S. Ikenaga, "Large scale integrated circuit having low internal operating voltage," US Patent No. 5179539, Jan. 1993.

[22] G. Kitsukawa, K. Itoh, R. Hori, Y. Kawajiri, T. Watanabe, T. Kawahara, T. Matsumoto and Y. Kobayashi, "A 1-Mbit BiCMOS DRAM using temperature-compensation circuit techniques," IEEE J. Solid-State Circuits, vol. 24, pp. 597–602, June 1989.

[23] S. Fujii, M. Ogihara, M. Shimizu, M. Yoshida, K. Numata, T. Hara, S. Watanabe, S. Sawada, T. Mizuno, J. Kumagai, S. Yoshikawa, S. Kaki, Y. Saito, H. Aochi, T. Hamamoto and K. Toita, "A 45-ns 16-Mbit DRAM with triple-well structure," IEEE J. Solid-State Circuits, vol. 24, pp. 1170–1175, Oct. 1989.

[24] M. Hiraki, K. Fukui and T. Ito, "A low-power microcontroller having a 0.5-μA standby current on-chip regulator with dual-reference scheme," IEEE J. Solid-State Circuits, vol. 39, pp. 661–666, Apr. 2004.

[25] P. Hazucha, G. Schrom, J.-H. Hahn, B. Bloechel, P. Hack, G. Dermer, S. Narendra, D. Gardner, T. Karnik, V. De and S. Borkar, "A 233MHz, 80–87% efficient, integrated, 4-phase dc-dc converter in 90nm CMOS," in Symp. VLSI Circuits Dig. Tech. Papers, June 2004, pp. 256–257.

[26] S. Abedinpour, B. Bakkaloglu and S. Kiaei, "A multi-stage interleaved synchronous buck converter with integrated output filter in a 0.18um SiGe process," in ISSCC Dig. Tech. Papers, Feb. 2006, pp. 356–357.

[27] S. Sakiyama, J. Kajiwara, M. Kinoshita, K. Satomi, K. Ohtani and A. Matsuzawa, "An on-chip high-efficiency a dn low-noise DC/DC converter using divided switches with current control technique," in ISSCC Dig. Tech. Papers, Feb. 1999, pp. 156–157.

[28] M. Hiraki, T. Ito, A. Fujiwara, T. Ohashi, T. Hamano and T. Noda, "A 63-uW standby power microcontroller with on-chip hybrid regulator scheme," IEEE J. Solid-State Circuits, vol. 37, pp. 605–611, May 2002.

[29] I. Oota, T. Inoue and F. Ueno, "A realization of low-power supplies using switched-capacitor transformers and its analysis," IEICE Trans. Electron., vol. J66-C, pp. 576–583, Aug. 1983.

[30] F. Ueno, T. Inoue, I. Oota and T. Umeno, "Analysis and applications of switched capacitor transformers by its formulation," IEICE Trans. Electron., vol. J73-C-II, pp. 66–73, Feb. 1990.

[31] N. Bansal and A. Katyal, "A switched-cap regulator for SoC applications," in Proc. 17th International Conference on VLSI Design, Jan. 2004, pp. 168–173.

[32] K. Yamada, N. Fujii and S. Takagi, "Capacitance value free switched capacitor DC-DC voltage converter realizing arbitrary rational conversion ratio," IEICE Trans. Fundamentals, vol. E87-A, pp. 344–349, Feb. 2004.

[33] S. Fujii, S. Saito, Y. Okada, M. Sato, S. Sawada, S. Shinozaki, K. Natori and O. Ozawa, "A 50-uA standby 1M x 1/256K x 4 CMOS DRAM with high-speed sense amplifier," IEEE J. Solid-State Circuits, vol. SC-21, pp. 643–648, Oct. 1986.

[34] Y. Nakagome, H. Tanaka, K. Takeuchi, E. Kume, Y. Watanabe, T. Kaga, Y. Kawamoto, F. Murai, R. Izawa, D. Hisamoto, T. Kisu, T. Nishida, E. Takeda and K. Itoh, "An experimental 1.5-V 64-Mb DRAM," IEEE J. Solid-State Circuits, vol. 26, pp. 465–472, Apr. 1991.

[35] H. Nakano, Y. Watanabe and S. Watanabe, "Consideration on a half Vcc generator," in IEICE 1990 Autumn National Convention Record, Oct. 1990, p. 5–252 (in Japanese).

[36] H. Tanaka, "Semiconductor integrated circuit device," US Patent No. 6339318, Jan. 2002.

[37] K. Itoh, VLSI Memory Chip Design, Springer-Verlag, NY, 2001, Chapter 5.

8
Voltage Up-Converters and Negative Voltage Generators

8.1. Introduction

Voltage up-converters and negative-voltage generators, which internally generate boosted supply and negative supply voltages, such as V_{PP} (i.e., V_{DH}) and V_{BB}, using charge pump circuits from the external supply voltage (V_{DD}), have been indispensable to ensure successful operations of memory LSIs. In fact, the resultant internal supply voltages have enabled stable and reliable operations for modern DRAM cells and flash memory cells. For DRAM cells, they eliminate the V_t-drop of cell transistors and prevent nMOSTs in the cell from being forward-biased [1]. In addition, they have been essential to program and erase a flash memory cell, as explained in Chapter 1. Here, the resultant internal voltages and their requirements are quite different between DRAM and flash memory cells. For example, V_{PP} necessary for DRAM cells is much lower than for flash memory. Instead, the cycle time (t_{cyc}) of the V_{PP}-pulse for DRAMs is much faster: In modern DRAM design with V_{DD} (maximum data-line voltage) = 1.8 V, V_{PP} is 3.8 V with a boost ratio (V_{PP}/V_{DD}) of 2.1, while in flash memory with V_{DD} = 1.8 V, V_{PP} is 20 V with a boost ratio of 11. The cycle time is about 100 ns for DRAMs while for flash memories it is 1 μs at most, even for V_t-verify operation. Thus, despite a lower boost ratio, the V_{PP}-generator for DRAMs must provide more charges to compensate for the charge lost at the load every cycle, if the same load capacitance and conversion efficiency of the V_{PP}–generator are assumed. In the near future, without these converters, all ultra-low voltage nano-scale LSIs could not be designed successfully. Even SRAM cells are expected to use voltage up-converters to boost the cell power supply, as explained in Chapter 3. In addition to the memory cells, peripheral circuits in memory LSIs and logic gates in logic LSIs will require the converters to help limit the subthreshold current and the speed variation, as mentioned in Chapters 1–5. The results from applying boosted supply voltages and negative supply voltage to gates, source, or substrate of nMOST and pMOSTs.

Key design issues for the converters are the voltage conversion efficiency, degree of voltage setting accuracy, load-current delivering capability and stability of the output voltage, power consumed by the converter itself (especially during the standby period), and cost of implementation. Additional design issue is the output level monitor that is indispensable for stabilizing the output voltage and protecting scaled devices from excessively high voltages. Unfortunately, even for a given boost ratio, each design issue becomes more difficult with device and voltage scaling. For example, instability problems caused by the floating output of voltage converters using charge pump circuits are more prominent. Low-V_t MOSTs in the load may collapse to even lower V_t or even to a depleted state, causing unexpectedly large subthreshold currents and fatal operations throughout the chip, if the current delivering capability of the converter is not sufficient. In unusual conditions such as power on/off and burn-in test with high stress voltage and high temperature, such detrimental effects become worse despite a normal low-V_t in the normal operation [1, 2]. Thus, in-depth understanding of voltage converters themselves and their load characteristics is required for successful designs of ultra-low voltage nano-scale RAMs. Some of load characteristics are discussed in detail in Ref. [1].

Unfortunately, the boost ratio increases with voltage scaling, which makes converter designs more difficult. This is because, for example, the necessary V_t and thus necessary boosted word voltage of DRAM cells must be kept high to ensure the data retention characteristics, while the voltage of the peripheral logic circuits continues to be lower. Thus, the boost ratio for DRAM cells increases from 2.1 at 1.8-VV_{DD} (V_{DD}: maximum data-line voltage) to 3.6 at 0.5-VV_{DD}, as discussed in Chapter 2. Such is the case for logic circuits. The V_t of MOSTs in logic circuits must be quite high and almost constant (e.g., > 0.3 V) during inactive periods to minimize subthreshold current, independent of V_{DD} scaling. On the contrary, V_t during active periods must be decreased for high speed operations. The high V_t is effectively realized by applying gate-source backbiasing or substrate biasing (V_{PP} or V_{BB}) to low-actual V_t MOSTs for the active periods, as mentioned in Chapter 4. Thus, V_{PP} or V_{BB} must be increased with V_{DD} scaling, implying an ever-larger boost ratio (V_{PP}/V_{DD} or V_{BB}/V_{DD}). In any event, the increased boost ratio necessitates voltage multipliers for large output voltages (i.e., V_{PP} and V_{BB}) by increasing the number of stages of the charge pump circuit, causing degradation of the voltage conversion efficiency. Moreover, it dramatically reduces the load-current delivering capability because the high V_t of transfer MOSTs in the converter must be maintained to preserve a low subthreshold current. The increased boost ratio also calls for an increase in power supply current, as mentioned later, which increases the power of the converter itself. Therefore, new circuits to cope with these issues are needed.

Both converters, the voltage up-converter and negative-voltage generator, share similar circuit configurations. The voltage doubler using capacitors is the well-known converter that forms the basis for the voltage multiplier. The voltage multiplier is categorized as a capacitor-based circuit and an inductor-based circuit, as discussed later (see Table 8.2). The capacitor-based circuit requires few off-chip components, but its power efficiency decreases as the output voltage falls down with increasing load current. In contrast, the inductor-based circuit has a high power efficiency of 80–90%, independent of the output voltage, but 3–5 off-chip components including an inductor and switching devices are needed. Therefore, the capacitor-based circuit is low cost and suitable for a small load current, while the inductor-based circuit is suitable for a large load current, but expensive. The capacitor-based circuit is further categorized as the

Dickson-type converter and the switched-capacitor-type converter, as discussed later (Table 8.1). The Dickson-type converter needs a long time to reach the final voltage of nV_{DD} (n: the number of the boost stages) as a result of repetitive boosting operations, while the switched-capacitor-type converter boosts quickly to nV_{DD} at a time. The maximum voltage applied to the switching MOSTs is $2V_{DD}$ for the Dickson-type converter, while it is nV_{DD} for the switched-capacitor-type converter. On the other hand, the maximum voltage applied to the boost capacitors is nV_{DD} for the Dickson-type circuit, while it is V_{DD} for the switched-capacitor type circuit.

In this chapter, voltage converters for boosted voltages and negative voltages are described in terms of the circuit configuration and operating principle, and the above-described design issues from the viewpoint of memory designers. First, basic voltage converters using capacitors and their applications are described, exemplified by the voltage doubler. Second, Dickson-type and switched-capacitor-type voltage multipliers are discussed independently, and then compared in terms of the conversion efficiency and the influence of parasitic capacitances on the efficiency. Third, converters using inductors are described. Finally, a level monitor is briefly explained.

8.2. Basic Voltage Converters with Capacitor

8.2.1. Voltage Doubler

Figure 8.1 shows a voltage doubler using switches, which is a basic circuit for voltage up-converters. Here, C_0 is a smoothing capacitor which includes the load capacitor, and C_1 is the boost capacitor. At first, when switches SW_1 and SW_4 are closed and Φ_1 is at V_{SS} level, C_1 is charged up to $C_1 V_{DD}$. Next, when SW_1 and SW_4 are opened, SW_2 and SW_3 are closed and Φ_1 is raised to V_{DD} level, the C_0 potential rises to V_{PP} expressed as

$$V_{PP} = \frac{2C_1}{C_1 + C_0} V_{DD}. \tag{8.1}$$

The charge in C_1 is discharged slowly through the load. The output voltage V_{PP}, however, is maintained around $2V_{DD}$ by repeating these operations if the load impedance is sufficiently high, as shown in Fig. 8.1(b).

Let us analyze the characteristics of the voltage doubler using an equivalent circuit shown in Fig. 8.1(c). A cycle is divided into pre-charge and boost stages (8.1(b)). At first, during the pre-charge period, the charge stored on C_0 is discharged by the load. At the same time, the charge flows into C_1 from the V_{DD} line. The relationships between the output voltage, the load current and the potential of N_1 are expressed as

$$C_0 \frac{dV_{PP}}{dt} = -I_L, \quad V_{N1} = V_{DD}. \tag{8.2}$$

In addition, the relationship between the V_{PP} level just before boost, $V_{PP\,min}$, and that just after boost, $V_{PP\,max}$, is given, using the law of charge conservation between C_0 and C_1, as

$$C_1 \cdot 2V_{DD} + C_0 \cdot V_{PP\,min} = (C_0 + C_1) \cdot V_{PP\,max}.$$

The output voltage after boost is also expressed as

$$(C_0 + C_1) \frac{dV_{PP}}{dt} = -I_L. \qquad (8.3)$$

As shown in Figure 8.1(b), the output voltage takes a maximum $V_{PP\,max}$ just after boost ($t = T_1$). It falls with time and becomes a minimum $V_{PP\,min}$ after

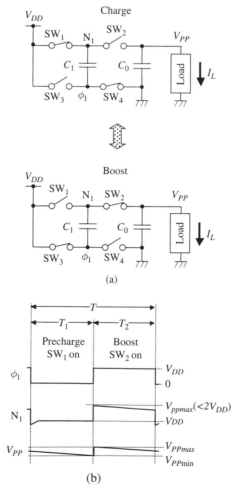

FIGURE 8.1. Principle of voltage doubler; circuit diagram (a), operating waveform (b), operating mode (c), equivalent circuit (d).

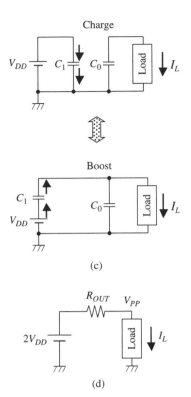

FIGURE 8.1. (Continued)

$t = T_1 + T_2$. The difference between $V_{PP\,\max}$ and $V_{PP\,\min}$ is the so-called ripple. The smaller the ripple, the more stably the load circuits operate. Using Eqs. (8.2) and (8.3), the average output voltage $\overline{V_{PP}}$, ripple v and conversion efficiency η of the doubler are given as

$$\overline{V_{PP}} = 2V_{DD} - I_L \cdot (T_1 + \frac{T_2 C_0}{C_0 + C_1})(\frac{1}{2C_0} + \frac{1}{C_1}) \qquad (8.4)$$

$$\cong 2V_{DD} - \frac{T}{C_1} \cdot I_L, (C_0 >> C_1),$$

$$v = \frac{I_L}{C_0} \cdot (T_1 + \frac{T_2 C_0}{C_0 + C_1}) \cong \frac{I_L T}{C_0}, (C_0 >> C_1), \qquad (8.5)$$

$$\eta = \frac{V_{PP}}{2V_{DD}}. \qquad (8.6)$$

Figure 8.1(d) shows the equivalent circuit of a voltage doubler consisting of a voltage source of $2V_{DD}$ and an output resistance R_{OUT}. It is derived from Eq. (8.4), in which the first term and second term express the voltage of the voltage source and the output resistance, respectively. This equation shows that to enhance the current

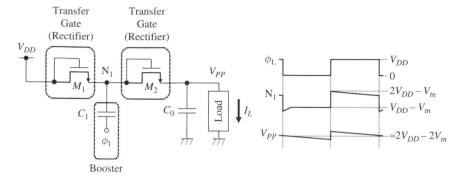

FIGURE 8.2. Example of voltage doubler circuit.

delivering capability of the circuit, the output resistance should be made small by reducing the period T and using a large capacitor C_1. In addition, for a smaller ripple, a shorter period or a larger smoothing capacitor is desired. Moreover, the conversion efficiency deteriorates as V_{PP} becomes low. This means that the power loss due to the equivalent output resistance increases as the load current increases.

In practice, there exists no perfect switch. Figure 8.2 shows a conventional charge pump circuit [2]. A clock Φ_1 in this figure works as switches SW_3 and SW_4 in Fig. 8.1 while a precharge nMOST M_1 works as the switch SW_1. A transfer nMOST M_2 acts as the series switch SW_2. In this circuit configuration, however, the V_t-drops at M_1 and M_2 and the substrate-bias effect of M_2 reduce V_{PP}, causing degradation of the boost ratio V_{PP}/V_{DD}, especially at low voltages. Note that a low enough V_t creates a subthreshold current, causing a malfunction at the floating node. A feedback charge-pump circuit [3] shown in Fig. 8.3 solves the problem. By utilizing a feedback MOST, node N_1 is precharged at V_{DD} instead

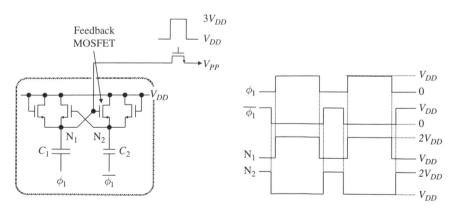

FIGURE 8.3. Feed-back type voltage doubler for low voltage operation. Reproduced from [3] with permission; © 2006 IEEE.

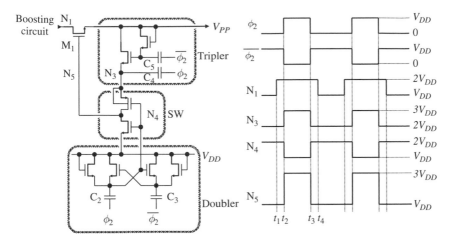

FIGURE 8.4. Transfer switch of voltage doubler for low voltage operation. Reproduced from [3] with permission; © 2006 IEEE.

of $V_{DD} - V_T$ while Φ_1 is "high". Thus, node N_1 is boosted to $2V_{DD}$ when $\overline{\Phi}_1$ goes "high" and a full V_{DD} is also stored at node N_2 while Φ_1 is "low", and node N_2 is boosted to $2V_{DD}$ when Φ_1 goes "high". This feedback technique can eliminate a voltage loss due to the threshold voltage drop of the nMOSTs. Figure 8.4 shows the whole V_{PP}-converter. The major components of the converter are the primary charge-pump circuit shown in Fig. 8.3, a transmission gate M_1, and a control pulse generator for the transmission gate. The control pulse generator consists of a the $3V_{DD}$-driver to drive the gate of the transmission gate, a $3V_{DD}$-booster to generate $3V_{DD}$ voltage and to feed it to the $3V_{DD}$-driver, and the other $2V_{DD}$-booster, which generates a gate control signal for the $3V_{DD}$-driver. A timing diagram for the drive pulse is shown in Fig. 8.4. Node N_3 is boosted from V_{DD} to $2V_{DD}$ at t_1. Before t_1, node N_6 stays at V_{DD} since both node N_4 and node N_5 voltages are $2V_{DD}$. When Φ_2 goes "high" and $\overline{\Phi}_2$ goes "low" at t_2, node N_4 is boosted from $2V_{DD}$ to $3V_{DD}$, and node N_5 is pulled down from $2V_{DD}$ to V_{DD}. Then, node N_6 goes from V_{DD} to $3V_{DD}$, and the transmission gate M_1 turns on, and $2V_{DD}$ is obtained at the output of the V_{PP} converter. In practice, the boost ratio is degraded by parasitic capacitances. However, an almost constant boost ratio as large as 1.8 was obtained at $V_{DD} = 0.5$–2V for the V_{PP} converter with feedback while a decrease in boost ratio was observed when V_{DD} is less than 1.5V for the conventional circuit without feedback [3], as shown in Fig. 8.5. A drawback is that the transfer nMOST M_1 needs a highly boosted gate voltage to eliminate the V_t drop.

Figure 8.6 shows another voltage doubler [4] using pMOST transfer gates. Although the boosted gate voltage is not needed, the reverse bias of the junctions must be ensured by keeping the n-well (i.e. substrate of the pMOST) voltage always higher than the source voltage. This is done using two sets of cross-coupled pMOSTs. M_3 and M_4 are the transfer gates that alternately pump the load. M_1 and M_2 quickly give the highest voltage to either of n-wells of M_3

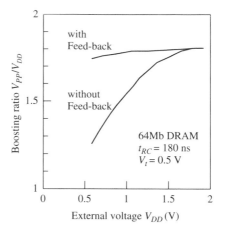

FIGURE 8.5. External voltage vs. boosting ratio. Reproduced from [3] with permission; © 2006 IEEE.

and M_4 to ensure the reverse bias of the junctions. This circuit does not need a gate boosting circuit for the charge transfer gate. Therefore, a timing margin is unnecessary between Φ_1 and $\overline{\Phi_1}$. This simple circuit configuration enables high-speed operation. On this account, this circuit has the advantage of requiring smaller capacitors.

8.2.2. Negative Voltage Generator

A charge pump circuit can generate not only a positive boosted voltage described above, but also a negative voltage. Figure 8.7 shows the principle of a negative voltage generator. At first, SW_1 and SW_3 are closed and inverter output Φ_1 is at V_{DD}. C_1 is charged, so that its upper electrode is negative and its lower electrode is positive. Next, SW_1 and SW_3 are opened, SW_2 and SW_4 are closed and Φ_1

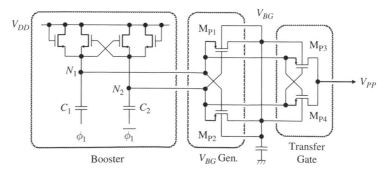

FIGURE 8.6. Voltage doubler using pMOST transfer switch for low voltage operation. Reproduced from [4] with permission; © 2006 IEEE.

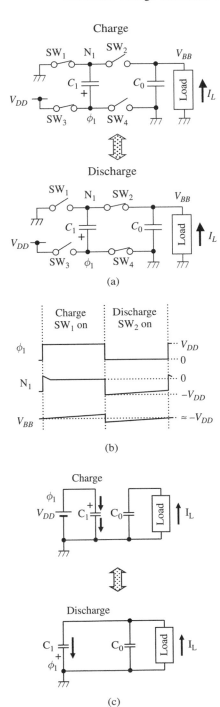

FIGURE 8.7. Principle of negative voltage generator; simplified circuit (a), operating waveform (b), and equivalent circuit (c).

goes to the V_{SS} level. Node N_1 becomes negative. The voltage at the top electrode of C_0 becomes a constant V_{BB} after repetitive operations. This is the principle of the negative voltage generator. Figure 8.8 shows the circuit which utilizes MOST diodes instead of SW_1 and SW_2. C_1 is charged with $V_{DD} - V_t$ when Φ_1 is at V_{DD}. When Φ_1 becomes V_{SS} level, the N_1 voltage becomes negative, and the current flows through M_2 from the load. The output level finally becomes V_{BB} after repetitive operations. When $I_L = 0$, V_{BB} becomes $-V_{DD} + 2V_{tn}$ which is higher by the V_{tn} of M_2 than the voltage of N_1, which is $-V_{DD} + V_{tn}$. In this way, a voltage loss of $2V_{tn}$ occurs as long as MOST diodes are used as switching elements, and the circuit cannot output a low enough voltage with voltage scaling. This is the same problem faced by the positive voltage doubler described above.

Figure 8.9 shows an example of a negative voltage generator for low voltage operation. The first stage is the negative charge pump consisting of the same cross-coupled pMOSTs configuration as in Fig. 8.3 to output full V_{DD} swing (from 0 to $-V_{DD}$) pulses with opposite phases from the nodes N_1 and N_2. The second stage is the V_{BG} generator that prevents M_{N3} and M_{N4} from forward biasing. The final stage is the charge transfer gate to alternately pump the load. Because M_{N3} (or M_{N4}) is turned on when the gate and source voltages of M_{N3} (or M_{N4}) are 0 and $-V_{DD}$, respectively, the negative voltage at N_1 is transferred to the output V_{BB} without a V_t loss. However, if the back gate (i.e., substrate or body) potentials of M_{N3} and M_{N4} rise over their source potentials, the parasitic diode between the source and back-gate turns on and may cause a latch-up. Thus, the back-gate potential must be kept always lower than the source potential. M_{N1} and M_{N2} meet the requirement because the gate voltage of one of M_{N1} and M_{N2} is 0 when its source potential is $-V_{DD}$, the MOST is turned on, and M_{N1} and M_{N2} thus supply the voltage $(-V_{DD})$ to the back

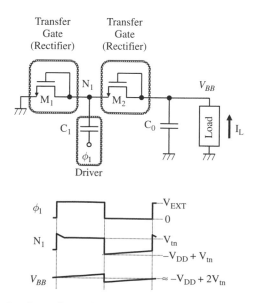

FIGURE 8.8. Example of negative voltage generator.

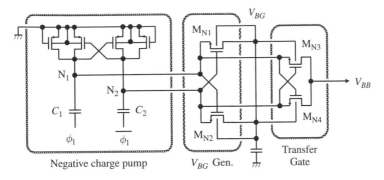

FIGURE 8.9. Negative voltage generator for low-voltage operation.

gate of M_{N3} and M_{N4}. The back gate is always kept at the minimum potential of this circuit, preventing latch up. This circuit also has an advantage of smaller capacitors in the same way as the circuit in Fig. 8.6.

8.2.3. Applications to Memories

Voltage Up-Converter

Figure 8.10 shows a V_{PP}-converter with a level monitor [10] detecting the V_{PP} level. When the level is low, a ring oscillator starts to operate so that a pumping capacitor C_P is driven and thus V_{PP} is raised. When V_{PP} reaches a certain level, the level monitor turns the ring oscillator off so that charge injection to the V_{PP}-load stops. The whole circuit is activated by ENABLE, which is a slow-cycle pulse such as the refresh control pulse with 16-μs cycle time during the standby period, or a random pulse generated synchronously with activation of a chip-enable signal. Here, it is assumed that the V_T of the level detecting MOST M_M is the same as that of memory cell-MOST. First, the operation during the standby period is explained for the case of a slow-cycle refresh-control (ENABLE) pulse. The input makes the M_M-gate high because node a rises to a high level. When refreshing the cells, a word-line activation lowers V_{PP} by drawing current from the supply. As long as the resultant V_{PP} is higher than $V_{DD} + V_t$, however, M_M continue to be on, allowing M_1 and M_2 to form a current mirror. Thus, node b is held at a high level if the transconductance of M_2 is larger enough than that of M_0 or M'_0, and the detecting speed(i.e. the time until M_2 detects a high level at node a) is fast enough. Thus, the ring-oscillator enable signal(OSCEN) is kept to a low level, turning the oscillator off. If V_{PP} is lower than $V_{DD} + V_t$, however, node b is discharged by M_0 and M'_0 because M_2 is off, allowing the oscillating to start oscillation with a high-level OSCEN. The resultant low-OSC makes not only nodes c, d and e high, but also nodes f and g change from V_{DD} and V_{PP} to 0V and $V_{DD} - V_t$, respectively. Even when ENABLE is turned off, the input NOR gate allows node a and OSCEN to maintain the same levels, allowing the oscillator to continue oscillation and the M_M-gate to remain at V_{PP}. Thus, the level monitor is ready for the V_{PP}-level detection.

296 Voltage Up-Converters and Negative Voltage Generators

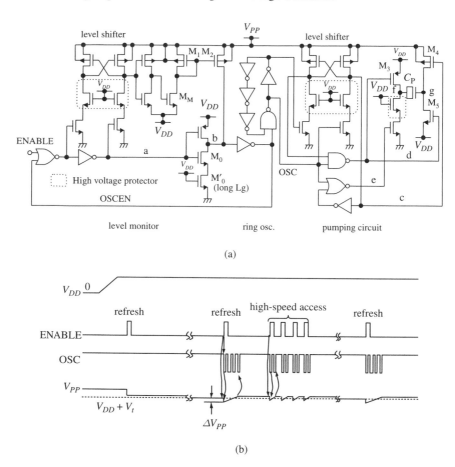

FIGURE 8.10. V_{PP} generator with level monitor; schematic of circuit (a), and timing diagram (b). Reproduced from [10] with permission; © 2006 IEEE.

The succeeding high-OSC eventually makes node f change from 0V to V_{DD}. Since M_4 is on while M_5 is off at this moment, the voltage change by V_{DD} that is developed across C_P injects charges to the V_{PP}-load so that V_{PP} is slightly raised. Subsequent repetitive OSC-pulses gradually raise V_{PP} until V_{PP} reaches $V_{DD} + V_t$. As soon as V_{PP} exceeds $V_{DD} + V_t$, M_M is turned on so that the oscillation stops and thus V_{PP} stays at $V_{DD} + V_t$. Thus, V_{PP} is maintained constant by application of the refresh-control pulses. Even when the chip is randomly-accessed, the V_{PP} degradation is compensated in the same manner by application of an ENABLE pulse generated synchronously with a chip-enable pulse. Here, the nMOSTs with their gates biased to V_{DD} surrounded by dashed lines are inserted at the nodes at V_{PP} to protect the nMOSTs from a high stress voltage, as discussed in Chapter 9, and M'_0 has enough channel length to reduce V_{PP} current.

Figure 8.11 shows another V_{PP} generator [11]. It features the use of two kinds of charge-pump circuits, a main pump and an active kicker, to provide charges for

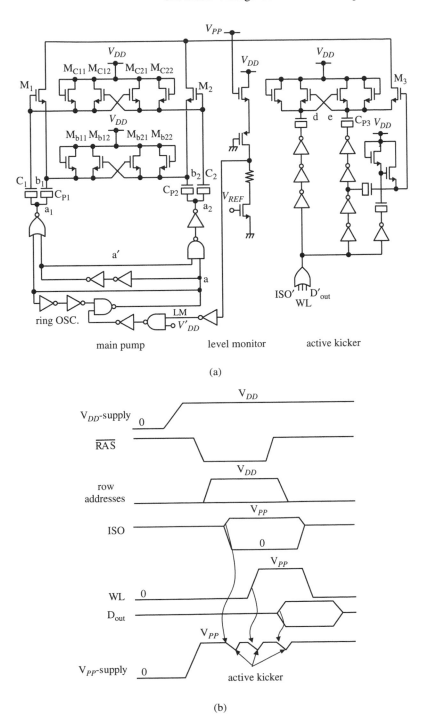

(a)

(b)

FIGURE 8.11. V_{PP} generator with two sets of pump circuits and level monitor; schematic of circuit (a), and timing diagram (b). Reproduced from [11] with permission; © 2006 IEEE.

a purely capacitive output load. The main pump compensates for a small charge loss due to the leakage current of the load. It is driven by a ring oscillator which is activated when V_{PP} is lower than the level determined by the level monitor. The active kicker operates synchronously with the load-circuit operation such as ISO driving (Fig. 2.26), word-line driving, and output buffer driving. This circuit compensates for a large charge loss due to load circuit operations. Non operation of the main pump and extremely slow cycle operation of the active kicker in the date-retention mode provide a minimized retention current.

Negative Voltage Generator.

Figure 8.12 shows the V_{BB} generator featuring two sets of charge pump circuits [12]: a slow cycle ring oscillator 1 for supplying a small current during retention and stand-by modes and a fast cycle ring oscillator 2 for supplying a sufficiently larger current during the active cycle or when the level monitor detects the V_{BB} level is high. Thus, it minimizes the retention current by shutting down the fast cycle circuit. Here, CMOS inverters (IV_1, IV_2) are waveform shapers, and inverter IV_3 is a buffer to drive a heavy capacitance (C_2). To suppress the current flowing into the substrate, M_3 must be small (i.e. small W/L).

Figure 8.13 shows another alternative which stops the oscillation of the V_{BB} generator while the DRAM chip is not in an active cycle [13]. In normal mode the ring oscillator provides an enough I_{CP} at a high frequency, while in battery-backup mode it oscillates intermittently only during burst requests of refresh operation. A negative V_{BB} is realized by extracting holes from the p-substrate

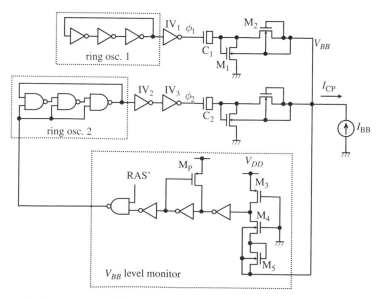

FIGURE 8.12. V_{BB} generator with mode switching and level monitor. Reproduced from [12] with permission; © 2006 IEEE.

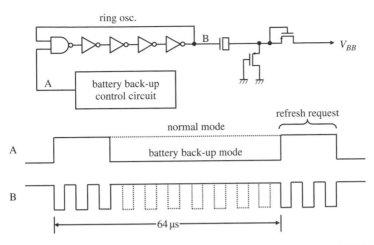

FIGURE 8.13. V_{BB} generator with pump-frequency control and output pMOS diodes. Reproduced from [13] with permission; © 2006 IEEE.

by pMOSTs at the output stage. The use of pMOSTs there prevents minority-carriers (electrons) injection from the diode in the generator to the nMOST memory cells, which differs from the use of nMOSTs described before.

Figure 8.14 shows another V_{BB} generator [1] in which charges are alternately pumped twice a cycle by using pMOSTs (M_2, M_4). In addition, the raised gate voltages of the cross-coupled M_1 and M_3 eliminate their V_t loss, offering a deeper V_{BB} of - $V_{DD} + V_t$(M_2, M_4). The V_t drop is still an obstacle to obtaining a deep V_{BB}.

Figure 8.15 shows a V_{BB} generator suitable for a low V_{DD} operation [14]. All MOSTs except M_1 are pMOSTs. When clock Φ goes down to 0V, the N_1-voltage also goes down to a negative voltage of - $V_{DD} + |V_{tP}|$ (V_{tP} : V_t of M_3), and thus M_2 is turned on so that the N_2-voltage is clamped to 0V, and M_1 is cut off. When Φ goes up to V_{DD}, the N_1-voltage is clamped to $|V_{tP}|$ and thus M_2 is cut off. At the same time, N_2-voltage instantaneously drops to -V_{DD}

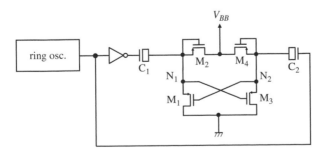

FIGURE 8.14. pMOST cross-coupled V_{BB} generator. Reproduced from [1] with permission of Springer.

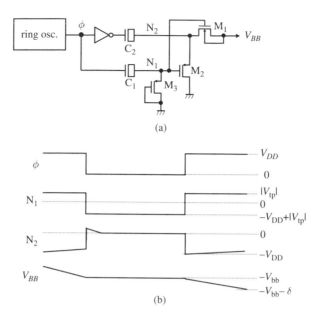

(a)

(b)

FIGURE 8.15. Hybrid pumping V_{BB} generator; schematic of circuit (a), and timing diagram (b). Reproduced from [14] with permission; © 2006 IEEE.

by capacitive coupling. Then, N_2 and the substrate continue to be charged up and discharged, respectively, until both voltages are equilibrated, allowing the charge at N_2 to be pumped to the substrate. As a result, one clock cycle deepens V_{BB} by δ. Here, the pumping is performed without a V_t drop at M_1 because the M_1-gate voltage is higher enough than the N_2 and substrate voltages during the equilibration process. By repetitive clock-applications the substrate is gradually discharged and the pumping ceases when V_{BB} reaches as low as $-V_{DD}$. Thus, perfect conversion of the power-supply voltage is established. In fact, a V_{BB} of -1.44V at $V_{DD} = 1.5$V was experimentally obtained.

8.3. Dickson-Type Voltage Multiplier

8.3.1. Voltage Up-Converter

This section describes voltage up-converters with a boost ratio of more than two which is becoming increasingly important in the nano-meter, as mentioned in Chapter 1. Figure 8.16 shows the so-called Dickson type charge pump that is an n-time voltage multiplier [5] achieved by an n-series-connected circuit of the above doubler. The advantage is that any multiple of the supply voltage can be theoretically generated using only two kinds of pulses, while the disadvantage is the need for high-breakdown-voltage boost capacitors in the latter stages. In the circuit, C_0 is the sum of load capacitance and smoothing capacitance,

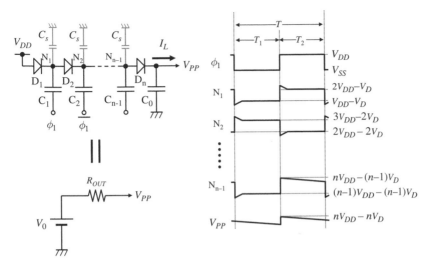

FIGURE 8.16. Dickson type voltage multiplier. Reproduced from [5] with permission; © 2006 IEEE.

C_1–C_{n-1} are boost capacitors, C_s's are parasitic capacitance, and D_1–D_n are diodes. Complementary pulses Φ_1 and $\overline{\Phi}_1$ are alternately supplied. When Φ_1 is at ground level, C_1 is charged to $V_{DD}-V_D$ through D_1 from V_{DD}. Here, V_D is a forward voltage drop of D_1. When Φ_1 is raised to V_{DD}, node N_1 is boosted to $2V_{DD}-V_D$ if $C_S \ll C_1$. Since $\overline{\Phi}_1$ is at ground level at this time, the charge in C_1 flows through D_2 into C_2. By this operation, the voltage across C_2 rises until the potential difference between N_1 and N_2 becomes V_D, and the voltage of N_2 thus reaches $2(V_{DD}-V_D)$. Then, $\overline{\Phi}_1$ becomes V_{DD}, and node N_2 is boosted to $2(V_{DD}-V_D)+V_{DD}$. Since Φ_1 is V_{SS} at this time, the charge in C_2 is transferred to C_3. When the voltage difference between N_2 and N_3 becomes V_D, in other words, when the voltage of N_3 becomes $2(V_{DD}-V_D)+V_{DD}-V_D = 3(V_{DD}-V_D)$, the charge flow stops. After that, Φ_1 becomes V_{DD}, and node N_3 is boosted to $3(V_{DD}-V_D)+V_{DD}$. Since $\overline{\Phi}_1$ is V_{SS} at this time, the charge in C_3 is transferred to C_4. When the voltage difference between N_3 and N_4 becomes V_D, that is, when the voltage of N_4 becomes $3(V_{DD}-V_D)+V_{DD}-V_D = 4(V_{DD}-V_D)$, the charge flow stops. This operation is repeated likewise, and each node voltage is boosted. Consequently, when the voltage difference between N_{n-1} and the output V_{PP} at the final stage becomes V_D, that is, when V_{PP} becomes $(n-1)(V_{DD}-V_D)+V_{DD}-V_D = n(V_{DD}-V_D)$, the charging of C_0 stops. In this way, V_{PP} finally becomes $n(V_{DD}-V_D)$.

If a load current exists at the final stage, the voltage drop occurs at the output, like for the voltage doubler. It also results from the parasitic capacitance C_S at each stage in the charge pump. Including these influences in the above equation, the relation between average output voltage $\overline{V_{PP}}$ and the output current I_L is expressed as follows [5].

$$\overline{V_{PP}} = \frac{nC + C_S}{C + C_S} \cdot V_{DD} - nV_D - \frac{(n-1)T}{C + C_S} \cdot I_L \qquad (8.7)$$

Here, C is the boost capacitance, C_S is the parasitic capacitance, T is the period of boosting clock, and I_L is the load current. This equation can be expressed as a series circuit of voltage source V_0 and a resistance R_{OUT}, as shown in the figure. Here, the total of the first and second terms of the right-hand of Eq. (8.7) corresponds to V_0, and the coefficient of I_L of the third term expresses R_{OUT}. Obviously, the resistance increases in proportion to the number of the stages n. In addition, the parasitic capacitance C_S lowers the output voltage, while it suppresses a rate of change of the output voltage for the load current with reduced R_{OUT}. The ripple is approximately expressed by the same Eq. (8.5) as that of a voltage doubler.

Figure 8.17 shows a Dickson-type voltage quintupler (5-time voltage multiplier) using nMOST diodes as rectification elements. Two-phase signals shifted by 180 degrees are alternately applied to the terminals of the capacitors, enabling a V_{PP} of $5(V_{DD} - V_t)$. The advantage is that the circuit can be easily constituted by using a standard CMOS process, while the disadvantage is the large voltage loss of $5V_t$. Furthermore, the V_t increases due to the body effect as the voltage difference between source and bulk becomes large. In addition, the voltage drop caused by the on- resistances of MOSTs and the load current must be taken into account.

With increasing the stage count, n, the resultant voltage drop (loss) nV_t is fatal for low-voltage LSIs. Figure 8.18 is a voltage multiplier to solve the problem [6]. There is no voltage loss when charging the boosting capacitors. For example, at the first stage, N_1 is raised to $2V_{DD}$ when Φ_1 becomes V_{DD}. The nMOST in the inverter thus turns on, and node G_1 becomes V_{DD}. It prevents the charge from flowing backward from N_1 to V_{DD}. On the other hand, the pMOST in the inverter becomes off because N_2 is at $2V_{DD}$. When Φ_1 becomes V_{SS}, the nMOST in the inverter is off because node N_1 becomes at V_{DD}. Because N_2 is raised to $3V_{DD}$, the pMOST in the inverter turns on, and the voltage of G_1 becomes $3V_{DD}$. The charge flows through M_1 from V_{DD} to node N_1, and C_1 is charged without a voltage loss. In this way, each boosting capacitor is charged without a voltage

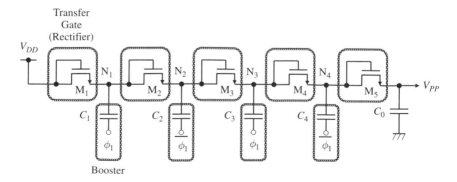

FIGURE 8.17. Example circuit of Dickson type voltage multiplier.

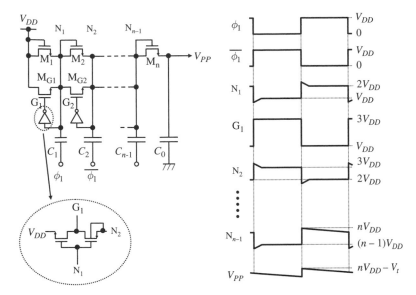

FIGURE 8.18. Dickson type voltage multiplier for low-voltage operation. Reproduced from [6] with permission; © 2006 IEEE.

loss, and a high output voltage can be generated with fewer stages even at low supply voltage. This is because the gate voltage of the transfer nMOST of each stage is controlled by the voltage of nV_{DD} and $(n+2)V_{DD}$. Since the last stage consists of only a MOST, the output voltage V_{PP} drops by the threshold voltage and becomes $nV_{DD} - V_t$.

Figure 8.19 shows the four-stage multiplier to further reduce the voltage loss [7]. This circuit can generate $4V_{DD}$. Figures 8.19(a) and (b) show the main part of the pump circuit, and level shifters for controlling the gate voltage, respectively. It uses both pMOSTs and nMOSTs as transfer MOSTs. The nMOSTs are used for the first half, while the pMOSTs are used for the second half. The voltage boosted in the second half is supplied to nMOSTs, and the voltage boosted in the first half is supplied to pMOSTs. These voltages are supplied through the level shifter S's that are controlled by the boosting clocks. By this configuration, a voltage that is higher by $2V_{DD}$ than the transferred voltage is supplied to the gate of each nMOST, while a voltage that is lower by $2V_{DD}$ than the transferred voltage is supplied to the gate of each pMOST. Therefore, each MOST can transfer the charge without any V_t loss. In addition, the on-resistance of transfer each MOST is low enough because the gate-source voltage becomes $2V_{DD}$. Thus, a multiplier with a low voltage loss even at a low supply voltage can be realized.

The current supplying capability of the circuit is proportional to frequency $(1/T)$, as expressed by Eq. (8.7) and is thus increased by a high-frequency clock. However, the circuit is difficult to implement because of a high voltage of $2V_{DD}$ supplied between terminals of charge transfer MOSTs. A thicker gate oxide and a longer channel length necessary for relevant MOSTs, compared with those of the drive circuit where only V_{DD} is involved, making high-speed operation impossible.

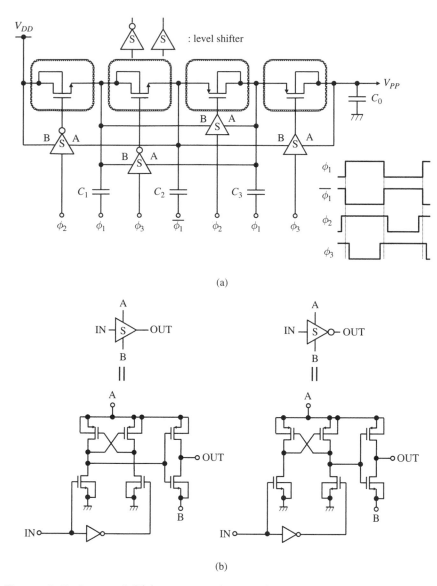

FIGURE 8.19. Improved Dickson type voltage multiplier for low-voltage operation; schematic of charge pump circuit (a), and level shift circuit (b). Reproduced from [7] with permission; © 2006 IEICE.

The circuit shown in Fig. 8.20 enables a high-speed operation [8] thanks to having only V_{DD}-tolerant short-channel MOSTs and smaller boosting capacitors. This circuit features two boosting capacitors, which are driven by complementary signals, Φ_1 and $\overline{\Phi}_1$, in pumping unit PUMP. Node N_X between the units is always kept at a boosted voltage V_H by this configuration. This enables the drain-source voltage of M_{02} (or M_{01}) to be applied V_{DD} at most when node N_{22}

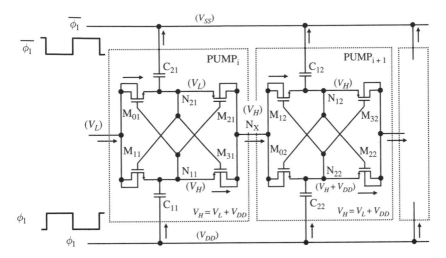

FIGURE 8.20. High efficient two-phase charge pump. Reproduced from [8] with permission; © 2006 IEEE.

(or N_{12}) in the next stage is boosted. The arrows in the figure show current flows when Φ_1 is V_{DD} and $\overline{\Phi}_1$ is V_{SS}. The voltage of each node at this time is shown in parentheses. Here, V_L is the voltage before boosting and V_H (= $V_L + V_{DD}$) is the voltage after boosting. Clearly, the voltage difference between each electrode of MOST is V_{DD} at most. A five-stage charge pump fabricated with a $0.18 - \mu$m process [8] has been reported to operate at a 100MHz at $V_{DD} = 1.8$V.

8.3.2. Negative Voltage Generator

A negative voltage multiplier can be composed of similar circuits to the multipliers discussed previously, with inverted polarities and a slight modification of the voltage relations of the rectifying devices. For instance, $-nV_{DD}$ is generated by an n-stages negative voltage multiplier, while $(n+1)V_{DD}$ is generated by a positive voltage multiplier. Figure 8.21 shows an example of the negative voltage multiplier of the Dickson type using diodes. The direction of the diodes is opposite to that in Fig. 8.16 and the voltage of the cathode of the first diode is ground instead. The voltages of the intermediate nodes $N_1 - N_n$ are gradually deeper from the left to the right. The output voltage is $-nV_{DD} + (n+1)V_D$, where V_D is the forward bias voltage of the diodes. Figure 8.22 shows another negative voltage multiplier without a V_t drop that is suitable for low voltage operations. The circuit is configured by inverting the polarities and voltage relations of the circuit in Fig. 8.19, and the principle of operation is the same. In this circuit, a voltage of $-3V_{DD}$ is generated from V_{DD} at the maximum. nMOSTs and pMOSTs are used for charge transfer in the first half and the second half, respectively. The voltages of the intermediate nodes in the second half are applied to the

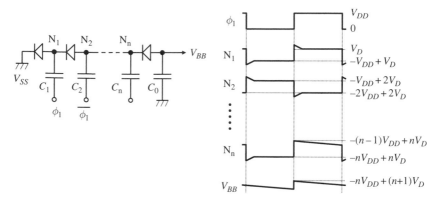

FIGURE 8.21. Dickson type negative voltage multiplier using diodes.

pMOSTs, and the voltages in the intermediate nodes in the first half are applied to the nMOSTs. These voltages are applied to the gates of the transfer MOSTs through level shifters S controlled by boosting clocks. The charge is transferred without a V_t loss, since a voltage lower than the transferred voltage by $2V_{DD}$ is applied to each pMOST gate, while a voltage higher than the transferred voltage by $2V_{DD}$ is applied to each nMOST gate. Moreover, because the gate-sources voltage is $2V_{DD}$ at the maximum, the on-resistance of each transfer MOST is low enough. Therefore, a negative voltage multiplier without any voltage loss is realized even if a large load current flows.

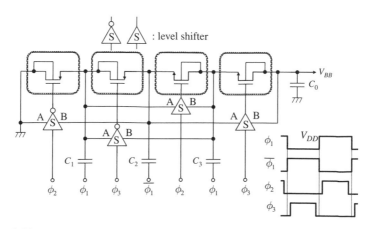

FIGURE 8.22. Negative voltage tripler for low-voltage operation. Reproduced from [7] with permission; © 2006 IEICE.

8.4. Switched-Capacitor (SC)-Type Voltage Multipliers

8.4.1. Voltage Up-Converter and Negative Voltage Generator

In addition to the Dickson-type multiplier, the switched capacitor generates boosted voltages. Figure 8.23 shows the principle of a switched-capacitor multiplier with $n-1$ stages. First, $C_1 - C_{n-1}$ are connected in parallel, so they are simultaneously charged to V_{DD}. Next, they are serially connected, and the bottom terminal of C_1 is also connected to the V_{DD} line. Then, the voltages of $C_1 - C_{n-1}$ and the V_{DD} are added, so a voltage of nV_{DD} is obtained at the output terminal. Figure 8.24 shows the principle of a switched-capacitor negative voltage multiplier with n stages. $C_1 - C_n$ are connected in parallel, so they are charged at the same time first. Next, they are switched to the series connection. Here, they are connected so that the output may be negative. The terminal of C_1 is connected to ground at the same time, so the output voltage becomes $-nV_{DD}$.

8.4.2. Fractional Voltage Up-Converters

The previous section describes how the power efficiency of the voltage up-converter using capacitors decreases with an increase in the load current. Thus, if a fractional voltage converters, such as for 3/2 or 4/3 of the input voltage, using capacitors is realized, the power efficiency can be improved. For example,

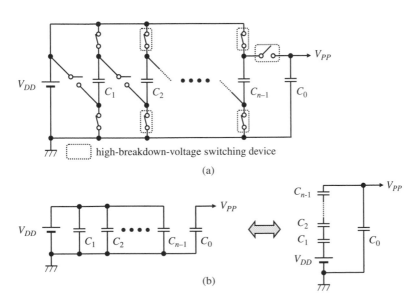

FIGURE 8.23. Switched capacitor type positive voltage multiplier; simplified circuit (a), and equivalent circuit (b).

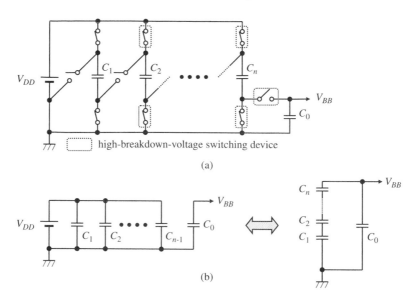

FIGURE 8.24. Switched capacitor type negative voltage multiplier; simplified circuit (a), and equivalent circuit (b).

the efficiency for an ideal 2-time charge pump becomes $V_{PP}/(2V_{DD})$. On the other hand, the efficiency for an ideal 1.5-time charge pump is expressed as $V_{PP}/(1.5V_{DD})$. Figure 8.25 shows the circuits to achieve this. For instance, two capacitors with the same capacitance are used to generate $3/2V_{DD}$. First, they are connected in series between V_{DD} and V_{SS}. As a result, C_1 and C_2 are charged

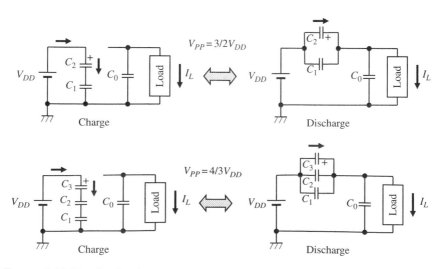

FIGURE 8.25. Fractional voltage up converter.

with $V_{DD}/2$, respectively. Next, they are connected in parallel and their lower electrodes are connected to the V_{DD} line and the upper electrodes are connected to the load. After the repetitive operations, a voltage of $V_{DD} + V_{DD}/2 = 3/2V_{DD}$ appears at C_0 in the case of no load current. Three capacitors are used to generate $4/3V_{DD}$. We should connect them to parallel after connecting three capacitors to series between V_{DD} and V_{SS} as mentioned above. Note that the on resistances of the switching devices should be minimized, as described in Section 7.3, because the output voltage is lowered due to the load current and the on resistances.

8.5. Comparisons between Dickson-Type and SC-Type Multipliers

8.5.1. Influences of Parasitic Capacitances

In this section a more detailed analysis including the influence of parasitic capacitance on the efficiency is performed. Figure 8.26 shows the equivalent circuits of the Dickson-type converter (a) and the switched-capacitor-type converter (b). The critical parameter for the efficiency is the parasitic capacitance C_P for both circuits. In the Dickson type, the capacitance C_P of the signal wiring that drives the boost capacitors is dominant. This consists of the substrate capacitance of

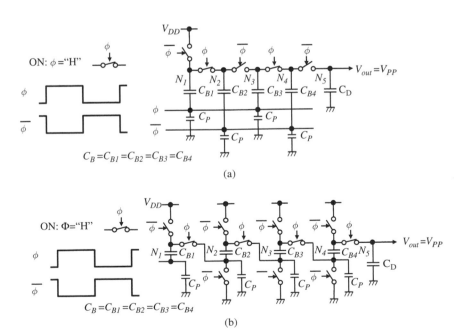

FIGURE 8.26. Model for efficiency analysis of charge pump circuits; model for Dickson type charge pump (a), and model for switched capacitor type charge pump (b).

the boost capacitors and the wiring capacitance. This capacitance wastes some of the charging and discharging current between V_{DD} and V_{SS} while no direct influence on the boost level. On the other hand, in the switched-capacitor type, the parasitic capacitance at the boost nodes is dominant. Although the quantity of charge that comes from each boost node is different, a charge of $(n-1)V_{DD} \, C_P$ is charged and discharged at the maximum. Therefore, the efficiency is expected to be lower than for the Dickson type.

The capacitance between two poly-silicon layers or between the gate and channel of a MOST is usually used for the boosting capacitor. The former has the advantage that there is no charge loss due to the threshold voltage, and the boost efficiency can be higher because the parasitic capacitance is small. Therefore, it is often used for small-scale on-chip EEPROMs. On the other hand, the latter is often used for large capacity DRAMs and the flash memories, where the cost has the first priority. This is because no additional layers are necessary, and capacitance per unit area is large. However, because the parasitic capacitance of the latter, which includes drain-substrate and source-substrate junction capacitances (or the well-substrate junction capacitance), is larger than that of the former, the boosting efficiency is not so high as that of the former. The C_S, which is the parasitic capacitance on the boosted nodes, is disregarded here because it is generally small enough compared with the substrate capacitance of the boosting capacitor though it is considered in Subsection 8.3.1.

Figure 8.27 compares the parasitic-capacitance dependency of the switched-capacitor-type and the Dickson-type quintuplers. Both are calculated based on

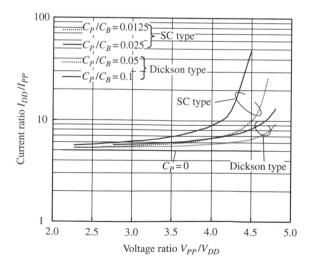

FIGURE 8.27. Dependency of parasitic capacitance of input/output current ratio in voltage multiplier (voltage quintupler).

the equivalent circuits in Fig. 8.30. The smaller the I_{DD}/I_{PP} is, the higher the efficiency is. If the parasitic capacitance is not considered, $I_{DD}/I_{PP} = 5$ for both. If it is included, I_{DD}/I_{PP} increases with V_{PP}/V_{DD}. However, the increase rate of the Dickson type is smaller than that of the switched-capacitor type, implying that the Dickson type has a higher conversion efficiency. In addition, it is found that the dependency on C_P/C_B, which is the ratio of the parasitic capacitance to the boost capacitance, of the Dickson type is smaller than that of the switched-capacitor type. This is because the amount of charging and discharging of the parasitic capacitance that occurs for the former is smaller than for the latter, as noted above. Here, the ratio C_P/C_B is assumed to be 0.05 and 0.0125 for the Dickson type, and 0.1 and 0.025 for the switched-capacitor type, because the latter can use larger boost capacitors with a thinner gate oxide (in principle, 1/4 in a voltage quintupler). If both ratios can be equal with a thick film for all the boost capacitors, the difference further grows. The deriving process of the analytical solution of these characteristics is shown in Appendix A8.1.

8.5.2. Charge Recycling Multiplier

The conversion efficiency of voltage multipliers decreases due to the influence of parasitic capacitance, as explained in Section 8.5.1. A method to reduce this influence is shown in Fig. 8.28. Figure 8.28(a) is for the Dickson type [9], and (b) is for the switched-capacitor type [15]. Both are fundamentally the same though the latter circuit is more complex. The feature is that a period when both Φ and $\overline{\Phi}$ are "low" is provided. During this period, both of two driving signal lines N_L and $N_L{}'$ are floating. At this time, both voltages are equalized by Φ_S. Since one of the signal lines that was at V_{SS} is raised to $V_{DD}/2$ by this equalization, it is raised only from $V_{DD}/2$ to V_{DD} in the next cycle. Thus, the charge that flows from the V_{DD} line, $V_{DD}/2 \times C_P$, is halved. Therefore, this can be called a charge-recycling method. On the other hand, since $V_{DD}/2$ is replaced by $nV_{DD}/2$ (n: the number of boosting stages) in the switched-capacitor type, a greater effect can be expected. The derivation process for the analytical solution of these characteristics is shown in Appendix A8.1.

Figure 8.29 shows the result of the calculation from the analytical solution for both types of quintuplers. The ratio C_P/C_B, is assumed to be 0.1 for both circuits. It is shown that there is an improvement of several times in the switched-capacitor type, while about 10–30% in the Dickson type. This is because the quantity for charging the parasitic capacitance is larger in the switched-capacitor type.

Table 8.1 compares the Dickson-type converter and the switched-capacitor-type converter. They have the pros and cons. The former needs a long time to reach the final voltage of nV_{DD} as a result of repetitive boosting operations,

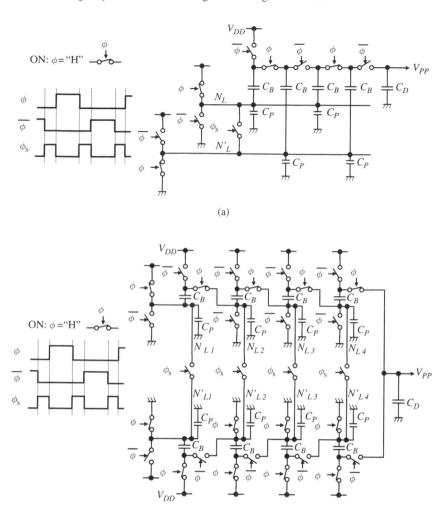

FIGURE 8.28. Model for efficiency analysis of charge pump circuit using charge recycle technique; model for Dickson type charge pump(a) [9], and model for switched capacitor type charge pump (b) [17].

while the latter boosts quickly to nV_{DD} at a time. The maximum voltage applied to the switching MOSTs is $2V_{DD}$ for the former, while it is nV_{DD} for the latter. On the contrary, the maximum voltage applied to the boost capacitors is nV_{DD} for the former, while it is V_{DD} for the latter. The influence of parasitic capacitance on the efficiency is small for the former, while it is large for the latter, although the difference in the influence can be reduced by charge-recycling.

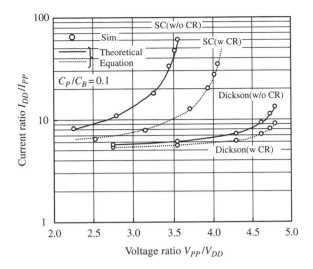

FIGURE 8.29. Efficiency of voltage multiplier (voltage quintupler) using charge recycle technology.

TABLE 8.1. Comparison of Dickcson type and switched capacitor type converter.

Type	Dickson	Switched cap.
Time to final level	$V_{DD} \to 2V_{DD} \to$ $\ldots \to nV_{DD}$	$V_{DD} \to nV_{DD}$
Power efficiency	$< \frac{|V_{out}|}{nV_{DD}}$	$< \frac{|V_{out}|}{nV_{DD}}$
Max. voltage to MOST	$2V_{DD}$	nV_{DD}
Max. voltage to capacitor	nV_{DD}	V_{DD} (except smoothing cap.)
Influence of parasitic capacitances	Small	Large

Dickson pump $\approx$ switched cap. pump for voltage doubler
n: Number of boosting stages

8.6. Voltage Converters with an Inductor

The voltage up-converter can also be implemented with an inductor. Figure 8.30 shows an example of a voltage up-converter using an inductor. In general, it is difficult to form an on-chip inductor because of the following reasons. First, it is difficult to form a large inductor of several μH because a large area is necessary to make spiral-shaped wiring, and the number of wiring layers is limited from three to about six. Second, a current is generated in the substrate and wiring by the electromagnetic induction, causing a power loss. Therefore, an off-chip

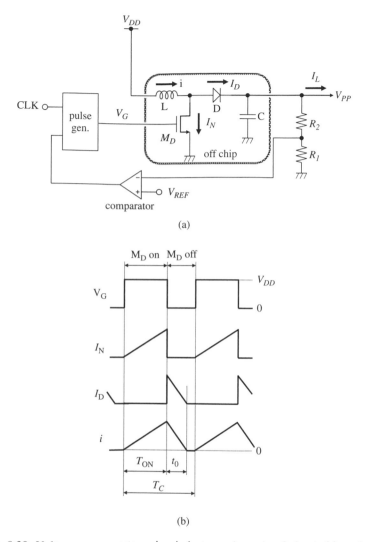

FIGURE 8.30. Voltage up converter using inductor; schematic of circuit (a), and timing diagram (b).

inductor is usually used. Moreover, since it is difficult to form a transistor and a diode with a high breakdown voltage and low on-resistance on a chip, these necessary elements are often put outside along with the inductor. The operation can be explained as follows (see Fig. 8.30(b)), ignoring the comparator for a while. The turn-on period of drive transistor M_D and the cycle time are denoted T_{ON} and T_C, respectively. First, the gate of drive transistor M_D is forced to "H" to turn it on. Then, the following current flows through inductor L.

$$i = I_N = \frac{V_{DD}}{L} \cdot t.$$

Next, the current of the inductor flows to the load through the diode when turning M_D off after T_{ON}. This current and the time until this current becomes 0 are given as

$$i = \frac{V_{DD}}{L} \cdot T_{ON} - \frac{V_{DD} - V_{PP} - V_D}{L} \cdot t.$$

$$t_0 = \frac{V_{DD}}{V_{DD} - V_{PP} - V_D} \cdot T_{ON}.$$

Here, V_D is the voltage drop with the diode. Since the load current continues to flow during the period of t_0 and the cycle time of the switching is T_C, the average load current is obtained by dividing the quantity of charge that flows during t_0 by T_C. This is expressed as:

$$I_L = \frac{V_{DD}}{V_{DD} - V_{PP} - V_D} \cdot T_{ON} \times \frac{V_{DD}}{L} \cdot T_{ON} \times \frac{1}{2} \div T_C$$

Solving for V_{PP} results in the following equation:

$$V_{PP} = \frac{(V_{DD} \cdot T_{ON})^2}{2 \cdot T_C \cdot I_L \cdot L} + V_{DD} - V_D \tag{8.8}$$

This is the maximum output voltage of this circuit. It should be noted that the voltage V_{PP} is controlled by T_{ON}. Here, the comparator compares $R_1/(R_1 + R_2)V_{PP}$ and V_{REF}, and turns M_D off if $R_1/(R_1 + R_2)V_{PP}$ exceeds V_{REF}. As a result, the output voltage V_{PP} is determined only by the resistance ratio and the reference voltage as

$$V_{PP} = \frac{R_1 + R_2}{R_1} \cdot V_{REF}. \tag{8.9}$$

Figure 8.31 shows an example of the negative voltage generator using an inductor. In this circuit, the inductor, a drive transistor, and a diode, etc. are also put external to the chip. In the same manner, the following equations are provided for the output voltage V_{BB} of this circuit.

$$V_{BB} = -\frac{(V_{DD} \cdot T_{ON})^2}{2 \cdot T \cdot I_L \cdot L} + V_D. \tag{8.10}$$

This is the maximum output voltage of this circuit. Here, the magnitude of V_{BB} can be controlled by controlling T_{ON} as well as the voltage up converter. The comparator circuit compares $R_1/(R_1 + R_2) V_{BB}$ and V_{REF} as well as the voltage up converter, and turns M_D off when $R_1/(R_1 + R_2)V_{BB}$ exceeds V_{REF}. However, V_{REF} must be negative. As a result, the output voltage V_{BB} is determined only by the resistance ratio and the reference voltage as follow

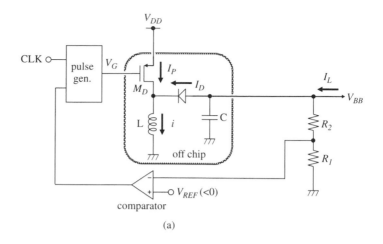

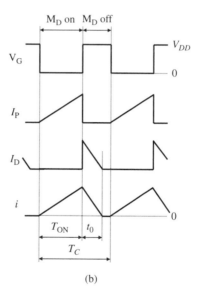

FIGURE 8.31. Negative voltage generator using inductor; schematic of circuit (a), and timing diagram (b).

$$V_{BB} = \frac{R_1 + R_2}{R_1} \cdot V_{REF}, \quad V_{REF} < 0 \qquad (8.11)$$

If the level monitor shown in Figure 8.33(a) is used, V_{BB} can be controlled by a positive reference voltage.

Table 8.2 compares the converters that use a capacitor and an inductor. The capacitor-type converter(with no load current) has the voltage conversion ratio of an integer determined by the number of boosting stages, while the inductor-type

TABLE 8.2. Comparison of voltage converter with capacitor and voltage converter with inductor.

Type	Capacitor	Inductor
Voltage conversion ratio	n	any
Power efficiency	$< \frac{|V_{out}|}{nV_{DD}}$	$> 80\text{–}90\%$
Off-chip components	0–1	3–5
Extra pins	0–1	≥ 2

n: Number of boosting stages

converter features a non-integer voltage conversion ratio achieved by controlling the duty cycle of the control pulse. The former has a power efficiency that decreases as the output voltage falls down when the load current increases. The latter has a power efficiency independent of the output voltage and that remains always as high as 80–90% for the same reason as the voltage down-converter with inductor, as discussed in Chapter 7. The former needs few off-chip components, while the latter needs 3–5 components including an inductor and switching devices as well as two or more extra pins for the connection with the off-chip components. Therefore, the capacitor-based circuit is suitable for a small load current and low cost design, while the inductor-based circuit is suitable for a large load current, but expensive design.

8.7. Level Monitor

8.7.1. Level Monitor for Voltage Up-Converter

In order to protect scaled devices from an excessively raised voltage, the level monitor is a key to designing a voltage up-converter. Figure 8.32 shows a level monitor for voltage multiplier. Although we explained the monitor circuits in Figs. 8.10 and 8.11, the circuit shown here, similar to those described in Chapter 7, is more general one. The comparator, Comp, compares the reference voltage V_{REF} with the voltage V'_{PP} generated by dividing V_{PP} by resistors, and stops voltage-up operation of the charge pump when V'_{PP} is higher than V_{REF}. The output voltage is expressed as

$$V_{PP} = (R_1 + R_2)/R_2 \cdot V_{REF}.$$

Here, V_{PP} level can precisely be set by dividing the resistances. However, the layout area of the resistors is large in comparison with that of MOSTs because the sheet resistance of a resistor is generally low. Figure 8.32(b) shows another division using MOSTs for a smaller layout area. A reference voltage is input to the gate of one of the two dividing MOSTs. This method features more flexible output-voltage setting than using n cascaded MOSTs with diode

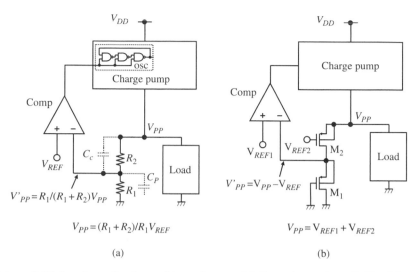

FIGURE 8.32. Level monitor for positive voltage multiplier; using resistor divider (a), and using MOST divider (b).

connection. Assuming that W/L of M_1 and M_2 are equal, the output voltage V_{PP} is expressed as:

$$V_{PP} = V_{REF1} + V_{REF2}.$$

In general, a higher V_{PP} requires larger resistors so as to reduce the current consumption in the voltage divider. Therefore, the parasitic capacitance (C_P in the figure) of resistors grows large, causing a delay of the feedback signal. If the delay is large, the ripple of V_{PP} increases and may cause unstable operation of the load circuit of V_{PP}, as discussed in Chapter 7. It is effective to connect bypass capacitor C_C in parallel with R_2 to reduce the delay. A large C_C is better because it works as a speed-up capacitor.

8.7.2. Level Monitor for Negative Voltage Multiplier

Figure 8.33 shows level monitors for negative voltage multipliers. The circuits need an extra positive reference voltage, which differs from those in Fig. 8.32. This is because the output of the voltage dividing circuit should be positive to make the comparator operate at a positive single-power supply. For the resistor dividing type, the reference voltage V_{REF2} is applied to R_1, while V_{REF1} is applied to the comparator. Output voltage V_{BB} is expressed as $V_{BB} = (R_1 + R_2)/R_1 V_{REF1} - R_2/R_1 V_{REF2}$, and $V_{REF1} < V_{REF2}$. For the MOS driving type, three reference voltages are needed. The output voltage, assuming that W/L of M_1 and M_2 are equal, is expressed as $V_{BB} = V_{REF3} + V_{REF1} - V_{REF2}$, and $V_{REF2} - V_{REF1} > V_{tn}$.

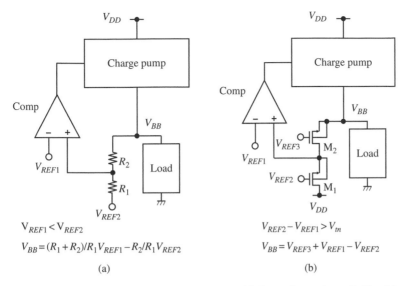

$V_{REF1} < V_{REF2}$

$V_{BB} = (R_1 + R_2)/R_1 V_{REF1} - R_2/R_1 V_{REF2}$

(a)

$V_{REF2} - V_{REF1} > V_{tn}$

$V_{BB} = V_{REF3} + V_{REF1} - V_{REF2}$

(b)

FIGURE 8.33. Level monitor for negative voltage multiplier; using resistor divider (a), and using MOST divider (b).

A8.1. Efficiency Analysis of Voltage Up-Converters

A8.1.1. Dickson-Type Charge Pump Circuit

Let us derive the equation for the voltage quintuplers shown in Figs. 8.26(a) and 8.28(a).

(1) Case 1: parasitic capacitance ignored.

In Figure 8.26(a), let us denote the voltages of nodes $N_1 - N_4$ after charging and before charging by (V_{10}, V_1), (V_{20}, V_2), (V_{30}, V_3), and (V_{40}, V_4), respectively. Then, the following equations are obtained for each node.

Node N_4, N_5: Because node N_4 becomes $V_{40} + V_{DD}$ just after boost and becomes V_{out} after time T, we obtain the following equation:

$$(V_{40} + V_{DD} - V_{out}) C_B = I_{out} T. \qquad (A8.1)$$

Here, the lefthand side shows the charge that flows out of C_{B4}, and the righthand side shows the charge that flows to the load.

Node N_3, N_4: The voltage of node N_3 is V_{30} just before boost, the voltage of node N_4 is V_4 just before charging. The voltages of nodes N_3 and N_4 are equalized to V_{40} after boost. Therefore,

$$V_4 C_B + V_{30} C_B = (V_{40} - V_{DD}) C_B + V_{40} C_B, \qquad (A8.2)$$

$$V_4 = V_{out} - V_{DD}. \qquad (A8.3)$$

Node N_2, N_3, and N_1 and N_2: We obtain the following equations in the same manner:

$$V_3 C_B + V_{20} C_B = (V_{30} - V_{DD}) C_B + V_{30} C_B \tag{A8.4}$$

$$V_3 = V_{40} - V_{DD}. \tag{A8.5}$$

$$V_2 C_B + V_{10} C_B = (V_{20} - V_{DD}) C_B + V_{20} C_B, \tag{A8.6}$$

$$V_2 = V_{30} - V_{DD}, \tag{A8.7}$$

$$V_{10} = V_{DD}. \tag{A8.8}$$

Here, T is the cycle time of clock □. Thus, from Eqs. (A8.6)–(A8.8), Eqs. (A8.4), (A8.5) and (A8.9), and Eqs. (A8.2), (A8.3) and (A8.11) we obtain

$$V_{20} = \frac{1}{2} (V_{DD} + V_{30}). \tag{A8.9}$$

$$V_{30} = \frac{1}{3} (V_{DD} + 2V_{40}). \tag{A8.10}$$

$$V_{40} = \frac{1}{4} (V_{DD} + 3V_{out}). \tag{A8.11}$$

Finally, substituting Eq. (A8.11) into Eq. (A8.1) and solving for V_{out} result in

$$V_{out} = 5V_{DD} - 4\frac{I_{out} T}{C_B}. \tag{A8.12}$$

Generalizing the above equation results in the following equation:

$$V_{out} = nV_{DD} - (n-1)\frac{I_{out} T}{C_B}. \tag{A8.13}$$

Next, let us find the input current that flows from V_{DD} line, Φ, and $\overline{\Phi}$. The charging current I_{i1} ($V_{DD} \to N_1$) to node N_1 is derived as follows. Because the voltage of node N_1 becomes V_{20} after boost, the voltage just before charging, V_1, is $V_{20} - V_{DD}$. Therefore,

$$V_1 = V_{20} - V_{DD} \tag{A8.14}$$

$$(V_{DD} - V_1) C_B = I_{i1} T. \tag{A8.15}$$

Eqs. (A8.14), (A8.15), and (A8.1)–(A8.9) result in $I_{i1} = I_{out}$. The charging currents I_{i2} ($\Phi \to N_1 \to N_2$) to node N_2, I_i ($\overline{\Phi} \to N_2 \to N_3$) to node N_3, and I_{i4} ($\Phi \to N_3 \to N_4$) to node N_4 are expressed as

$$(V_{20} - V_2) C_B = I_{i2} T. \tag{A8.16}$$

$$(V_{30} - V_3) C_B = I_{i3} T. \tag{A8.17}$$

$$(V_{40} - V_4) C_B = I_{i4} T. \tag{A8.18}$$

Using Eq. (A8.4), the charging current I_{i5} $(\overline{\Phi}N_4N_5)$ to node N_5 is expressed as $I_{i5} = I_{out}$. Substituting Eqs. (A8.1) – (A8.9) into Eqs. (A8.16) – (A8.18) and solving I_{i2}, I_{i3}, I_{i4}, and I_{out} results in $I_{i2} = I_{i3} = I_{i4} = I_{out}$. Here, $I_{i1} - I_{i5}$ are current that flow from V_{DD} line in one cycle. Thus, the total input current I_i and the generalized current are given as

$$I_i = I_{DD} = \sum_{n=1}^{5} I_{in} = 5I_{out}. \qquad (A8.19)$$

$$I_{DD} = nI_{out}. \qquad (A8.20)$$

(2) Case 2: parasitic capacitance included.

In the Dickson type circuit, the voltage swing between both ends of parasitic capacitance C_P is V_{DD}. Therefore, the following current I_P due to parasitic capacitance C_P is added to Eq. (A8.20).

$$I_p = \frac{\sum C_p V_{DD}}{T} = \frac{(n-1)\kappa C_B V_{DD}}{T} \qquad (A8.21)$$

Here, n is the number of boost stages and κ is a ratio of parasitic capacitance and boosting capacitance, C_P/C_B. Therefore, the input current is expressed as

$$I_{DD} = nI_{out} + \frac{(n-1)\kappa C_B V_{DD}}{T}. \qquad (A8.22)$$

Solving Eq. (A8.13) for C_B and substituting it into Eq. (A8.22) give

$$I_{DD} = nI_{out} + \frac{(n-1)^2 \kappa I_{out} V_{DD}}{nV_{DD} - V_{out}} = \left\{ n + \frac{(n-1)^2 \kappa V_{DD}}{nV_{DD} - V_{out}} \right\} I_{out} \qquad (A8.23)$$

$$\frac{I_{DD}}{I_{out}} = n + \frac{(n-1)^2 \kappa}{n - V_{out}/V_{DD}} V_{DD}. \qquad (A8.24)$$

This is the equation that expresses the I/O current ratio of the charge pump circuit.

(3) Case 3: using charge recycle.

Figure 8.28 (a) shows a simplified model for the charge recycling. Due to the charge recycling, the voltage swing between both ends of parasitic capacitance, C_P, is halved. Therefore, the current that flows to C_P (the second term of Eq. (A8.23)) is also halved. Therefore, the I/O current ratio is given as

$$\frac{I_{DD}}{I_{out}} = n + \frac{(n-1)^2 \kappa}{2\left(n - V_{out}/V_{DD}\right)}. \qquad (A8.25)$$

A8.1.2. Switched-Capacitor-Type Charge Pump Circuit

Let us derive the equation for the voltage quintuplers shown in Figs. 8.26(b) and 8.28(b).

(1) Case 1: parasitic capacitance ignored.

In this method, all $n - 1$ boosting capacitors are connected in series after V_{DD} is charged in them, and the voltage of nV_{DD} is obtained for no load current. If load current I_{out} flows, the potential difference between the electrodes of each boosting capacitor equally decreases. This is expressed by the following equation:

$$\frac{(nV_{DD} - V_{out})}{n-1} C_B = I_{out}T. \tag{A8.26}$$

Solving this equation for V_{out} results in the following equation:

$$V_{out} = nV_{DD} - (n-1)\frac{I_{out}T}{C_B}. \tag{A8.27}$$

It is obvious that this is the same as Eq. (A8.13) of the Dickson charge pump circuit. Next, let us consider the relation between the input current and the output current. During boosting the charge of each boost capacitor has decreased by $I_{out}T$, so it should be charged by the same. Because $(n - 1)$ capacitors are simultaneously charged, the relation between input current I_{i1} and the charge is given as

$$(nV_{DD} - V_{out}) C_B = I_{i1}T \tag{A8.28}$$

Substituting this equation into Eq. (A8.26) results in the following equation:

$$I_{i1} = (n-1) I_{out}. \tag{A8.29}$$

Moreover, the input current also flows during boosting because one of the terminals of C_{B1} is connected with V_{DD} line, and this is equal to I_{out}. Therefore, the total input current is given by adding this current and the current of Eq. (A8.29) as

$$I_{DD} = I_{out} + (n-1) I_{out} = nI_{out}. \tag{A8.30}$$

This is also equal to Eq. (A8.20) for the Dickson method.

(2) Case 2: parasitic capacitance included.

It is difficult to derive a general formula for the switched-capacitor circuit because the output voltage changes due to the charge sharing with parasitic capacitance in addition to the load current. Even if it can be derived, the equation is complex and not comprehensive. Therefore, we show only the deriving method and an equation under a specific condition here. Here, we assume that the decrease of the charge of each boost capacitor C_{Bi} is equal to the output charge. $V_1 - V_n$ in the following equations are the voltages of node $N_1 - N_n$ just after boosting.

$$V_{DD}C_B - (V_n - V_{n-1})C_B = I_{out}T \text{ for node } N_n. \tag{A8.31}$$

$$V_{DD}C_B - (V_{n-1} - V_{n-2})C_B = I_{out}T + V_{n-1}C_p \text{ for node } N_{n-1}. \tag{A8.32}$$

$$V_{DD}C_B - (V_3 - V_2)C_B = I_{out}T + (V_{n-1} + V_{n-2} + \cdots + V_3)C_p \text{ for node } N_3. \tag{A8.33}$$

$$V_{DD}C_B - (V_2 - V_1)C_B = I_{out}T + (V_{n-1} + V_{n-2} + \cdots + V_3 + V_2)C_p \text{ for node } N_2. \tag{A8.34}$$

Adding both sides of the equations from node N_2 to N_n results in the following equation:

$$(n-1)V_{DD}C_B - (V_n - V_1)C_B = (n-1)I_{out}T + \{(n-2)V_{n-1} + (n-3)V_{n-2} + \cdots$$
$$+2V_3 + V_2\}C_p \tag{A8.35}$$

Taking the difference of the equations of node N_i and $N_{i-1}(i = 3, \ldots, n)$, $n-2$ simultaneous equations that contain n unknown variable from V_n to V_1 are obtained. Here, because V_n and V_1 are constant ($V_n = V_{out}$, $V_1 = V_{DD}$), and V_i ($i = 2 - (n-1)$) are linear functions of V_{out} and V_{DD}, Eq.(A8.35) is simplified as follows and V_{out} is given as

$$(n-1)V_{DD}C_B - (V_{out} - V_{DD})C_B = (n-1)I_{out}T + (aV_{DD} + bV_{out})C_p \tag{A8.36}$$

$$\therefore V_{out} = \frac{n - \kappa a}{1 + \kappa b}V_{DD} - \frac{(n-1)T}{C_B(1+\kappa b)}I_{out}. \tag{A8.37}$$

Here, $\kappa = C_p/C_B$, and a and b are constants determined by the number of boosting stages, as explained later for $n = 5$.

Then, we derive the relation between the input current I_{DD} and the output current I_{out}. Here, the input current is the sum of the current flowing to the load and parasitic capacitance during boosting and the current flowing to each boosting capacitor during charging. Hence,

$$I_{DD} = fC_p \sum_{i=1}^{n-1} V_i + I_{out} + \{(V_{DD} - V_{out} + V_{n-1})C_Bf + (V_{DD} - V_{n-1} + V_{n-2})C_Bf$$

$$+ \cdots + (V_{DD} - V_2 + V_1)C_Bf\}$$

$$= fC_p \sum_{i=1}^{n-1} V_i + I_{out} + (nV_{DD} - V_{out})C_Bf, \quad V_1 = V_{DD}, \quad f = 1/T$$

Here, because $V_1 - V_n$ are linear functions of V_{out} and V_{DD}, as mentioned previously, I_{DD} is given as

$$I_{DD} = fC_P(cV_{DD} + dV_{out}) + I_{out} + (nV_{DD} - V_{out})C_Bf$$
$$= \{(cV_{DD} + dV_{out})\kappa + nV_{DD} - V_{out}\}C_Bf + I_{out}, \quad \kappa = C_P/C_B.$$

Here, c and d are constants determined by the number of boosting stages. Solving Eq. (A8.37) for C_B results in the following equation:

$$\frac{I_{DD}}{I_{out}} = 1 + \frac{(n-1)\left\{(c\kappa + n) + (d\kappa - 1)\frac{V_{out}}{V_{DD}}\right\}}{(n - \kappa a) - (1 + \kappa b)\frac{V_{out}}{V_{DD}}}. \tag{A8.38}$$

Here, constants a, b, c, and d are obtained for $n = 5$ as follows. Equations of the charge that flows to each node are expressed as:

$$V_{DD}C_B - (V_5 - V_4)C_B = I_{out}T, \tag{A8.39}$$

$$V_{DD}C_B - (V_4 - V_3)C_B = I_{out}T + V_4C_p, \tag{A8.40}$$

$$V_{DD}C_B - (V_3 - V_2)C_B = I_{out}T + (V_4 + V_3)C_p, \tag{A8.41}$$

$$V_{DD}C_B - (V_2 - V_1)C_B = I_{out}T + (V_4 + V_3 + V_2)C_p. \tag{A8.42}$$

Adding from Eqs. (8.39) to (8.42) results in the following equation:

$$(4V_{DD} - V_5 + V_1)C_B = 4I_{out}T + (3V_4 + 2V_3 + V_2)C_p. \tag{A8.43}$$

$$\therefore V_{out} = 5V_{DD} - \frac{4I_{out}T}{C_B} - (3V_4 + 2V_3 + V_2)\frac{C_p}{C_B}. \tag{A8.44}$$

Subtracting Eq. (A8.40) from Eq. (A8.39), Eq. (A8.41) from Eq. (A8.40), and Eq. (A8. 42) from Eq. (A8.41) result in

$$-V_3 + \left(2 + \frac{C_p}{C_B}\right)V_4 - V_5 = 0, \tag{A8.45}$$

$$-V_2 + \left(2 + \frac{C_p}{C_B}\right)V_3 - V_4 = 0, \tag{A8.46}$$

$$-V_1 + \left(2 + \frac{C_p}{C_B}\right)V_2 - V_3 = 0. \tag{A8.47}$$

By using Eqs. (A8.44)–(A8.47) the voltage relationship is obtained for $V_5 = V_{out}$, $V_1 = V_{DD}$, and $C_p/C_g = \kappa$ as

$$3V_4 + 2V_3 + V_2 = \frac{(10 + 6\kappa + \kappa^2)V_{DD} + (14 + 14\kappa + 3\kappa^2)V_{out}}{(2 + \kappa)(2 + 4\kappa + \kappa^2)} = aV_{DD} + bV_{out}, \tag{A8.48}$$

$$a = \frac{10 + 6\kappa + \kappa^2}{(2 + \kappa)(2 + 4\kappa + \kappa^2)}, \quad b = \frac{14 + 14\kappa + 3\kappa^2}{(2 + \kappa)(2 + 4\kappa + \kappa^2)}. \tag{A8.49}$$

On the other hand, the relationship between the input and output currents is given as

$$I_{DD} = fC_p \sum_{i=1}^{4} V_i + I_{out} + (5V_{DD} - V_{out}) C_B f. \tag{A8.50}$$

Substituting V_2, V_3, and V_4, which are derived from Eqs. (A8.45)–(A8.47), into the term of $\sum$ in the above equation gives

$$\sum_{i=1}^{4} V_i = V_{DD} + V_2 + V_3 + V_4 = \frac{\left(5 + 5\kappa + \kappa^2\right) V_{DD} + (3 + \kappa) V_{out}}{2 + 4\kappa + \kappa^2} = cV_{DD} + dV_{out}, \tag{A8.51}$$

$$c = \frac{5 + 5\kappa + \kappa^2}{2 + 4\kappa + \kappa^2}, \quad d = \frac{3 + \kappa}{2 + 4\kappa + \kappa^2} \tag{A8.52}$$

(3) Case 3: using charge recycling.

Using charge recycling, the voltage swing at the parasitic capacitance is halved. Therefore, it is equivalent to the replacement of C_P with $C_P/2$ in Eqs. (A8.32), (A8.33), and (A8.34). This is also equivalent to the replacement of κ with $\kappa/2$. If this replacement is applied to Eqs. (A8.38), (A8.49), and (A5.52), the I/O current ratio for $n = 5$ can be obtained.

References

[1] K. Itoh, *VLSI Memory Chip Design*, Springer-Verlag, NY, 2001.
[2] Y. Nakagome, M. Horiguchi, T. Kawahara, K. Itoh, "Review and prospects of low-voltage RAM circuits," IBM J. R & D, vol. 47, no. 5/6, pp. 525–552, Sep./Nov. 2003.
[3] Y. Nakagome, H. Tanaka, K. Takeuchi, E. Kume, Y. Watanabe, T. Kaga, Y. Kawamoto, F. Murai, R. Izawa, D. Hisamoto, T. Kisu, T. Nishida, E. Takeda and K. Itoh, "An experimental 1.5-V 64-Mb DRAM," IEEE J. Solid-State Circuits, vol. 26, pp. 465–472, Apr. 1991.
[4] P. Favrat, P. Deval and M. J. Declercq, "A high-efficiency CMOS voltage doubler," IEEE J. Solid-State Circuits, vol. 33, pp. 410–416, Mar. 1998.
[5] J. F. Dickson, "On-chip high-voltage generation in MNOS integrated circuits using an improved voltage multiplier technique," IEEE J. Solid-State Circuits, vol. SC-11, pp. 374–378, June 1976.
[6] J.-T. Wu and K.-L. Chang, "MOS charge pumps for low-voltage operation," IEEE J. Solid-State Circuits, vol. 33, pp. 592–597, Apr. 1998.
[7] T. Myono, A. Uemoto, S. Kawai, E. Nishibe, S. Kikuchi, T. Iijima and H. Kobayashi, "High-efficiency charge-pump circuits with large current output for mobile equipment applications," IEICE Trans. Electron., vol. E84-C, pp. 1602–1611, Oct. 2001.

[8] R. Pelliconi, D. Iezzi, A. Baroni, M. Pasotti and P. L. Rolandi, "Power efficient charge pump in deep submicron standard CMOS technology," IEEE J. Solid-State Circuits, vol. 38, pp. 1068–1071, June 2003.

[9] C. Lauterbach, W. Weber and D. Römer, "Improvement of boosted charge pumps," IEEE J. Solid-State Circuits, vol. 35, pp. 719–723, May 2000.

[10] R. C. Foss, G. Allan, P. Gillingham, F. Larochelle, V. Lines and G. Shimokura, "Application of a high-voltage pumped supply for low-power DRAM," in Symp. VLSI Circuits Dig. Tech. Papers, June 1992, pp. 106–107.

[11] D.-J. Lee, Y.-S. Seok, D.-C. Choi, J.-H. Lee, Y.-R. Kim, H.-S. Kim, D.-S. Jun and O.-H. Kwon, "A 35ns 64Mb DRAM using on-chip boosted power supply," in Symp. VLSI Circuits Dig. Tech. Papers, June 1992, pp. 64–65.

[12] K. Sato, H. Kawamoto, K. Yanagisawa, T. Matsumoto, S. Shimizu and R. Hori, "A 20ns static column 1Mb DRAM in CMOS technology," in ISSCC Dig. Tech. Papers, Feb. 1985, pp. 254–255.

[13] Y. Konishi, K. Dosaka, T. Komatsu, Y. Ionue, M. Kumanoya, Y. Tobita, H. Genjyo, M. Nagatomo and T. Yoshihara, "A 38-ns 4-Mb DRAM with a battery-backup (BBU) mode," IEEE J. Solid-State Circuits, vol. 25, pp. 1112–1117, Oct. 1990.

[14] Y. Tsukikawa, T. Kajimoto, Y. Okasaka, Y. Morooka, K. Furutani, H. Miyamoto and H. Ozaki, "An efficient back-bias generator with hybrid pumping circuit for 1.5-V DRAM's," IEEE J. Solid-State Circuits, vol. 29, pp. 534–538, Apr. 1994.

[15] H. Tanaka, M. Isoda and T. Kawahara, "Nonvolatile memory and processing system," US Patent No. 6,781,890, Aug. 2004.

9
High-Voltage Tolerant Circuits

9.1. Introduction

On-chip voltage converters are becoming increasingly important for ultra-low voltage nano-scale memories, as discussed in Chapter 1. They include the reference voltage generator, the voltage down-converters, the voltage up-converter and negative voltage generator with charge pump circuits, and level shifters to adjust resultant voltage differences between internal blocks, and between the internal core and I/O circuits. In the past, DRAMs and flash memories have required such voltage converters to ensure stable operations and retention characteristics of memory cells, and they will continue to need such converters. Even future SRAM cells may need such converters for ensuring low subthreshold current and stable operation, as discussed in Chapters 1–3. For example, the raised voltage necessary for such memory cells must be kept high, independent of device scaling, to ensure data-retention characteristics, although the operating voltage for peripheral logic circuits can be scaled down with device scaling. Consequently, the voltage difference between memory cells and peripheral circuits will grow with device scaling. In addition, interface circuits of chips must operate at quite a high external voltage, although some internal circuits using scaled devices can operate at another low external voltage. Moreover, in the near future, even some logic gates will have to operate at high voltages using raised (boosted) supply voltages and/or negative supply voltages, to manage subthreshold currents, as discussed in Chapters 2–5. These circumstances unavoidably call for stress voltage-immune circuits for the memory cell and its related circuits, interface-related circuits, and subthreshold-current sensitive circuits. This chapter describes high-voltage tolerant circuit techniques for such circuits.

9.2. Needs for High-Voltage Tolerant Circuits

Device scaling causes the reliability problem, such as device-parameter degradation due to hot carrier, breakdown of gate oxide due to high electric field, and punchthrough. Thus, releasing the stress-voltage is indispensable for MOSTs in circuits operating at high stress voltages. Figure 9.1 shows the mechanism of hot-carrier generation, as shown in Chapter 1. There exists a large electric field around the drain of a short-channel device. Carriers (electrons for nMOSTs and holes for pMOSTs) accelerated by the field collide with the lattice, generating

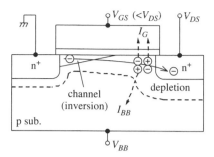

FIGURE 9.1. Hot-carrier injection mechanism [1].

electron-hole pairs (DAHC: Drain Avalanche Hot Carrier). In the case of an nMOST, the generated electrons with high energy reach the gate through gate oxide (gate current), while the holes flow to the substrate (substrate current). The CHE (Channel Hot Electron), where the accelerated electrons are directly injected into gate oxide, is smaller than DAHC. The substrate current raises the substrate potential and may cause latch-up. On the other hand, the electrons are trapped in the gate oxide if there are trap levels. The trapped electrons enhance the threshold voltage and decrease drain current. Thus, the device performance is degraded by hot carriers. In the case of a pMOST, the generated holes flow to the gate and the electrons flow to the substrate. However, the injection of holes into gate oxide is less than that of an nMOST because a hole has a larger effective mass. The electrons that flow to the substrate lower the substrate potential and may cause latch-up. However, since the latch-up can be avoided by careful layout design, the degradation of nMOSTs due to the electron injection into gate oxide is the most serious problem.

In addition, the gate-oxide thickness becomes smaller according to device scaling, causing breakdown and/or long-range reliability problem (TDDB: Time Dependent Dielectric Breakdown) with too large of an electric field across the gate oxide. Moreover, when the electric field between the drain and source of a short-channel device becomes large to the extent that current flow cannot be controlled by the gate voltage, punchthrough occurs, and the MOST is no more a switching device.

9.3. Concepts of High-Voltage Tolerant Circuits

The following circuit techniques are proposed for the above problems. Figure 9.2(a) shows a circuit to alleviate the hot-carrier effect. This circuit has a protection device only for an nMOST because of the above reasons. Here, M_{N2} is the protection device. The series connection of two nMOSTs halves the drain-source voltage of each nMOST and reduces hot-carrier injection. Fig. 9.2(b) shows the relationship between supply voltage V_{DD} and substrate current I_{BB},

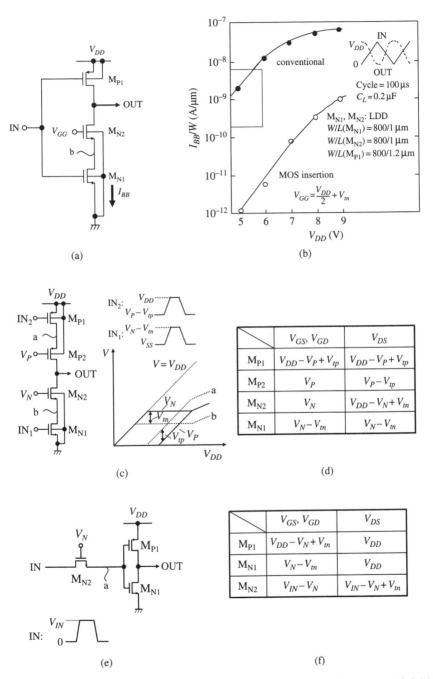

FIGURE 9.2. Concepts of high-voltage tolerant circuits; hot-carrier tolerant circuit [2] (a), relationship between substrate current and supply voltage [2] (b), drain-applied high-voltage tolerant circuit [3] (c), maximum voltage applied to each MOST (d), gate-applied high-voltage tolerant circuit (e), and maximum voltage applied to each MOST (f).

which is a good monitor of the amount of hot carriers [1]. Although I_{BB} exponentially increases with V_{DD}, as expressed in Eq. (1.16) in Chapter 1, the insertion of the protection device reduces I_{BB} by three orders of magnitude. The gate voltage of M_{N2} is $V_{DD}/2 + V_{tn}$ (V_{tn} is the threshold voltage of M_{N2}) to minimize I_{BB}. This technique is applied to the circuit shown in Fig. 8.11 in the previous chapter. The MOSTs surrounded by dashed lines are the protection devices, which protect nMOSTs from higher voltages than V_{DD}.

Circuit techniques for protecting the gate oxide from high voltages are classified into two types. One is for the case of a high voltage being applied to the drain terminal and the other is for the case of being applied to the gate terminal. Figure 9.2(c) shows the former, while (e) shows the latter. The circuit in Fig. 9.2(c) reduces both gate-source voltage and drain source voltage. This circuit has two protection devices M_{P2} and M_{N2}, which respectively protect M_{P1} and M_{N1} from high voltage. This is because the gate-source breakdown voltage is almost equal for pMOSTs and nMOSTs, unlike the unequal behavior for hot carrier injection. The gate voltages of the protection devices are between V_{DD} and ground (V_P for M_{P2} and V_N for M_{N2}). The minimum potential of node a is $V_{P-} + |V_{tp}|$, and the maximum potential of node b is $V_N - V_{tn}$, limiting the drain-source voltages of M_{P1} and M_{N1}. The characteristics of V_P and V_N are also shown in the figure. V_N and V_P are regulated based on ground and V_{DD}, respectively, so that sufficiently high voltages are applied to the protection devices at low V_{DD} in order to reduce the on resistances. At higher V_{DD}, $V_P = V_N = V_{DD}/2$. The voltage level of input signals IN_1 is between ($V_N - V_{tn}$) and V_{SS}, while that of IN_2 is between V_{DD} and ($V_P + |V_{tp}|$) to limit the gate-source voltages. The input signals are generated by the circuit described later. Fig. 9.2(d) shows the maximum voltages applied to each MOST. For example, when $V_{DD} = 3.5\,\text{V}$, $|V_{tp}| = V_{tn} = 0.5\,\text{V}$, $V_P = V_N = 1.75\,\text{V}$, the maximum voltage is $1.75\,\text{V}$, much lower than V_{DD}. The injection of hot-carriers is reduced because of the lower drain-source voltages. This circuit also features the full-swing ($V_{DD} - V_{SS}$) output signal. Figure 9.2(e) shows the technique for the case that a high voltage is applied to the gate. A protection device M_{N2} is inserted between the input and the gates of M_{P1} and M_{N1}. Even if a high voltage is applied to the input, the voltage of node a is limited to $V_N - V_{tn}$ (V_{tn} is the threshold voltage of M_{N2}). On the other hand, a low-level input voltage passes through M_{N2} as it is. Figure 9.2(f) shows the maximum voltages applied to each MOST.

9.4. Applications to Internal Circuits

9.4.1. Level Shifter

Figure 9.3 shows a level shifter circuit using the high-voltage tolerant technique in Fig. 9.2(c). The circuit diagram and the operating waveforms are shown in Fig. 9.3(a) and (b), respectively. This circuit converts the input signal with a voltage swing of V_{DD1} into the output signal with a larger swing of V_{DD2}. This circuit consists of two sets of the inverters shown in Fig. 9.2(c). The

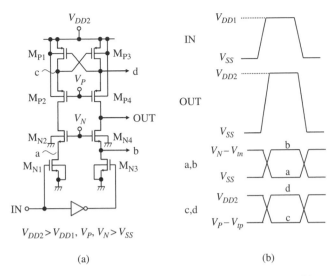

(a)   (b)

FIGURE 9.3. High-voltage tolerant level shifter(1) [3]; shematic diagram (a), and operating waveforms (b).

pMOSTs M_{P1} and M_{P3} are cross-coupled, while nMOSTs M_{N1} and M_{N3} receive complementary input signals. The voltages between terminals of MOSTs are the same as in Fig. 9.2(d). The voltage of nodes a and b is between $V_N - V_{tn}$ and ground, while that of nodes c and d is between V_{DD2} and $V_P + |V_{tp}|$. Therefore, they can be used as input signals IN_1 and IN_2 for the circuit in Fig. 9.2(c).

Figure 9.4 shows another example of a high-voltage tolerant level shifter circuit. The circuit diagram and the operating waveforms are shown in Fig. 9.4(a) and (b), respectively. This circuit is fundamentally the same as the circuit in Fig. 9.3. The differences include that thick gate-oxide MOSTs are used except for the input nMOSTs (M_{N1}, M_{N3}) and that the gates of M_{P2}, M_{P4}, M_{N2} and M_{N4} are supplied with the input signal instead of dc voltage. The thick gate-oxide MOSTs M_{N2} and M_{N4} are the protection devices for thin gate-oxide MOSTs M_{N1} and M_{N3}, respectively. When input signal IN does not change, V_{DD1} ($< V_{DD2}$) is applied to the gates of M_{N2} and M_{N4}. The source potentials of M_{N2} and M_{N4}, which are drain-source voltage of M_{N1} and M_{N3}, are limited to $V_{DD1} - V_{tn}$, even if one of the output terminals OUT and $\overline{OUT}$ is at V_{DD2} level. When IN becomes at V_{DD1} level, M_{N1} is turned on and M_{N3} is turned off. The voltage of node n_1 is boosted above V_{DD1} by capacitor C_1. Since the on resistance of M_{N2} becomes lower than constant-V_{DD1} case, the charge of $\overline{OUT}$ is quickly discharged. On the other hand, the gate voltage of M_{P4} changes from V_{DD1} level to V_{SS} level. Since the on resistance of M_{P4} becomes lower than constant-voltage case, output node OUT is quickly charged. Thus, this circuit converts a small-swing input signal into a large-swing output signal at a high speed.

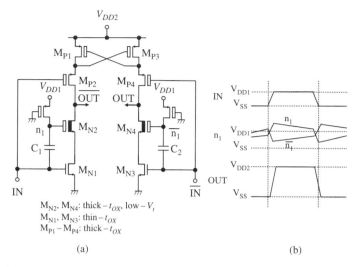

FIGURE 9.4. High-voltage tolerant level shifter(2); shematic diagram (a), and operating waveforms (b). Reproduced from [4] with permission; © 2006 IEEE.

9.4.2. Voltage Doubler

Fig. 9.5 shows a voltage doubler using the high-voltage tolerant circuit technique. The voltages of nodes A, B and C are $2V_{DD}$, $2V_{DD}$ and $3V_{DD}$, respectively, when Φ_1 is at V_{DD}. Therefore $2V_{DD}$ is applied between node B and ground, and between node C and V_{DD}. To tolerate the high voltage, this circuit utilizes the following techniques. First, two serially-connected MOSTs M_2 and M_3 are inserted between node B and ground. The gate of M_2 is connected to V_{DD}, while

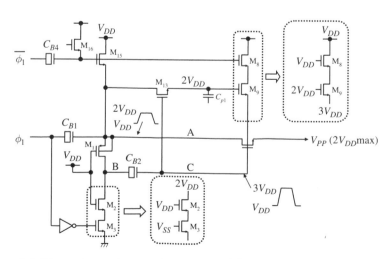

FIGURE 9.5. High-voltage tolerant voltage doubler [5].

that of M_3 receives a switching signal. Since the drain-source voltages of M_3 and M_2 are limited to $V_{DD} - V_t$ and $2V_{DD} - (V_{DD} - V_t) = V_{DD} + V_t$, respectively, MOSTs with ordinary breakdown voltage can be used as M_2 and M_3. Second, serially connected MOSTs M_8 and M_9 are inserted between node C and V_{DD}. The gate of M_9 is connected to $2V_{DD}$, which is supplied from node A through M_{13}. If the falling timing of node C is earlier than that of node A, the gate voltage $2V_{DD}$ does not flow back to node A. The source potential of M_8 is $2V_{DD} - V_t$. Since the drain-source voltages of M_8 and M_9 are $(2V_{DD} - V_t) - V_{DD} = V_{DD} - V_t$ and $3V_{DD} - (2V_{DD} - V_t) = V_{DD} + V_t$, respectively, MOSTs with ordinary breakdown voltage can be used as M_8 and M_9. The gate voltages of both M_8 and M_9 are $2V_{DD}$ during precharging period ($\Phi_1 = V_{SS}$ and $\overline{\Phi_1} = V_{DD}$). Thus the capacitor C_{B2} is precharged without having a V_t drop.

9.5. Applications to I/O Circuits

Input and output buffers of the chip are different from internal circuits in that a higher voltage than V_{DD} or a negative voltage may be applied to I/O terminals due to signal overshoots and undershoots. Therefore, more strict protection of devices against high voltage is required.

9.5.1. Output Buffers

Figure 9.6 shows a high-voltage tolerant output buffer [3] using the technique shown in Fig. 9.2(c). The MOSTs M_{P2} and M_{N2} are the protection devices for M_{P1} and M_{N1}, respectively. The input signal for M_{P1} is generated by the level shifter shown in Fig. 9.3. The delay circuit inserted at the gate of M_{N1} adjusts the driving timings of M_{N1} and M_{P1} to avoid an unfavorable DC-current flow, which occurs if both M_{N1} and M_{P1} turn on at the same time.

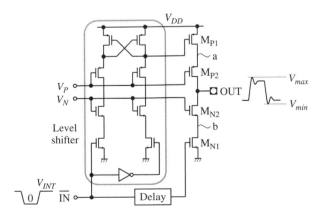

FIGURE 9.6. High-voltage tolerant output buffer. Reproduced from [3] with permission; © 2006 IEEE.

The gate voltages of M_{P2} and M_{N2}, V_P and V_N, are determined by the following criteria. Let us denote the maximum voltage that can be allowed (including design margin) to apply to the gate oxide as V_{safe}. First, from the gate-drain voltage of M_{N2}, the following inequality stands:

$$|V_{OUT} - V_N| < V_{safe}. \tag{9.1}$$

Second, from the gate-drain voltage of M_{N1},

$$V_N - V_{tn} < V_{safe} \tag{9.2}$$

stands because the minimum voltage of M_{N1}'s gate is zero and the maximum voltage of node b equals $V_N - V_{tn}$, where V_{tn} is the threshold voltage of an nMOST. Finally, since M_{N2} must be turned on,

$$V_N > V_{tn} \tag{9.3}$$

stands. From inequalities (1.1), (1.2) and (1.3), the allowable range of V_N is illustrated by the hexagon in Fig. 9.7(a). Here, V_{max} and V_{min} are the maximum and minimum voltages, respectively, which may be applied to the output terminal. Similarly the allowable range of V_P is expressed by the following inequalities:

$$|V_{OUT} - V_P| < V_{safe}, \tag{9.4}$$

$$V_P + |V_{tp}| > V_{DD} - V_{safe}, \tag{9.5}$$

$$V_{DD} - V_P > |V_{tp}|, \tag{9.6}$$

and is illustrated by the hexagon in Fig. 9.7(b). If $V_{safe} + V_{min} > V_{max} - V_{safe}$, that is,

$$V_{safe} > \frac{V_{max} - V_{min}}{2}, \tag{9.7}$$

both V_P and V_N may be DC voltages between $V_{safe} + V_{min}$ and $V_{max} - V_{safe}$. It is favorable for reducing on resistances that V_P is as low as possible and that V_N is as high as possible in the range. However, if inequality (1.7) does not stand, V_P and V_N must be dynamically changed according to V_{OUT} as described below.

Figure 9.8 shows a high-voltage tolerant output buffer using the dynamic changing of gate bias [6]. The circuit shown in Fig. 9.9 generates the gate voltages V_P and V_N. The output voltage V_{OUT} couples to V_P and V_N through the feedback circuit consisting of M_{P9}, M_{N9}, M_{P10} and M_{N10}. When V_{OUT} is at high level, a current flows through M_{N9} and M_{P8}. The voltage $V_P = V_N$ is determined by the source follower of M_{P8} as $V_P = V_N = V_1 - V_{tn} + |V_{tp}|$. On the other hand, when V_{OUT} is at low level, a current flows through M_{N8} and M_{P9}. The voltage $V_P = V_N$ is determined by the source follower of M_{N8} as $V_P = V_N = V_2 + |V_{tp}| - V_{tn}$. Thus, the voltage $V_P = V_N$ is limited between $V_1 - V_{tn} + |V_{tp}|$ and $V_2 + |V_{tp}| - V_{tn}$. The gate voltage of M_{N4} in the level shifter is also dynamically changed according to V_{OUT}. The pMOST M_{P4}, the gate of which is biased to a fixed voltage of 1.1 V,

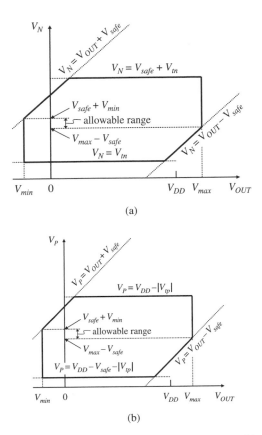

FIGURE 9.7. Determination of gate voltages; determining V_N (a), and determining V_P (b) [3].

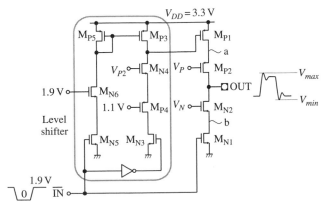

FIGURE 9.8. High-voltage tolerant output buffer. Reproduced from [6] with permission; © 2006 IEEE.

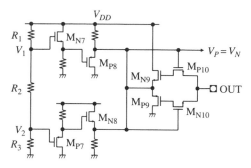

FIGURE 9.9. Circuit for generating dynamic voltages V_{P1} and V_N. Reproduced from [6] with permission; © 2006 IEEE.

prevents the gate of M_{P1} from being discharged below $1.1\,V + |V_{tp}|$. It is reported that this output buffer is tolerant to an overshoot of 4.1 V and undershoot of $-0.8\,V$ despite of the usage of MOSTs with $V_{safe} = 2.0\,V$.

9.5.2. Input Buffers

Figure 9.10 shows a high-voltage tolerant input buffer [7] using the technique shown in Fig. 9.2(e). When the input terminal is at high level (3.3 V), an nMOST M_1 gives a high level voltage of $V_N - V_{tn}$ on node a. Here V_N is a DC voltage of 2.5 V. Once the high input is received, the feedback pMOST M_3 holds node a at V_{INT} level (2.5 V). Thus, the MOSTs are protected from a high-level input voltage of 3.3 V. A capacitor C helps to speed up rising edges.

Figure 9.11 shows another high-voltage tolerant input buffer [8], which features dynamic changing of M_1's gate voltage V_N. When the input terminal IN is at low level, the node a follows to the input and node b transitions to $V_{INT}(2.5\,V)$. This turns M_7 and M_5 on and a DC current flows from V_{DD} to ground through $M_5 - M_8$. The conductance ratio among $M_5 - M_8$ is designed so that $V_N \sim 2.2\,V$. Thus, the gate oxide of M_1 is protected from undershoots. When IN goes high, nodes a and b go high and low, respectively. This cuts the DC conduction path, and V_N is raised to V_{DD}. Therefore M_1 passes a high enough level of $V_{DD} - V_{tn}$ to fully turn off the pMOST M_4, while protecting M_3 and M_4 from overshoots.

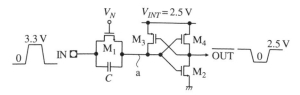

FIGURE 9.10. High-voltage tolerant input buffer [7].

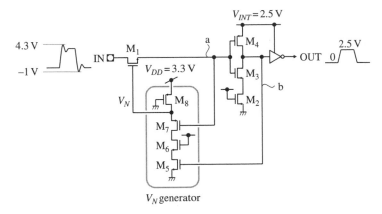

FIGURE 9.11. High-voltage tolerant input buffer [8].

References

[1] T. Sakurai, K. Nogami, M. Kakumu and T. Iizuka, "Hot-carrier generation in submicrometer VLSI environment," IEEE J. Solid-State Circuits, vol. SC-21, pp. 187–192, Feb. 1986.

[2] K. Nogami, K. Sawada, M. Kinugawa and T. Sakurai, "VLSI circuit reliability under AC hot-carrier stress," in Symp. VLSI Circuits Dig. Tech. Papers, May 1987, pp. 13–14.

[3] Y. Nakagome, K. Itoh, K. Takeuchi, E. Kume, H. Tanaka, M. Isoda, T. Mushya, T. Kaga, T. Kisu, T. Nishida, Y. Kawamoto and M. Aoki, "Circuit techniques for 1.5-3.6-V battery-operated 64-Mb DRAM," IEEE J. Solid-State Circuits, vol. 26, pp. 1003–1010, July 1991.

[4] Y. Kanno, H. Mizuno, N. Oodaira, Y. Yasu and K. Yanagisawa, "μI/O architecture for 0.13-um wide-voltage-range system-on-a-package (SoP) designs," in Symp. VLSI Circuits Dig. Tech. Papers, June 2002, pp. 168–169.

[5] H. Tanaka, "Charge pump with improved reliability," US Patent No. 6456152, Sep. 2002.

[6] G. Singh, "A high speed 3.3V IO buffer with 1.9V tolerant CMOS process," in Proc. ESSCIRC, Sep. 1997, pp. 128–131.

[7] D. Greenhill, E. Anderson, J. Bauman, A. Charnas, R. Cheeria, H. Chen, M. Doreswamy, P. Ferolito, S. Gopaladhine, K. Ho, W. Hsu, P. Kongetira, R. Melanson, V. Reddy, R. Salem, H. Sathianathan, S. Shah, K. Shin, C. Srivatsa and R. Weisenbach, "A 330MHz 4-way superscalar microprocessor," in ISSCC Dig. Tech. Papers, Feb. 1997, pp. 166–167.

[8] J. Connor, D. Evans, G. Braceras, J. Sousa, W. W. Abadeer, S. Hall and M. Robillard, "Dynamic dielectric protection for I/O circuits fabricated in a 2.5V CMOS technology interfacing to a 3.3V LVTTL bus," in Symp. VLSI Circuits Dig. Tech. Papers, June 1997, pp. 119–120.

Index

Printed in the United States of America.